AF576045

Innovative Agrifood Supply Chain in the Post-COVID 19 Era

Innovative Agrifood Supply Chain in the Post-COVID 19 Era

Editor

Dimitris Skalkos

MDPI • Basel • Beijing • Wuhan • Barcelona • Belgrade • Manchester • Tokyo • Cluj • Tianjin

Editor
Dimitris Skalkos
Laboratory of Food Chemistry
Department of Chemistry
University of Ioannina
Ioannina
Greece

Editorial Office
MDPI
St. Alban-Anlage 66
4052 Basel, Switzerland

This is a reprint of articles from the Special Issue published online in the open access journal *Sustainability* (ISSN 2071-1050) (available at: www.mdpi.com/journal/sustainability/special_issues/Supply_Chain_Network).

For citation purposes, cite each article independently as indicated on the article page online and as indicated below:

LastName, A.A.; LastName, B.B.; LastName, C.C. Article Title. *Journal Name* **Year**, *Volume Number*, Page Range.

ISBN 978-3-0365-4188-4 (Hbk)
ISBN 978-3-0365-4187-7 (PDF)

Contents

About the Editor

Dimitris Skalkos

Dr. Dimitris Skalkos is currently an associate professor in food business management of innovation, at the Laboratory of Food chemistry, Department of Chemistry, University of Ioannina, Greece. He has been assistant and associate professor at The University of The Aegean, Greece (Department of Food Science & Nutrition), and visiting assistant professor at The University of Toledo (Department of Chemistry), Ohio, USA. Dr. Skalkos career expands beyond academia, for more than 30 years, since he has worked as: 1) the founder, and first director of the Business Innovation Center (BIC) of Epirus-Greece, 2) the owner of Synthesis Consulting Ltd, 3) the co-owner of Paskal Herbal extracts S.A. and has worked as business consultant in more than 20 years with 50 national, and multi-national companies of various sectors of activity, mainly in the food sector. He has also worked as project manager or scientific director on more than 20 EU funding programs involved with the promotion of innovation, the innovative pilot actions, the establishment of academic spin off companies, the food products development, and the food business management. His research interests currently focus on the development of all innovative aspects within the food businesses including production, processing, organization, marketing, consumers' motives and knowledge transfer.

Preface to "Innovative Agrifood Supply Chain in the Post-COVID 19 Era"

The world is changing rapidly in the age beyond Coronavirus. The current period of deprivation and anxiety, together with the coming global economic crisis, will usher in new consumer attitudes and behaviors that will change the nature of today's capitalism. There are signs today of a growing anti-consuming movement with five types of anti-consumerists: life simplifiers, degrowth activists, climate activists, food choosers, and conservation activists. Citizens will reexamine what they eat, how much they eat, and how all this is influenced by class issues and inequality. They will reexamine their eating habits and emerge from this terrible period with a new, more equitable form of food consumption. Consequently, the food supply chain network will have to change dramatically, adjusting to the new attitudes, perceptions and preferences of the consumers of the post-COVID-19 era. Innovation will play a vital role in modernizing the food supply chain to meet the new challenges of the upcoming global economy. The process "from farm to fork" as the holistic approach to the production and consumption of food will become a key factor for the sustainability and the progress of the food industry. This Special Issue is focused on 11 selected topics from different parts of the agrifood supply chain in view of the post-COVID-19 era expanding from innovative scientific insights and technological advances of natural resources, organic pollutants identification, new food product development, traceability, and packaging, chain management, to consumer's attitudes, and eating motivations, aiming to tackle the foreseen changes of global economy, and society.

Dimitris Skalkos
Editor

Editorial

Innovative Agrifood Supply Chain in the Post-COVID 19 Era

Dimitris Skalkos

Laboratory of Food Chemistry, Department of Chemistry, University of Ioannina, 45110 Ioannina, Greece; dskalkos@uoi.gr; Tel.: +30-2651-0083-45

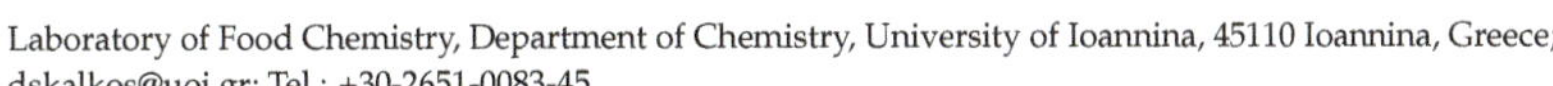

The world is changing rapidly in the age of Coronavirus [1]. The long period of deprivation, economic austerity and anxiety that is foreseen will usher new consumer attitudes and behaviors, which will change the nature of today's capitalism. There are signs of a growing anti-consuming movement at present [2], with five types of anti-consumerists: life-simplifiers, degrowth activists, climate activists, food-choosers, and conservation activists. Citizens will reexamine what they eat, how much they eat, and how all this is influenced by class issues and inequality. They will reexamine their eating habits and emerge from this terrible period with a new, more equitable form of food consumption [3]. Consequently, the agrifood supply chain network will have to change dramatically to adjust to the new attitudes, perceptions, and preferences of consumers in the post-COVID-19 era. Innovation will play a vital role in modernizing the agrifood supply chain to meet the new challenges of the upcoming new global economy [4]. The process "from farm to fork", the holistic approach to the production and consumption of food, will become a key factor in the sustainability and progress of the food industry. This Special Issue is focused on 11 selected topics from different parts of the agrifood supply chain in view of the post-COVID-19 era, expanding from the innovative scientific insights and technological advances of natural resources, organic pollutants' identification, new food product development, traceability, and packaging, and chain management, to consumer's attitudes, and eating motivations, aiming to tackle the changes that are foreseen for the global economy and society.

Citation: Skalkos, D. Innovative Agrifood Supply Chain in the Post-COVID 19 Era. *Sustainability* **2022**, *14*, 5359. https://doi.org/10.3390/su14095359

Received: 26 April 2022
Accepted: 27 April 2022
Published: 29 April 2022

Nature-based solutions (NbSs) encompass a broad range of practices that can be introduced in the agri-food supply chain and address multiple environmental challenges in the post-COVID-19 era, while providing economic and societal benefits [5]. The study of Takavakoglou et al. explores the potential role of constructed wetlands, revealing application opportunities in different segments of the supply chain, identifying linkages with societal challenges and EU policies, and discussing their potential limitations, as well as the future challenges and perspectives [6]. The pandemic opened an opportunity for the reformation of economies and the transition towards a greener model of development.

The ecosystems and public health of the agrifood supply chain are increasingly affected by the presence of pesticides and pharmaceuticals, which will become a major concern in the post COVID-19 era due to the increase in global public attention to health issues [7]. This condition is of major importance in regions with fish farms in their aquatic environment. The study of Martinaiou et al. developed a solid-phase extraction method to optimize and validate the analysis of 7 pesticides and 25 pharmaceuticals in seawater using LC-HR-LTQ/Orbitrap-MS [8]. The method was then successfully applied in seawater samples collected from an aquaculture farm to evaluate its validity. At the same time, passive sampling was conducted as an alternative screening technique, showing the presence of contaminants that were not detected with spot sampling.

The post COVID-19 customers are seeking quality, innovative, healthy foods with natural ingredients to protect themselves, the environment and provide sustainability to local economies [4,9]. The study of Karantonis et al. developed a series of functional baked goods, such as whole-meal sliced bread, chocolate cookies and breadsticks, which are rich in natural enrichment source of omega-3 fatty acids and fiber [10]. The source is the flaxseed (*Linum usitatissimu*) added to the foods. The final products were tested as sources of omega-3 fatty acids in terms of α-linolenic acid, as well as for their in vitro antithrombotic/anti-inflammatory affects. The results showed high omega-3 fatty acids concentrations (>0.6 g per 100 g of product) in all products, exerting higher in vitro antithrombotic/anti-inflammatory activity, which was in a different grade compared to the conventional products. In the same research field of healthy foods, the study of Papagianni et al. examined the benefits of novel functional cookies enriched with olive paste [11]. The production of these cookies and their antioxidant activities was reported earlier by the same authors [12]. Olive paste exerts bioactivity due to its richness in bioactive components, such as oleic acid and polyphenols. This interventional human study investigated whether the fortifications of cookies with olive paste and herbs may affect postprandial lipemia, oxidative stress, and other biomarkers in healthy volunteers [11]. Total plasma antioxidant capacity according to FRAP, ABTS, and resistance to copper-induced plasma oxidation, serum lipids, glucose, uric acid, and antithrombotic activity in platelet-rich plasma were determined at each timepoint 0.5 h, 1.5 h and 3 h after eating cookie meal alone compared with enhanced with 20% olive paste. There was a significant decrease in triglycerides' concentration in the last 1.5 h of the intervention as compared to the control group ($p < 0.05$). A tendency towards a decrease in glucose levels and an increase in the plasma antioxidant capacity was observed at 0.5 h and 1.5 h, respectively, in the intervention compared to the control group. The remaining biomarkers did not show statistically significant differences ($p > 0.05$).

Moving from the innovative food products to innovative packaging, the study of Tsironi et al. proposes an alternative form of food preservation and packaging for minced lamp meat, with the fewest possible preservatives and additives, as well as an extended shelf life [13]. Whey protein isolate (WPI) films, alone and with incorporated essential oils (WPI + EO) at different concentrations, were prepared and then examined for their possible delaying effect on the deterioration of minced lamb meat. The essential oils, as natural antimicrobials, were oils of ginger (Zingiber Officinale Roscoe) and rosemary (*Rosmarinus officinalis* L.). the results showed that films with 1% EO significantly improved the microbiological qualities of meat. Regarding physicochemical properties, the same pattern was observed for pH, while the oxidation degree was significantly reduced. Finally, color attribute measurements recorded fluctuations between samples; however, overall, no considerable discoloration was observed. Within the subject of packaging, the important issue examining societies' moving towards sustainable habits, questioning their actions and considering their impact on the environment, is evaluated [14]. Macena et al. examine, via an online questionnaire, the habits of Portuguese citizens concerning plastic food packaging, sustainability and recycling, as well as the knowledge effects of plastic materials or their residues on the environment, and focuses on aspects related to sustainability [15]. The participants tend to think about the negative impact of plastic packages on the environment; 39% sometimes do not buy plastic; and 30% try to look for alternatives, 81% support the avoidance of plastic utensils and reductions in the use of plastic bags, most participants have a good knowledge of recycling and strongly agree with the use of recycled materials, and 87% of respondents practice the separation of different types of waste for recycling.

Food traceability is another research area of major concern in the post-COVID-19 period since it tackles the selected concerns of "new" consumers [16]. It is an essential tool for both industry and consumers to confirm the characteristics of leading food products' industries to ensure the traceability of their merchandise. Dima et al. carried out an online market research survey to determine the significant concerns of the Greek customers regarding eating pork and pork products, their opinion on related traceability information, and their preferences regarding how they would like to receive this information [17]. Consumers expressed high interest in the expiry date of the meat (87.9%), followed by the means and conditions of transport of the meat products (79%), as well as a preference to buy traceable compared with untraceable pork (79%), and their belief that the quality and safety of pork products would be improved with traceability (70.1%), signifying the importance of traceability for consumers.

In the coming period, alternative solutions to feeding the growing world population with less stress on the planet will be investigated, contributing to the preservation of the environment, such as edible insects (EIs) [18]. Guine et al., using a questionnaire survey, explored the level of information that people in a traditionally non-insect-eating country have about the sustainability issues related to EIs, and some possible factors that could motivate their consumption [19]. It was found that the highest motivators to consume EIs are their contribution to preserving the environment and natural resources, followed by their being a more sustainable option (for 64.7% and 53.4% of participants, respectively).

Traditional foods (TFs) can also play a major role as the food of choice for the "new" consumers, the "anti-consumers" in the coming post-COVID-19 period worldwide [20]. Consumers' trust in TFs after COVID-19 was studied by Skalkos et al. [21], using the variables of safety, healthiness, sustainability, authenticity and taste, assessing consumers' confidence and satisfaction with the TFs, their raw materials, and the technologies used for their production. The results show that the participants trust Greek TFs because they "strongly agree", by an average of 20%, and "agree", by an average of 50%, that they are safe, healthy, sustainable, authentic and tasty. Furthermore, the second study of Skalkos et al. examined the consumers' perception of the traceability of TFs in the post-COVID-19 era. [22]. Traceability was tested using variables related to the package, product, quality, process, and personal information of these foods. The results show that the participants consider traceability regarding questions on package information to be "quite important" and "very important" by an average of 68%, on food information by 64%, on quality information by 69%, on production process information by 78%, and on personal information by 65%.

In the last part of this Special Issue, a measurement instrument model for sustainable supply chain management (SSCM) critical factors, practices and performance is developed and validated in the food industry by Mastos et al. [23]. SSCM is one of the key sustainability concepts, and received significant attention in the last two decades [24]. It involves the management of material, information, and capital flows, as well as the cooperation among all companies in the supply chain. The validity of the proposed instrument was confirmed through an e-mail questionnaire answered by 423 Greek food companies. The extracted SSCM critical factors were "firm-level sustainability critical factors" and "supply chain sustainability critical factors". The extracted SSCM practices were "supply chain collaboration" and "supply chain strategic orientation". The extracted SSCM performance factors were "economic performance", "social performance" and "environmental performance". The three developed constructs constitute a measurement instrument that can be used both by practitioners who desire to implement SSCM and researchers who can apply the proposed scales in other research projects, or use them as assessment tools.

Funding: This research received no external funding.

Acknowledgments: As Guest Editor of the Special Issue "Innovative Agrifood Supply Chain in the Post-COVID 19 Era", I would like to express my deep appreciation to all the authors whose valuable work was published under this issue, and thus contributed to the success of the edition.

Conflicts of Interest: The authors declare no conflict of interest.

References

1. Grossmann, I.; Twardus, O.; Varnum, M.E.W.; Jayawickreme, E.; McLevey, J. Expert Predictions of Societal Change: Insights from the World After COVID Project. *Am. Psychol.* **2022**, *77*, 276–290. [CrossRef] [PubMed]
2. Kotler, P. The consumer in the age of coronavirus. *J. Creat. Value* **2020**, *6*, 12–15. [CrossRef]
3. Caso, D.; Guidetti, M.; Capasso, M.; Cavazza, N. Finally, the chance to eat healthily: Longitudinal study about food consumption during and after the first COVID-19 lockdown in Italy. *Food Qual. Prefer.* **2022**, *95*, 104275. [CrossRef] [PubMed]
4. Galanakis, C.M.; Rizou, M.; Aldawoud, T.M.S.; Ucak, I.; Rowan, N.J. Innovations and technology disruptions in the food sector within the COVID-19 pandemic and post-lockdown era. *Trends Food Sci. Technol.* **2021**, *110*, 193–200. [CrossRef]

5. Kisser, J.; Wirth, M.; De Gusseme, B.; Van Eekert, M.; Zeeman, G.; Schoenborn, A.; Vinnerås, B.; Finger, D.C.; Kolbl Repinc, S.; Bulc, T.G.; et al. A review of nature-based solutions for resource recovery in cities. *Blue-Green Syst.* **2020**, *2*, 138–172. [CrossRef]
6. Takavakoglou, V.; Pana, E.; Skalkos, D. Constructed Wetlands as Nature-Based Solutions in the Post-COVID Agri-Food Supply Chain: Challenges and Opportunities. *Sustainability* **2022**, *14*, 3145. [CrossRef]
7. Othmani, W. Pesticide use and COVID-19: An Indictment of Global Agriculture and An Advocacy of the Mediterranean Food Culture. *Res. Sq.* **2020**, 1–15. [CrossRef]
8. Martinaiou, P.; Manoli, P.; Boti, V.; Hela, D.; Markou, E.; Albanis, T.; Konstantinou, I. Quality control of emerging contaminants in marine aquaculture systems by spot sampling-optimized solid phase extraction and passive sampling. *Sustainability* **2022**, *14*, 3452. [CrossRef]
9. Olaimat, A.N.; Shahbaz, H.M.; Fatima, N.; Munir, S.; Holley, R.A. Food Safety During and After the Era of COVID-19 Pandemic. *Front. Microbiol.* **2020**, *11*, 1854. [CrossRef]
10. Karantonis, H.C.; Nasopoulou, C.; Skalkos, D. Functional Bakery Snacks for the Post-COVID-19 Market, Fortified with Omega-3 Fatty Acids No Title. *Sustainability* **2022**, *14*, 4816. [CrossRef]
11. Papagianni, O.; Moulas, I.; Loukas, T.; Magkoutis, A.; Skalkos, D.; Kafetzopoulos, D.; Dimou, C.; Karantonis, H.C.; Koutelidakis, A.E. Trends in food innovation: An interventional study on the benefits of consuming novel functional cookies enriched with olive paste. *Sustainability* **2021**, *13*, 1472. [CrossRef]
12. Argyri, E.A.; Piromalis, S.P.; Koutelidakis, A.; Kafetzopoulos, D.; Petsas, A.S.; Skalkos, D.; Nasopoulou, C.; Dimou, C.; Karantonis, H.C. Olive paste-enriched cookies exert increased antioxidant activities. *Appl. Sci.* **2021**, *11*, 5515. [CrossRef]
13. Tsironi, M.; Kosma, I.S.; Badeka, A.V. The Effect of Whey Protein Films with Ginger and Rosemary Essential Oils on Microbiological Quality and Physicochemical Properties of Minced Lamb Meat. *Sustainability* **2022**, *14*, 3434. [CrossRef]
14. Sundqvist-Andberg, H.; Åkerman, M. Sustainability governance and contested plastic food packaging–An integrative review. *J. Clean. Prod.* **2021**, *306*, 127111. [CrossRef]
15. Macena, M.W.; Carvalho, R.; Cruz-Lopes, L.P.; Guiné, R.P.F. Plastic food packaging: Perceptions and attitudes of portuguese consumers about environmental impact and recycling. *Sustainability* **2021**, *13*, 9953. [CrossRef]
16. Sinha, A.; Priyadarshi, P.; Bhushan, M.; Debbarma, D. Worldwide trends in the scientific production of literature on traceability in food safety: A bibliometric analysis. *Artif. Intell. Agric.* **2021**, *5*, 252–261. [CrossRef]
17. Dima, A.; Arvaniti, E.; Stylios, C.; Kafetzopoulos, D.; Skalkos, D. Adapting Open Innovation Practices for the Creation of a Traceability System in a Meat-Producing Industry in Northwest Greece. *Sustainability* **2022**, *14*, 5111. [CrossRef]
18. Ordoñez-Araque, R.; Egas-Montenegro, E. Edible insects: A food alternative for the sustainable development of the planet. *Int. J. Gastron. Food Sci.* **2021**, *23*, 100304. [CrossRef]
19. Guiné, R.P.F.; Florença, S.G.; Anjos, O.; Correia, P.M.R.; Ferreira, B.M.; Costa, C.A. An insight into the level of information about sustainability of edible insects in a traditionally non-insect-eating country: Exploratory study. *Sustainability* **2021**, *13*, 2014. [CrossRef]
20. Skalkos, D.; Kosma, I.S.; Chasioti, E.; Skendi, A.; Papageorgiou, M.; Guiné, R.P.F. Consumers' Attitude and Perception toward Traditional Foods of Northwest Greece during the COVID-19 Pandemic. *Appl. Sci.* **2021**, *11*, 4080. [CrossRef]
21. Skalkos, D.; Kosma, I.S.; Vasiliou, A.; Guine, R.P.F. Consumers' trust in Greek traditional foods in the post COVID-19 era. *Sustainability* **2021**, *13*, 9975. [CrossRef]
22. Skalkos, D.; Kosma, I.S.; Chasioti, E.; Bintsis, T.; Karantonis, H.C. Consumers' perception on traceability of greek traditional foods in the post-COVID-19 era. *Sustainability* **2021**, *13*, 2687. [CrossRef]
23. Mastos, T.; Ckotzamani, K.; Kafetzopoulos, D. Development and validation of a measurement instrument for sustainability in food supply chains. *Sustainability* **2022**, *14*, 5203. [CrossRef]
24. Ansari, Z.N.; Kant, R. A state-of-art literature review reflecting 15 years of focus on sustainable supply chain management. *J. Clean. Prod.* **2017**, *142*, 2524–2543. [CrossRef]

 *sustainability*

Article

Plastic Food Packaging: Perceptions and Attitudes of Portuguese Consumers about Environmental Impact and Recycling

Morgana Weber Macena [1], Rita Carvalho [2], Luísa Paula Cruz-Lopes [1,3] and Raquel P. F. Guiné [2,3,*]

1 School of Technology and Management, Polytechnic Institute of Viseu, 3504-510 Viseu, Portugal; morganaweber.m@gmail.com (M.W.M.); lvalente@estgv.ipv.pt (L.P.C.-L.)
2 Agrarian School, Polytechnic Institute of Viseu, 3500-606 Viseu, Portugal; anaamaral_26@hotmail.com
3 CERNAS Research Centre, Polytechnic Institute of Viseu, 3504-510 Viseu, Portugal
* Correspondence: raquelguine@esav.ipv.pt

Abstract: The use of plastics for packaging has some advantages, since they are flexible and inexpensive. However, most plastics are of single use, which, combined with low recycling or reuse ratios, contributes substantially to environmental pollution. This work is part of a project studying the habits of Portuguese citizens concerning plastic food packaging and focuses on aspects related to sustainability. The survey was carried out via an online questionnaire about sustainability, recycling, and knowledge of the effects of plastic materials or their residues on the environment. The results were obtained based on a statistical analysis of the data. The participants tend to think about the negative impact of plastic packages on the environment; 39% sometimes do not buy plastic; and 30% try to look for alternatives. A substantial fraction, 81%, support the avoidance of plastic utensils and reduction in the use of plastic bags. Most participants have a good knowledge of recycling and strongly agree with the use of recycled materials, and 87% of respondents practice separation of different types of waste for recycling. Changing plastic consumption habits has not been an easy task. Nevertheless, it is expected that society will increasingly move toward sustainable habits, questioning its actions and considering their impact on the environment.

Keywords: food packaging; plastic; recycling; knowledge; impact; questionnaire survey

Citation: Weber Macena, M.; Carvalho, R.; Cruz-Lopes, L.P.; Guiné, R.P.F. Plastic Food Packaging: Perceptions and Attitudes of Portuguese Consumers about Environmental Impact and Recycling. *Sustainability* **2021**, *13*, 9953. https://doi.org/10.3390/su13179953

Academic Editors: Graeme Moad and Tony Robert Walker

Received: 6 July 2021
Accepted: 30 August 2021
Published: 4 September 2021

Publisher's Note: MDPI stays neutral with regard to jurisdictional claims in published maps and institutional affiliations.

1. Introduction

Food packaging is absolutely essential and modern food systems could not function properly without packaging. Todays' food chains are characterized by their vast geographical spread as well as by value chains at the global level [1]. The primary function of food packaging is to protect the product it contains, preserving its safety and organoleptic characteristics. Among these, properties such as flavour, colour and aroma are highly important for the consumer who will purchase and consume the product. Additionally, the package serves as a barrier for microorganisms and undesirable changes in temperature, light, and moisture, protecting the product during transport and storage against microbial spoilage, chemical modifications, or physical changes. [2]. The packaging functions required for a food package system are expressed as PC3, which stands for Protection, Containment, Communication and Convenience [3]. However, selecting an appropriate package is not the only factor that guarantees the product's shelf-life. In fact, besides selecting the proper material for packaging, which is crucial, the conditions under which the food is stored are equally important [4]. The package is the face of a product and is often the only experience consumers have before making a purchase [5]. Thus, it is essential that the package presents good aesthetics [4] to convince the consumers to buy the product. In this way, packaging can drive sales in a competitive market, as packaging can be designed to enhance the image or differentiate one product from others [6]. In addition, packages bring essential information about the product, such as a list of ingredients, nutritional composition, preparation instructions, brand identification, and prices [5].

Materials that have been commonly used in food packaging embrace glass [7,8], metals [9,10], paper [11,12], plastics [5,13–15], wood [16,17], textile and cork [4]. Modern packaging can encompass more than one type of material to explore and combine the functional or aesthetic properties of each one [5]. The kind of packaging applied varies according to the product characteristics, the level of protection required, the intended shelf-life, the target market, the distribution and the sales circuit [4]. Packaging production translates into a globalized industry characterized by its internal diversity, while on the other hand, each of its sectors individually influences the market [18].

The use of plastic bags to carry groceries and goods goes back to the 1970s [19], but plastic materials have been increasingly used for food since then. In the latest decades, the relative share of plastic on food packaging systems has been way too high due to the many advantages associated with the use of plastics for food packaging: they are fluid and mouldable, offering considerable design flexibility; they are inexpensive and lightweight; and they have a wide range of physical and optical properties [5]. However, they also have disadvantages, the most important ones being their permeability to light, gases, vapours, and low-weight molecules [5]. Plastics can be divided into two groups: thermoplastics and thermosets. Thermoplastics do not suffer chemical changes in their production so that they can be recycled. Contrarily, thermosets suffer chemical changes in their production, which preclude a new merge; therefore, they are not recyclable [2].

Most plastics are produced from petroleum and are discarded in the environment where they are not degradable, creating considerable environmental problems. The incorrect disposal of plastic bags, and other forms of plastics, has created a problem, since they have found a way to be everywhere, including the oceans, posing a threat to aquatic life, agricultural lands, and the environment in general [20]. Thus, alternatives to plastic bags are necessary, but these alternatives should be less harmful to the environment or have no impact at all [19]. The majority of plastics are of single-use; thus, a significant proportion of this material is lost each year. The immense production, combined with low levels of recycling or reuse, and insufficient sustainable policies to support the circular plastic economy, result in a large contribution of waste to the environment. The United States Environmental Protection Agency (EPA) estimated that 14.5 million tons of plastic containers and packaging were generated in 2018, corresponding approximately to 5% of municipal solid waste generation (in this analysis, the "plastic packaging" as a category excluded single-service plates and cups, as well as trash bags, which are classified as nondurable goods). In 2019, plastic packaging generated around 54% of the global anthropogenic waste [21]. According to EPA, the recycling rate of PET bottles and jars was 29.1 percent in 2018 (910,000 tons).

There are two ways to reduce the primary production of packaging, reuse, and recycling. In the reuse, the product is returned and reused in its original form. Another way to reuse is replacement; that is, containers which allow refilling. Examples of reuse are beverage packaging, such as returnable glass bottles, plastic packaging for personal care products, and cleaning products that would enable the use of refills, as well as refillable water bottles. Recycling involves converting the materials, involving reprocessing into new products [5]. Thus, to make recycling economically viable, the materials need to have a market. Recycling effectiveness is linked to several factors, such as the correct disposal of the material, the type of material, and its conditions after use. Materials such as paper and cardboard, metals, and glass have a more consolidated recycling market, unlike plastics, which have, however, gained more attention recently.

Plastic is not biochemically inert; thus, it can interact with the human body and the environment, causing negative impacts [21]. However, investment in truly sustainable innovations is still scarce. Industries that opt for sustainable packaging generally turn to the use of recycled materials, not considering the production of packaging which uses sustainable raw materials with a low degradation time [2]. Reducing the amount of packaging in food products represents an opportunity, as well as a challenge, for the food and beverage industry, as the main concern is related to food safety. Thus, finding ways to reduce its quantity and subsequent waste is a very challenging task [22]. The requirements for packaging and articles which remain in contact with food are becoming systematically more strict [18], as they can affect the health of consumers and the environment. Nevertheless, the criteria for packaging to produce the lowest environmental impact are difficult to define [22].

Recycled metal and glass materials are considered safe for use in packaging that remain in contact with food, as the heat used to melt and form the material is sufficient to kill microorganisms and pyrolyze organic contaminants. However, in the case of plastics, reprocessing uses enough heat to destroy microorganisms, but it is not enough to pyrolyze all organic contaminants. Thus, post-consumer recycled plastics are hardly used for food packaging [5]. In general, the smaller the number of polymeric components and complexity of plastic packaging, the greater is the recycling value, due to the reduction of steps and technological resources applied in the process [23]. The profitability of the package recycling market shows its attractive aspects for business initiatives in the sector. Still, the success of recycling is directly related to cultural, political, and socioeconomic factors, such as the implementation of recycling companies, the existence of selective collection, and the continuous availability of recyclable waste, incentive programs for recycling projects, encouraging the sale of recycled products, as well as actions in the production–use-consumption chain of packaging [23]. Understanding the profile of people who buy plastic is vital for planning future plastic reduction interventions, legislation, and campaigns [24].

The role of consumers is of most importance in order to help decision-making bodies and governmental regulators to successfully implement measures in order to reduce the use of plastic, and particularly those of single use, which have a high impact on the environment, as well as on human health, as final elements of the possible contamination chains. The study by Adam et al. [25] explored consumer's attitudes towards the single-use plastics in Ghana considering their effect on marine pollution. They found that while some consumers avoid the consumption of single-use plastics, others consume them without any restrictions. Nevertheless, there was a third group that, although also conscientious about the implications of single-use plastics, still sometimes use them. A study conducted with Canadian consumers [26] revealed that practically all of the participants (around 94%) felt motivated to reduce the consumption of foods packed using single-use plastic. In this study, the authors also said that environmental concerns were more critical than food safety from the point of view of consumers. On this point, it was an undeniable fact that the Covid-19 pandemic brought to light new challenges concerning food safety, and therefore the work by Kitz et al. investigated the consumer perception of food packaging with single-use plastics during the Covid-19 period. They found that the motivation to reduce plastics was not so strong as before the pandemic, but this decline was not so pronounced among women as it was among men.

Although there is vast information in the literature about the negative impact of plastics on human health, as well as for the environment at the global scale, the information about the consumer's perceptions and knowledge and to what extent this shapes their behaviour and food choices is scarcer. To the best of our knowledge, this has not yet been accomplished for Portuguese consumers. This work is part of a project studying plastic food packaging, including Portuguese citizens' practices, knowledge, and concerns, from different perspectives, namely the impact on human health and the environment. This particular work has focused on the aspects related to sustainability, including recycling practices and knowledge about the impact of plastics on the ecosystems on a global scale.

2. Materials and Methods

2.1. Research Questions

Having in mind the aim of this study to investigate the practices of Portuguese consumers towards the use of plastics for food packaging and recycling practices, as well as the degree of knowledge about their impact on the environment, our main research questions were:

RQ1: How do Portuguese consumers perceive the impact of plastic food packaging, including their negative impact on the environment?

RQ2: What are consumers' attitudes towards minimising the harmful impacts of plastic, including practicing recycling?

RQ3: What is the degree of knowledge of Portuguese consumers concerning recycling?

RQ4: What is the influence of the sociodemographic characteristics of the Portuguese consumers on eco-responsible behaviour towards plastic packaging and knowledge about recycling?

These research questions were assessed though a questionnaire survey, using an appropriate instrument for data collection.

2.2. Questionnaire Survey

The survey was done by a questionnaire that was designed purposely for this project. The instrument included six sections with questions to collect data for different goals: (I) Sociodemographic variables; (II) buying habits; (III) opinions about packaging; (IV) impact of packages in health and the environment; (V) recycling of plastic products; (VI) Education about plastic and recycling; (VII) knowledge about recycling; (VI) knowledge about the effects of plastic on health and the environment. This manuscript addressed the questions related to sustainability, including attitudes and recycling practices, as well as knowledge about the effects of plastic materials or their residues (such as microplastics) on the environment.

The survey was applied on a convenience sample due to the recruitment facility and considered the disposition to participate. Although the use of convenience samples has some drawbacks, they are extremely useful for research with an exploratory nature [27,28]. The sample size calculation, although not being applied directly to convenience samples, is also a helpful indicator for this type of research. In this case, the indicative sample size was calculated considering a 95% confidence interval, corresponding to a level of significance of 5% and a z score of 1.96 [29,30]. The Portuguese population in 2019 (the last year available) was 10.286 million people [31], of which about 80% are adults, aged 18 years old or over, and targeting half of the adult population. The sample should include 385 participants [32–34] in order to be representative.

The data collection took place by the internet platform Google Forms, and the invitation to participate in the survey was sent by email and social networks. The inclusion criteria were: • Portuguese citizens; • participants with 18 years or more, meaning they were old enough to legally self-authorize to take part in the survey; • access to the internet; • access to a computer or other device through which they could answer the questionnaire; • able to understand the questions and express their responses; and • a willingness to participate in the research voluntarily and anonymously.

Strict ethical principles were obeyed when formulating the questionnaire and collecting the data, according to international standards (Declaration of Helsinki). The questionnaire was approved by the Ethical Commission at the Polytechnic Institute of Viseu with reference 09/SUB/2021. To all participants it was guaranteed that the internet tool used for the questionnaire would not record any data from the participants, such as email, IP or other sensitive information. Each participant could only access the questionnaire after agreeing to participate and after expressing informed consent.

The number of participants in the study was 487, exceeding the calculated indicative number of 385 previously referred to.

2.3. Data Analysis

For exploratory analysis of the data, basic descriptive statistics were used. Additionally, to access the relations between some of the categorical variables under study, the crosstabs and the chi-square test were used. The values of the Cramer's V coefficient allowed analysing the strength of the relations between variables. This coefficient varies from 0 to 1, and its meaning is as follows: $V \approx 0.1$, the association is weak; $V \approx 0.3$, the association is moderate; and $V \approx 0.5$ or over, the association is strong [35].

To validate the results obtained for the mean values calculated, a comparison of means was done by the analysis of variance (ANOVA), with the Post-Hoc Tukey HSD (Honestly Significant Difference) test for identification of the differences between samples for variables with three or more groups. For variables with two groups, the T-test for independent samples was used.

The variable accounting for the level of knowledge about recycling was submitted to a tree classification analysis to assess the relative importance of the sociodemographic variables. The analysis followed the CRT (Classification and Regression Trees) algorithm with a cross-validation and a minimum change in improvement of 0.0001, considering a limit of 5 levels and a minimum number of cases for parent or child nodes equal to 20 and 15, respectively [36]. A level of significance of 5% was considered in all statistical analyses.

To test the influence of the sociodemographic variables on eco-responsible behaviour towards plastic package, chi-square tests were conducted, based on the following null and alternative hypothesis:

Null Hypothesis (H0). *There are NO significant differences between groups regarding the measured variable (ex: avoid plastic utensils);*

Alternative Hypothesis (H1). *The differences between groups are significant.*

Additionally, to test the influence of the sociodemographic variables on the perception of the negative impact of plastics, ANOVA tests were conducted, based on the following null and alternative hypothesis:

Null Hypothesis (H0). *There are NO significant differences between groups regarding the perception of the negative impact of plastics;*

Alternative Hypothesis (H1). *The differences between groups are significant.*

In all cases, H0 was accepted if the *p*-value of the test was higher than 0.05, which was the level of significance established, while for values of *p* under 0.05, H0 was rejected and H1 was accepted.

3. Results

3.1. Sociodemographic Characterization of the Sample

Table 1 shows the sociodemographic characteristics of the sample at study. Most participants were female (70.4%) and resided in the central region of Portugal (64.9%). The participants' ages varied from 18 to 88 years old, the average age being equal to 37.7 ± 14.4 years. The variable age was categorized into young adults (aged between 18 and 30 years) corresponding to 41.1%, middle-aged adults (between 31 and 50 years) accounting for 35.5% and senior adults (51 years or older) representing 23.4%. The majority had completed university graduation (69.8%) and were currently employed (55.4%).

Table 1. Sociodemographic characterization of the sample (n = 487).

Variable	Group	n	%
Sex	Female	343	70.4
	Male	144	29.6
Residence	North	75	15.4
	Centre	316	64.9
	South and Islands	96	19.7
Age	Young adults (18–30 years)	200	41.1
	Middle-aged adults (31–50 years)	173	35.5
	Senior adults (≥51 years)	114	23.4
Education level	Up to secondary school or CET	147	30.2
	University Degree	340	69.8
Professional status	Employed	270	55.4
	Unemployed	24	4.9
	Student	134	27.5
	Retired	21	4.3
	Working-student	38	7.8

3.2. Attitudes and Perceptions Regarding Plastic Food Packaging

The majority of participants were responsible for buying the foods they consume (n = 338), while some only buy their food sometimes (n = 132). In the case of 17 participants, someone else buys their food.

At the moment of purchasing, participants tend to think about the negative impact of the plastic package on the environment, as seen in Figure 1. However, only 30% try to look for alternatives.

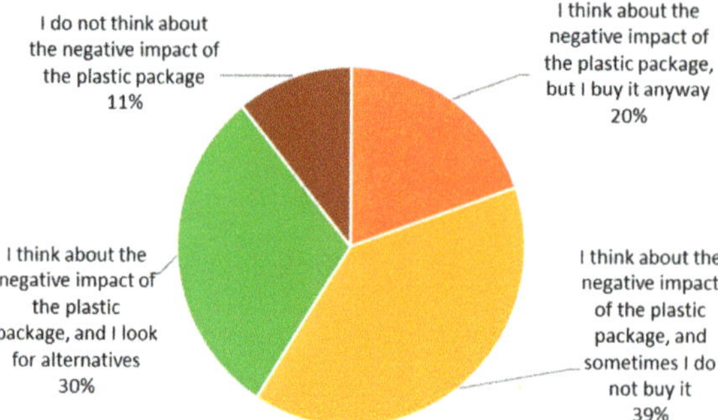

Figure 1. Thoughts regarding the impact of plastic food packaging at the moment of purchase.

The research also included a question aimed to evaluate how each participant classified on a scale from 1 (no impact) to 10 (maximum impact) concerning the negative impact of plastics on different elements of the environment, and the results are presented in Figure 2. In general, the participants classified the impacts into levels 7 to 10. The maximum impact (corresponding to score 10) was always the score that got the most answers, with particular relevance for the impact of plastics on the seas/oceans (attributed by 344 participants) followed by rivers (n = 298) and ecosystems (n = 292). The negative impact of plastics is a little less perceived on agricultural soils or forests than other elements of the environment.

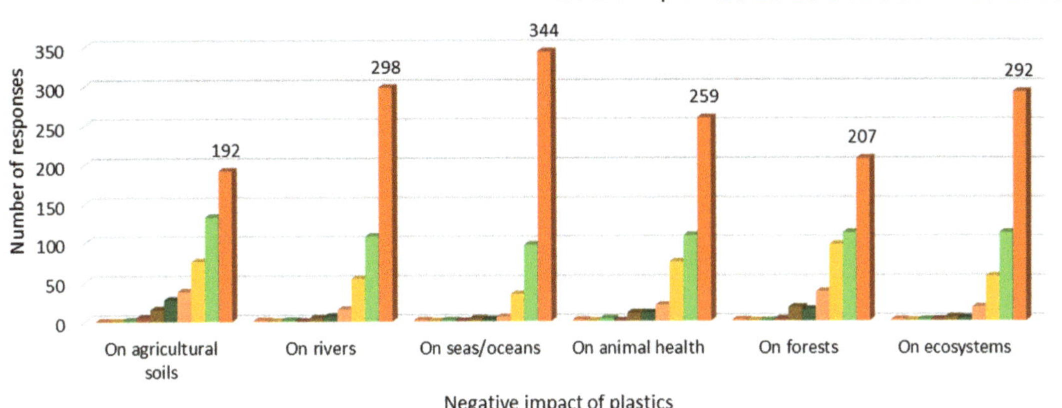

Figure 2. Perceived negative impact of plastics on the environment.

The scores given by the participants were used to calculate the indices that measure the perceived negative impact of plastic on the environment. These were calculated as the mean value and are, by decreasing order: impact on seas/oceans (9.55 ± 0.93), impact on rivers (9.34 ± 1.09), impact on ecosystems (9.32 ± 1.09), impact on animals (9.10 ± 1.09), impact on forests (8.81 ± 1.40) and impact on agricultural soils (8.73 ± 1.46), all measured in the scale from 1 (minimum impact) to 10 (maximum impact).

Table 2 presents some of the aspects investigated relating to the measures adopted by the participants to minimize the harmful impacts of plastic materials and their recycling. A very expressive majority of the participants separate the plastic residues for recycling (87%), avoid plastic utensils and reduce the use of plastic bags (81% for both options). However, when it comes to avoiding purchasing products with excessive plastic, only about half of the participants adopt this measure (55%). The results in Table 2 also show that most participants recycle as a usual practice (74%), and 59% admit doing it always. Most participants have containers for recycling in their homes (81%), and they are located essentially in the kitchen (60%). In comparison, at the working place, less participants have recycling containers (62%), and these are situated in the corridors (31%) or in the bar (29%). Public containers for recycling, particularly plastic, are usually present near the workplace (52%) and the house (87%) of the participants. On average, in a week, the participants deposit waste into public bins only once (for 55% of the participants), and regarding the plastic residues, they produce one bag per week (50%) or two to five bags (44% of participants).

Table 2. Attitudes towards minimising the harmful impacts of plastic and recycling.

Measures Adopted to Minimize the Harmful Impact of Plastics	n	%
Separate plastic residues for recycling	422	87%
Avoid consume plastic utensils, such as dishes, straws, glasses, silverware	393	81%
Reduce the use of plastic bags	393	81%
Use cloth bags to transport foods and other goods	299	61%
Avoid purchasing products with excessive plastic	266	55%
Recycling of Plastic Materials	**n**	**%**
Do you usually recycle?		
Yes	362	74%
No	30	6%
Sometimes	95	20%

Table 2. *Cont.*

Recycling of Plastic Materials	n	%
Do you usually select and separate plastic food packages for recycling?		
Never	5	1%
Sometimes	54	12%
Frequently	128	28%
Always	270	59%
In your house do you have containers for recycling?		
Yes	370	81%
No	87	19%
Where are the containers, in your house?		
In the kitchen	294	60%
In the attic	1	0%
In the basement	7	1%
In the garage	30	6%
Other	65	13%
On average, per week, how many times do you deposit residues in the public waste bin?		
Once	247	55%
2 times	125	28%
3–4 times	53	12%
More than 4 times	28	6%
On average, per week, how many bags full of plastic residues do you produce?		
One bag	226	50%
2–5 bags	212	47%
6–10 bags	13	3%
More than 10 bags	3	1%
In your area of residence, which type of public containers do you have to deposit residues?		
Common waste	452	93%
Plastic	425	87%
Glass	425	87%
Paper	422	87%
Oil	148	30%
Batteries	129	26%
In your workplace do you have containers for recycling?		
Yes	282	62%
No	175	38%
Where are the containers, in your workplace?		
In the office	60	12%
In the meeting rooms	15	3%
In the bar	139	29%
In the copies room	32	7%
In the corridors	150	31%
Others	124	25%
In your area of work which type of public containers do you have to deposit residues?		
Common waste	377	77%
Plastic	252	52%
Glass	192	39%
Paper	282	58%
Oil	37	8%
Batteries	82	17%

3.3. Information and Knowledge about Recycling

Practically, all participants (n = 478) refer that they believe school should have a more important role in the awareness about the harmful effects of plastics as well as about the recycling practices. They also believe that some aspects should be addressed at schools, like those which were presented to the respondents:

- Right attitudes about recycling food packaging (selected by 455 participants)
- Behaviours to have when using food packaging (n = 448)
- Packaging constituents and their decomposition (n = 405)
- Forms of degradation of plastic packaging in soil and water (n = 431)
- Risks to public health due to inappropriate recycling practices (n = 466)

Other topics also referred to by some participants on open question include: sustainable alternatives to the use of plastic; the various options available on the market regarding plastic replacements; schools setting examples; the problem of micro- and nano-plastics and their influence on human health and ecosystems; citizenship and responsibility in eco-sustainability; the principle of respect for ourselves and others; the awareness of the importance of reduction and recovery, rather than encouraging recycling; circular economy and in particular of plastic packaging; ways to replace plastic in the consumer society and what this transition would represent, opportunities and challenges, global impact on all ecosystems, on planet earth and future generations; the importance of avoiding endocrine disrupting plastics like bisphenol A (BPA); plastic particles, for example, released when the package is heated, and which can contaminate food.

Moreover, 99% of the participants (n = 481) agree that schools should have a more critical role in teaching about the sustainability of natural resources. They believe that the most appropriate means to receive information/alerts on good recycling practices are social media (405 participants agree on this) followed by email (n = 252) or text messages on the mobile phone (n = 191).

The frequency with which the participants obtain information about recycling through several ways is indicated in Figure 3. The results reveal that only the internet is referred to as a frequent source of information for a relevant percentage of participants (about 30%). In contrast, the other forms of obtaining information are used only sporadically or sometimes.

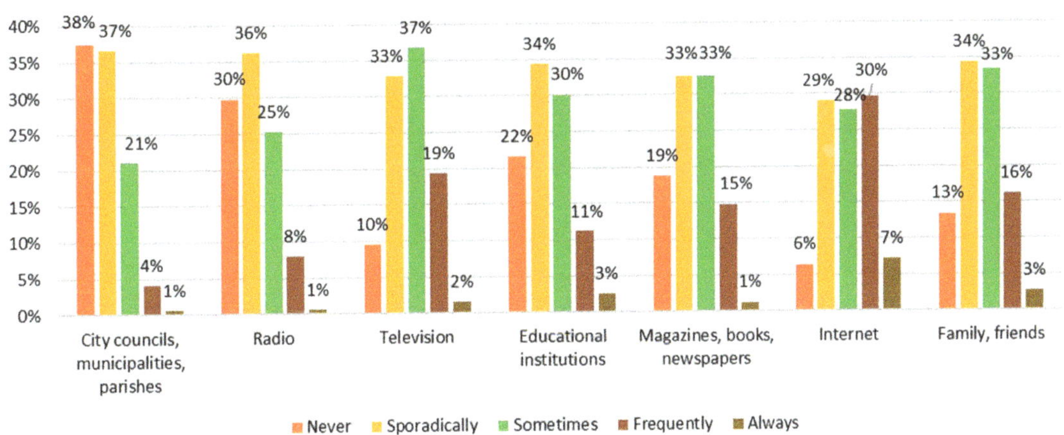

Figure 3. Frequency of obtaining information about recycling.

The knowledge about some facts related to recycling and plastics was assessed through a number of statements and the participants were asked to indicate their level of agreement on a five-point Likert scale from totally disagree (1) to totally agree (5) (Table 3). Most participants agreed (40.9%) or totally agreed (40.9%) with item 1, about the use of recycled materials. Regarding the fact that recycling of packaging materials originates new raw materials (item 2) 45.6% agreed, and 22.2% totally agreed with it. Similar percentages were obtained for item 3, which refers to the operations that plastic undergoes when being recycled. The lower impact of glass over plastic was also acknowledged by many participants (40.2% agree and 27.1% totally agree). Item 7 was a false statement included to evaluate if the participants could distinguish this false fact, and, although there were still many participants revealing an incorrect agreement with the item, nearly 30% totally disagreed and about 13% disagreed, which indicates that nearly half of the participants had a proper knowledge of the fact that not all glass materials are placed into the green recycling bin. The last two items were more difficult for the participants to express an opinion, as high percentages of participants did not express an opinion (56.3% and 61.6%, respectively, replied neither agree nor disagree).

Table 3. Knowledge about recycling.

Items	Totally Disagree (1)	Disagree (2)	Neither Agree Nor Disagree (3)	Agree (4)	Totally Agree (5)
1. It is better for the environment to buy products with packaging made from recycled materials	1.0%	2.3%	15.0%	40.9%	40.9%
2. If all plastic packaging is recycled, we will have new raw materials again	2.1%	11.9%	18.3%	45.6%	22.2%
3. When plastic arrives at the sorting stations, it is washed, crushed and processed, transforming itself, and giving rise to urban furniture, clothing, tubes, vases, etc ...	1.4%	5.1%	21.8%	47.2%	24.4%
4. When going to the supermarket and the same product is available in glass and plastic packaging, it is better to choose the glass one in view of the comparative impact of these materials	3.9%	8.6%	20.1%	40.2%	27.1%
5. Broken dishes and glasses must be placed in the green recycling bin (false statement)	29.8%	12.9%	12.9%	23.2%	21.1%
6. Portugal in 2019 met the plastic recycling target	11.1%	18.9%	56.3%	9.9%	3.9%
7. In just over two decades, Portugal has separated and sent for recycling more than 7 million tons of packaging waste	3.7%	5.1%	61.6%	23.0%	6.6%

3.4. Influence of Sociodemographic Factors on Eco-Responsible Behaviour towards Plastic Package

Table 4 presents the cross-tabulation between the sociodemographic variables measuring the attitudes of the participants regarding plastic packages. Concerning the question of whether the participants think about the impact of the plastic package in the moment of purchase, significant differences were found between groups for sex and education (p-value of 0.013 and 0.010, respectively). Still, the associations were weak in both cases ($V = 0.152$ and $V = 0.156$). The participants who tried to adapt their purchases more according to the minimization of the negative impacts of plastic were women with an under-university level of education.

Concerning the separation of residues for recycling, significant differences were found between groups for age and profession (p-value < 0.0005 in both cases), with moderate associations ($V = 0.263$ and $V = 0.241$). The separation of residues increased as age increased, and those who recycled more were retired (100%) or employed (92.6%).

Table 4. Association between sociodemographic variables and attitudes towards sustainability of plastics.

Attitudes	Sex [1]		Age [1]			Region [1]			Education [1]		Professional Status [1]				
	Female	Male	Young Adults	Middle-Aged Adults	Senior Adults	North	Centre	South & Islands	Under-University	University Degree	Employed	Unemployed	Student	Retired	Working-Student
Reflects on impact of plastic package in moment of purchase:															
Yes but buy anyway	18.7%	21.9%	20.5%	15.6%	24.3%	19.1%	19.6%	20.0%	16.5%	20.8%	18.4%	13.0%	24.2%	15.0%	19.4%
Yes, sometimes does not buy	42.2%	32.8%	38.4%	42.2%	36.9%	47.1%	36.6%	43.2%	30.8%	42.9%	41.7%	52.5%	35.5%	30.0%	33.3%
Yes, finds alternatives	31.0%	27.7%	27.0%	30.1%	35.1%	20.6%	32.0%	30.5%	36.8%	27.4%	29.3%	21.7%	28.2%	50.0%	36.1%
No	8.1%	17.5%	14.1%	12.1%	3.6%	13.2%	11.8%	6.3%	15.8%	8.9%	10.5%	13.0%	12.1%	5.0%	11.1%
p-value [2]	0.013			0.058			0.341		0.010				0.654		
Cramer's coefficient, *V*	0.152			0.114			0.085		0.156				0.082		
Separate residues to recycle:															
Yes	87.2%	85.4%	76.0%	93.1%	95.6%	78.7%	87.3%	90.6%	83.7%	87.9%	92.6%	83.3%	75.4%	100.0%	78.9%
No	12.8%	14.6%	24.0%	6.9%	4.4%	21.3%	12.7%	9.4%	16.3%	12.1%	7.4%	16.7%	24.6%	0.0%	21.1%
p-value [2]	0.603			<0.0005			0.062		0.204				< 0.0005		
Cramer's coefficient, *V*	0.024			0.263			0.107		0.058				0.241		
Avoid plastic utensils:															
Yes	84.0%	72.9%	78.0%	78.0%	89.5%	80.0%	80.4%	82.3%	79.6%	81.2%	78.1%	87.5%	82.8%	100.0%	76.3%
No	16.0%	27.1%	22.0%	22.0%	10.5%	20.0%	19.6%	17.7%	20.4%	18.8%	21.9%	12.5%	17.2%	0.0%	23.7%
p-value [2]	0.005			0.025			0.905		0.684				0.102		
Cramer's coefficient, *V*	0.128			0.123			0.020		0.018				0.128		
Avoid products with excessive plastic:															
Yes	57.4%	47.9%	49.0%	53.8%	65.8%	50.7%	52.8%	63.5%	48.3%	57.4%	58.1%	54.2%	47.8%	71.4%	44.7%
No	42.6%	52.1%	51.0%	46.2%	34.2%	49.3%	47.2%	36.5%	51.7%	42.6%	41.9%	45.8%	52.2%	28.6%	55.3%
p-value [2]	0.054			0.015			0.138		0.065				0.099		
Cramer's coefficient, *V*	0.087			0.131			0.090		0.083				0.126		
Reduce the use of plastic bags:															
Yes	80.5%	81.3%	79.0%	80.3%	84.2%	89.3%	76.6%	87.5%	74.8%	83.2%	83.0%	83.3%	76.1%	76.2%	81.6%
No	19.5%	18.8%	21.0%	19.7%	15.8%	10.7%	23.4%	12.5%	25.2%	16.8%	17.0%	16.7%	23.9%	23.8%	18.4%
p-value [2]	0.842			0.525			0.007		0.031				0.542		
Cramer's coefficient, *V*	0.009			0.051			0.142		0.098				0.080		

Table 4. *Cont.*

Attitudes	Sex [1]		Age [1]			Region [1]			Education [1]		Professional Status [1]				
	Female	Male	Young Adults	Middle-Aged Adults	Senior Adults	North	Centre	South & Islands	Under-University	University Degree	Employed	Unemployed	Student	Retired	Working-Student
Use cloth bags to transport foods and other goods:															
Yes	63.8%	55.6%	64.0%	61.8%	56.1%	57.3%	59.5%	70.8%	63.9%	60.3%	58.1%	41.7%	67.9%	71.4%	68.4%
No	36.2%	44.4%	36.0%	38.2%	43.9%	42.7%	40.5%	29.2%	36.1%	39.7%	41.9%	58.3%	32.1%	28.6%	31.6%
p-value [2]	0.086			0.384			0.100		0.447				0.056		
Cramer's coefficient, *V*	0.078			0.063			0.097		0.034				0.138		
Frequency of separation of plastic for recycling:															
Never	1.2%	0.8%	0.6%	2.4%	0.0%	1.5%	1.3%	0.0%	0.0%	1.6%	1.5%	0.0%	0.8%	0.0%	0.0%
Sometimes	12.3%	10.5%	20.2%	6.6%	6.3%	23.5%	11.7%	3.3%	17.6%	9.3%	8.0%	17.4%	21.0%	4.8%	9.1%
Frequently	25.9%	33.1%	30.9%	29.3%	21.4%	19.1%	29.4%	30.0%	35.3%	24.9%	27.6%	34.8%	29.4%	19.0%	27.3%
Always	60.5%	55.6%	48.3%	61.7%	72.3%	55.9%	57.5%	66.7%	47.1%	64.2%	62.8%	47.8%	48.7%	76.2%	63.6%
p-value [2]	0.464			<0.0005			0.006		0.001				0.063		
Cramer's coefficient, *V*	0.075			0.184			0.140		0.188				0.121		

[1] Percentages in Column. [2] Chi-square test *p*-value at a level of significance of 5%.

The avoidance of plastic utensils is stronger for women and for senior adults, with significant differences between groups ($p = 0.005$ and $p = 0.025$, respectively). However, the associations in both cases are weak ($V = 0.128$ and $V = 0.123$, respectively). Regarding avoidance of products with excessive plastic, only significant differences were found for groups of age ($p = 0.015$), with a higher avoidance rate for increasing age, although this association is weak ($V = 0.131$).

Reducing the use of plastic bags is less prevalent for participants from the central region of Portugal, with significant differences and a low association ($p = 0.007$ and $V = 0.142$), while it is significantly more prevalent in people with a university degree ($p = 0.031$ and $V = 0.098$). On the other hand, no significant differences were found for any of the sociodemographic variables studied for the use of cloth bags to transport foods and other goods.

Finally, concerning the frequency of separation of plastic for recycling, significant differences were found between groups of age, region and education level ($p < 0.0005$, $p = 0.006$ and $p = 0.001$, respectively), but the associations were weak in all cases ($V = 0.184$, $V = 0.140$ and $V = 0.188$).

3.5. Influence of Sociodemographic and Behavioural Factors on Perceptions of the Impact of Plastics and Knowledge about Recycling

For each participant, the six variables accounting for the negative impact of plastics (soils, rivers, oceans, animals, forest, ecosystems) were used to calculate an average perception of the negative impact of plastics on the environment and possible significant differences between groups of sociodemographic variables were tested as shown in Table 5. Only for sex was there found significant differences ($p = 0.018$), with women revealing a higher level of perception about the negative impact of plastics (9.21 ± 1.00) as compared with men (8.97 ± 1.10).

Table 5. Perception of the negative impact of plastics according to sociodemographic groups.

Variable	Group	Perception of the Negative Impact of Plastics on the Environment [1]
Sex	Female	9.21 ± 1.00
	Male	8.97 ± 1.10
	p-value [2]	0.018
Age	Young adults	9.25 ± 0.88 [a]
	Middle-aged adults	9.03 ± 1.01 [a]
	Senior adults	9.10 ± 1.28 [a]
	p-value [3]	0.137
Residence	North	9.18 ± 0.97 [a]
	Centre	9.20 ± 1.03 [a]
	South & Islands	8.91 ± 1.07 [a]
	p-value [3]	0.093
Education level	Under-university	9.26 ± 1.08
	University Degree	9.09 ± 1.01
	p-value [2]	0.088
Professional status	Employed	9.09 ± 1.12 [a]
	Unemployed	8.57 ± 1.09 [a]
	Student	9.26 ± 0.90 [a]
	Retired	9.18 ± 0.77 [a]
	Working-student	9.14 ± 0.99 [a]
	p-value [3]	0.866
Global sample		9.14 ± 1.03

[1] Mean value $\pm$ standard deviation (scale from 1—minimum impact to 10—maximum impact). [2] T-test for independent samples, level of significance of 5%. [3] ANOVA with Post-Hoc Tukey test, level of significance of 5%. Values with the same letter are not significantly different.

The seven items used to measure knowledge about recycling were used, after reversing the negative item (number 5), to assess an average level of knowledge for each participant, computed as the mean value, varying in the scale from −2 to +2. These values were then categorized into: very low knowledge—mean ∈[−2;−1[, low knowledge—mean ∈[−1;0 [, high knowledge—mean ∈[0;1[and very high knowledge—mean ∈[1;2], and this variable was used for the tree classification considering the sociodemographic variables studied, as shown in Figure 4. The obtained tree is five levels deep, with 21 nodes, from which 11 are terminal. The risk estimates were 0.267 for resubstitution and cross-validation, with standard errors of 0.020 in both cases. According to the results obtained, the first discriminant variable was profession, separating people employed from those with other job situations. Among the employed, the percentage of participants with a very high knowledge was higher (23.3%) than for other groups. For the employed, the next discriminant was residence, with those living in the North showing more people with a very high level of knowledge. For participants living in the centre and the South and Islands, the following discriminant variable was age, and for the next level, age separated again the young adults from the middle-aged adults (lower percentage of very high knowledge, 24.2%). Sex was the final discriminant for this group, for which women showed a higher percentage in the category of very high knowledge (26.6%).

For participants with other professional status than the employed, age was the second discriminating variable, separating young adults for which the percentage of very high knowledge was lower (10.2%) from the middle aged or senior adults. The discriminating variable for the next level in these two groups was education, and for the young adults with a level of education under university, the last discriminant was sex, with women showing a higher percentage for very high knowledge (9.1%).

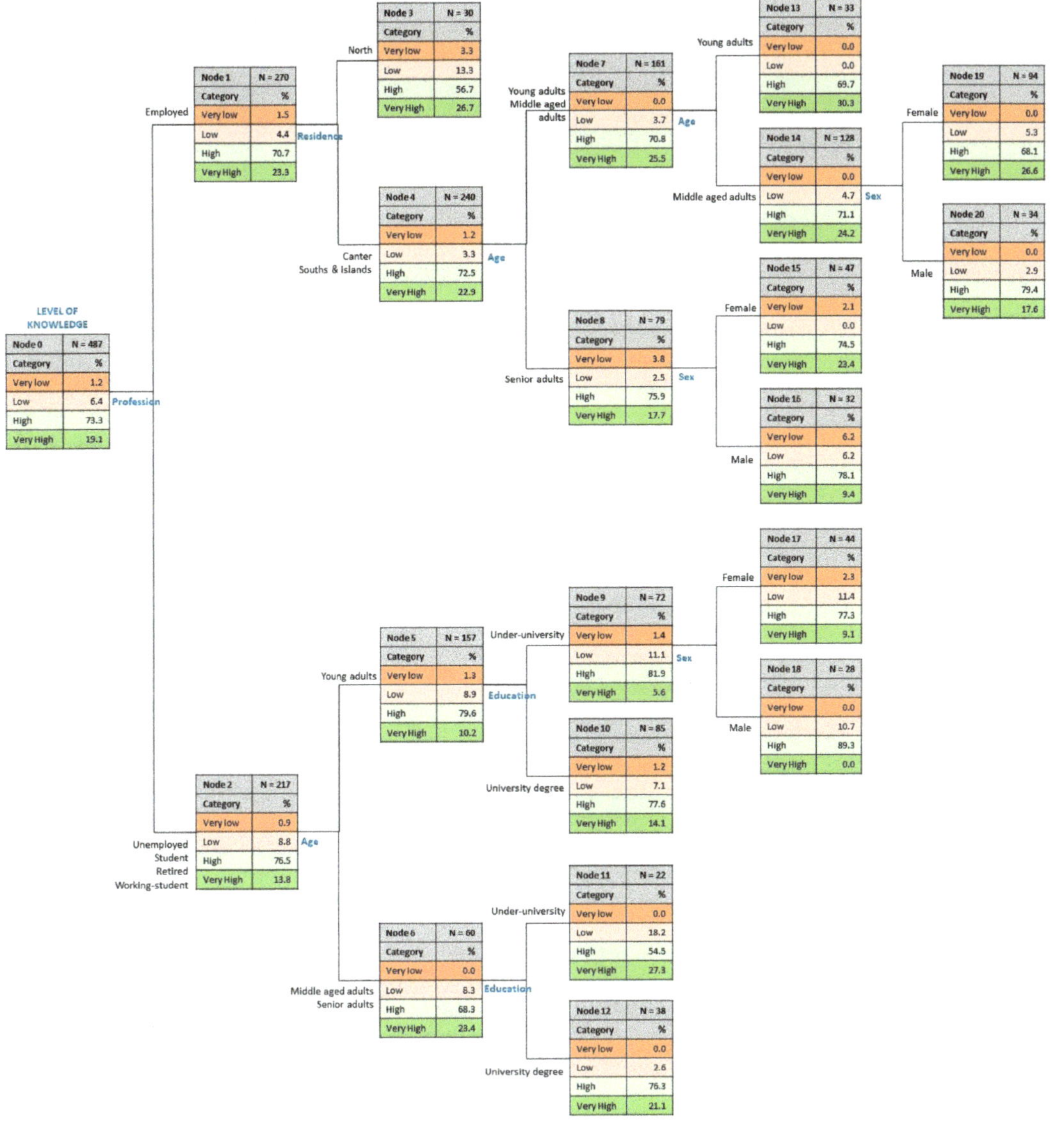

Figure 4. Tree classification for variable level of knowledge about recycling.

4. Discussion

Understanding the different perceptions of the public can allow government authorities to make informed decisions about funding and management priorities, promoting cooperation between society, institutions, and governments [37]. Therefore, knowing the consumers' awareness about the use of plastics and their effects on human health and for the environment can be a trigger for governmental authorities, as well as for industries, to actively promote the shift towards more sustainable packaging systems. Plastics are a part of many items present in our daily lives in many sectors, but packaging is one of the areas that highly contribute to the use of plastics, and in many cases single-use plastics. The increase in the use of plastic in various sectors has caused concern regarding the usage of natural resources for its production, the toxicity associated with its manufacture and use, and the environmental impacts generated by its disposal [20]. To positively contribute to sustainability, packages should be made from environmentally adequate sources, applying clean production technologies with the possibility of being recovered or recycled after being used. Sustainability also depends on consumers, and if the product is not correctly discharged, the sustainability is compromised [2].

The recycling of plastic packaging worries society due to the growing use of these materials and the environmental implications inherent to their non-rational post-consumer disposal [23]. In this work, it was observed that this concern is present since the great majority of participants practice recycling. A similar result was found by Forleo and Romagnoli [38] in Italy, with 87% of respondents always following the separate disposal of plastics. Several factors may have an impact on waste disposal and recycling [5]. Among these stand, for example, the presence of other materials (combined packaging), labels, dirt, damage, or food residues, which remain in post-consumer packaging [23]. Moreover, the economic feasibility of recycling, including the costs of collecting, separating, cleaning or reprocessing and transporting waste [5], highly influences the recycling of plastic materials. In this study, most participants agreed or totally agreed with the use of recycled materials, and demonstrated a good knowledge about recycling, and the higher negative impact of plastics over glass. Plastic is known as the most difficult household waste to degrade. Its degradation releases toxic residues that pollute soil, air, and water [39]. However, people are aware of the negative impact of plastic waste on the environment, and this study confirmed it, with the oceans/seas as the natural sites of greatest concern.

Problems related to its use in food packaging often result from the release of non-plastic components. When exposed to high temperatures, some plastics decompose or oxidize, producing low molecular weight substances that can be toxic. Another problem is related to the ingestion of nano, micro, or macro plastics by animals. Thousands of plastic bags are ingested by animals annually. A study of blue petrel chicks on South Africa found that 90% of them had plastic in their stomachs [20]. These facts contribute to a higher perception in society about the adverse effects of plastics in the oceans/seas. In this sense, measures have been adopted to reduce plastic consumption. China has restricted the use of plastic bags in retail since 2008 and a similar policy was implemented in Malaysia in 2011, in England in 2015 and in Indonesia in 2016 [39]. In Portugal, the plastic bags to carry groceries and other goods were free before, but presently are only provided against payment, encouraging the utilization of reusable bags, and the customers need to bring their own bags or containers. Moreover, in restaurants are prohibited the use of any plastic disposable utensils [40]. This methodology was implemented in Portugal some years ago as a preparatory way for the limitations that the European Union would demand following the regulations approved in 2019, according to which there would be a measurable reduction in the consumption of single-use plastic products in the EU until 2026 [41]. In Portugal, the decrease in the use of disposable plastics in the restoration was expected to start in 2020, before the deadlines established by the European directive. However, due to the Covid-19 pandemic, the measure was postponed. Decree-Law No. 22-A/2021 [42] "is postponed to 1 July 2021 the obligation of catering and beverage service providers to adapt to the provisions of Law no. 76/2019" [43], which determines "the non-use and non-availability of single-use plastic tableware, referring to "activities in the restaurant and/or beverage sector and in the retail trade". As an alternative to disposable plastic, the law defines that "reusable utensils must be used, or, alternatively, utensils made of biodegradable material". Uganda and South Africa have also banned single-use plastic bags. Other countries such as Kenya are considering implementing taxes on plastic bags, or even banning their use [20]. In the current study, 81% of the interviewees committed to avoiding plastic utensils and to reduce their use of plastic bags. The specific recycling rate for plastic packaging in Portugal reached 44% in 2018, surpassing the European targets, which stood at 22.5%. The collection of these packages, which are mostly placed in the yellow recycling bin, totalled 72,000 tons in 2018. In the first half of 2019, there was a 5% increase in the amount of plastic packaging waste sent for recycling. During this period, around 30 thousand tons of plastic were collected in the yellow recycling bin [44]. According to the Portuguese organism for recycling *Sociedade Ponto Verde* (Green Point society), plastic will continue to be part of the consumption cycle, so it is important that all agents have an active contribution in terms of the circularity, sustainability and recyclability of this material. Hence, their compromise is to promote development, knowledge and innovation, investing in valuing and promoting gains from an economic, environmental and positive reputation point of view of a brand, product or company. It is envisaged that Portugal will continue to meet the targets set by the European Union, which stand at 50% in 2025 and 55% in 2030. It is important to emphasize that this requires a joint commitment from all of society, including the citizens, the Government, national and local entities, the industry and the academic community [44].

Six months after implementing the charge for plastic bags in England, it was verified that the number of disposable plastic bags used dropped by more than 85%, around 500 million units. Likewise, there was an increase in the awareness of the environmental impact of household plastic waste and the population's support for the issue [24]. Studies carried out reveal that the reduction policy is effective, instigating the consumer to avoid the use of plastic bags across 52.3% a year [39]. In the current survey, the majority of participants admit to generating one bag of plastic waste per week, or two to five, which represents a great volume of residues.

A survey on marine pollution carried out in Greece by Gkargkavouzi et al. [37] indicated that, in general, respondents showed positive attitudes and a moderate knowledge about the theme of marine pollution and that they value the marine environment due to the ecosystem services provided. Among the main threats identified, garbage and industries were considered the most important, followed by fishing and agriculture. A study carried out in Italy by Forleo and Romagnoli [38] identified a low involvement of people regarding changes in their purchasing behavior to reduce the amount of plastic packaging. When the people were asked if, in the period of six months before the questionnaire was applied, they had adopted purchasing choices aimed at preventing the use of plastic waste, only 16% stated that it reduced a lot, and 24% slightly reduced the purchase. Changing plastic consumption habits has not been an easy task, as it directly depends on the change in the way individuals consume [38]. This could also be verified by the present survey, in which the participants tended to think about the negative impact of the plastic package on the environment. Therefore, the willingness to adopt plastic waste reduction should be strengthened and stimulated, especially among those individuals who are not at all committed or are not often aware of their purchasing and waste behavior. Because of this, research has been carried out to investigate bioplastics, which are polymers from renewable and/or biodegradable resources [45–48]. Degradable biopolymers are an alternative to traditional plastics, especially when recycling is not economically viable, or when the environmental impact must be minimized [49]. Bioplastics can be defined as plastics based on renewable resources, or plastics that are biodegradable and/or compostable. The use of bioplastics as food packaging materials has limitations, such as higher prices compared to conventional plastics and concerns about availability as well as land for its production [50].

Consumers are increasingly concerned about the safety offered by the products which they consume, such as food, water, health-related products such as medicines and other goods used in everyday life [20]. Lavelle-Hill et al. [24] verified that people more concerned with environmental issues are currently younger, female, have more money and a higher education. They found that young adults are more concerned with the environment, but older adults adopt more pro-environmental behaviors. Therefore, specific actions such as purchasing plastic bags may be less motivated by environmental factors and more by economic ones [24]. Nevertheless, this ecological conscientiousness may help increase the adoption of alternative biodegradable materials similar to plastics and bioplastics, many of them obtained from industrial agro-food wastes, as a replacement of traditional plastic materials [14,51]. Social awareness, education and public pressure play key roles in shaping and encouraging consumer behavioral changes towards a more environmentally friendly responsibility. Nevertheless, correct habits involve more than just motivation, but also self-discipline and a belief in the positive impact of behavior change [24].

5. Conclusions

This work investigates the habits related to food packaging in a sample of Portuguese citizens and their knowledge and concerns about the use of plastics. Regarding research question 1 (RQ1), it was observed that people are more aware of the environmental issues, with 89% confirming they think about the negative impact of plastic packages. Regarding this, the main concern relates to the impact on seas/oceans (maximum score attributed by 344 participants).

Concerning the RQ2, it was concluded that consumers have a conscience that recycling is a means to reduce environmental pollution and promote the sustainability of the packaging chain. In the same way, they are getting more informed and having a better attitude towards it, with 87% separating plastic materials for recycling. Most of the interviewees had concerns about the use of plastic packaging, and 55% are trying to change their habits so as to avoid the use of plastics in this context. Additionally, they know how to separate waste types so that they can be efficiently recycled and have been doing this where possible.

Regarding RQ3, it was concluded that participants know very well about aspects such as the lower environmental impact of recycled materials or the way materials are handled for recycling, but are less informed about the Portuguese effectiveness in meeting recycling targets.

Concerning RQ4, it was concluded that there are significant differences between women and men for the thought about the impact of plastic at the moment of purchase, for separating residues to recycle, for the reduction in plastic bags' usage, as well as for the perception of the negative impact of plastics on the environment. Differences between groups for other sociodemographic variables were, in general, not significant.

In this way, it is expected that society will increasingly move towards sustainable habits, questioning its actions and the impact they have on the environment. To measure this evolution, this study and other related studies might be implemented as a follow-up strategy (through longitudinal studies) to evaluate the real impact of the legislation presently available to minimize the use of plastics. It might also be important to replicate this study in other countries and compare results.

Author Contributions: Conceptualization, R.P.F.G.; methodology, R.P.F.G.; software, R.P.F.G.; validation, R.P.F.G.; formal analysis, R.P.F.G.; investigation, R.C., L.P.C.-L. and R.P.F.G.; resources, L.P.C.-L. and R.P.F.G.; data curation, R.P.F.G.; writing—original draft preparation, M.W.M., L.P.C.-L. and R.P.F.G.; writing—review and editing, L.P.C.-L. and R.P.F.G.; visualization, R.P.F.G.; supervision, R.P.F.G.; project administration, R.P.F.G.; funding acquisition, L.P.C.-L. and R.P.F.G. All authors have read and agreed to the published version of the manuscript.

Funding: This research was funded by FCT—Foundation for Science and Technology, I.P., Portugal, within the scope of the project Ref. UIDB/00681/2020. The APC was funded by FCT—Ref. UIDB/00681/2020.

Institutional Review Board Statement: The study was conducted according to the guidelines of the Declaration of Helsinki, and approved by the Ethics Committee of Polytechnic Institute of Viseu (reference 09/SUB/2021).

Informed Consent Statement: Informed consent was obtained from all subjects involved in the study.

Data Availability Statement: The data are available from the author Raquel Guiné, upon request.

Acknowledgments: This work is funded by National Funds through the FCT—Foundation for Science and Technology, I.P., within the scope of the project Ref. UIDB/00681/2020. Furthermore, we would like to thank the CERNAS Research Centre and the Polytechnic Institutes of Viseu and Coimbra for their support.

Conflicts of Interest: The authors declare no conflict of interest.

References

1. Sundqvist-Andberg, H.; Åkerman, M. Sustainability governance and contested plastic food packaging–An integrative review. *J. Clean. Prod.* **2021**, *306*, 127111. [CrossRef]
2. Landim, A.P.M.; Bernardo, C.O.; Martins, I.B.A.; Francisco, M.R.; Santos, M.B.; De Melo, N.R. Sustentabilidade quanto às embalagens de alimentos no Brasil. *Polímeros* **2016**, *26*, 82–92. [CrossRef]
3. Aggarwal, A.; Langowski, H.-C. Packaging functions and their role in technical development of food packaging systems: Functional equivalence in yoghurt packaging. *Procedia CIRP* **2020**, *90*, 405–410. [CrossRef]
4. De Souza, L.B.; Moura, A.A.C. Embalagens para alimentos: Tendências e inovações. *Hig. Aliment* **2017**, *5*, 25–29.
5. Marsh, K.; Bugusu, B. Food packaging? Roles, materials, and environmental issues. *J. Food Sci.* **2007**, *72*, R39–R55. [CrossRef]
6. Farmer, N. 6-Packaging and marketing. In *Packaging Technology*; Emblem, A., Emblem, H., Eds.; Woodhead Publishing: Cambridge, England, 2012; pp. 87–105.
7. Balzarotti, S.; Maviglia, B.; Biassoni, F.; Ciceri, M.R. Glass Vs. Plastic: Affective judgments of food packages after visual and haptic exploration. *Procedia Manuf.* **2015**, *3*, 2251–2258. [CrossRef]
8. Kim, H.J.; Kim, S.J.; An, D.S.; Lee, D.S. Monitoring and modelling of headspace-gas concentration changes for shelf life control of a glass packaged perishable food. *LWT Food Sci. Technol.* **2013**, *55*, 685–689. [CrossRef]
9. Vaireanu, D.-I.; Cojocaru, A.; Maior, I.; Ciobotaru, I.-A. Food packaging interactions in metal cans. In *Reference Module in Food Science*; Elsevier: Amsterdam, The Netherlands, 2018.
10. Pasias, I.N.; Raptopoulou, K.G.; Proestos, C. Migration from metal packaging into food. In *Reference Module in Food Science*; Elsevier: Amsterdam, The Netherlands, 2018.
11. Sapozhnikova, Y. Non-targeted screening of chemicals migrating from paper-based food packaging by GC-Orbitrap mass spectrometry. *Talanta* **2021**, *226*, 122120. [CrossRef]
12. He, Y.; Li, H.; Fei, X.; Peng, L. Carboxymethyl cellulose/cellulose nanocrystals immobilized silver nanoparticles as an effective coating to improve barrier and antibacterial properties of paper for food packaging applications. *Carbohydr. Polym.* **2020**, *252*, 117156. [CrossRef]
13. Testa, F.; Di Iorio, V.; Cerri, J.; Pretner, G. Five shades of plastic in food: Which potentially circular packaging solutions are Italian consumers more sensitive to. *Resour. Conserv. Recycl.* **2021**, *173*, 105726. [CrossRef]
14. Montoille, L.; Vicencio, C.M.; Fontalba, D.; Ortiz, J.A.; Moreno-Serna, V.; Peponi, L.; Matiacevich, S.; Zapata, P.A. Study of the effect of the addition of plasticizers on the physical properties of biodegradable films based on kefiran for potential application as food packaging. *Food Chem.* **2021**, *360*, 129966. [CrossRef]
15. Kwon, C.W.; Chang, P.-S. Influence of alkyl chain length on the action of acetylated monoglycerides as plasticizers for poly (vinyl chloride) food packaging film. *Food Packag. Shelf Life* **2021**, *27*, 100619. [CrossRef]
16. Van Hai, L.; Muthoka, R.M.; Panicker, P.S.; Agumba, D.O.; Pham, H.D.; Kim, J. All-biobased transparent-wood: A new approach and its environmental-friendly packaging application. *Carbohydr. Polym.* **2021**, *264*, 118012. [CrossRef] [PubMed]
17. Hossain, R.; Tajvidi, M.; Bousfield, D.; Gardner, D.J. Multi-layer oil-resistant food serving containers made using cellulose nanofiber coated wood flour composites. *Carbohydr. Polym.* **2021**, *267*, 118221. [CrossRef]
18. Wyrwa, J.; Barska, A. Innovations in the food packaging market: Active packaging. *Eur. Food Res. Technol.* **2017**, *243*, 1681–1692. [CrossRef]
19. Jalil, A.; Mian, N.; Rahman, M.K. Using plastic bags and its damaging impact on environment and agriculture: An alternative proposal. *Int. J. Learn. Dev.* **2013**, *3*, 1–14. [CrossRef]
20. Bashir, N.H.H. Plastic problem in Africa. *Jpn. J. Vet. Res.* **2013**, *61* (Suppl. 2013), S1–S11.
21. Rodrigues, M.; Abrantes, N.; Gonçalves, F.; Nogueira, H.S.; Marques, J.; Gonçalves, A.M. Impacts of plastic products used in daily life on the environment and human health: What is known? *Environ. Toxicol. Pharmacol.* **2019**, *72*, 103239. [CrossRef]
22. Henningsson, S.; Hyde, K.; Smith, A.; Campbell, M. The value of resource efficiency in the food industry: A waste minimisation project in East Anglia, UK. *J. Clean. Prod.* **2004**, *12*, 505–512. [CrossRef]
23. Forlin, F.J.; Faria, J.D.A.F. Considerações sobre a reciclagem de embalagens plásticas. *Polímeros* **2002**, *12*, 1–10. [CrossRef]
24. Lavelle-Hill, R.; Goulding, J.; Smith, G.; Clarke, D.D.; Bibby, P.A. Psychological and demographic predictors of plastic bag consumption in transaction data. *J. Environ. Psychol.* **2020**, *72*, 101473. [CrossRef]
25. Adam, I.; Walker, T.R.; Clayton, C.A.; Bezerra, J.C. Attitudinal and behavioural segments on single-use plastics in Ghana: Implications for reducing marine plastic pollution. *Environ. Chall.* **2021**, *4*, 100185. [CrossRef]
26. Walker, T.R.; McGuinty, E.; Charlebois, S.; Music, J. Single-use plastic packaging in the Canadian food industry: Consumer behavior and perceptions. *Humanit. Soc. Sci. Commun.* **2021**, *8*, 80. [CrossRef]
27. Guiné, R.P.F.; Florença, S.G.; Moya, K.V.; Anjos, O. Edible flowers, old tradition or new gastronomic trend: A first look at consumption in portugal versus Costa Rica. *Foods* **2020**, *9*, 977. [CrossRef]
28. Guiné, R.P.F.; Florença, S.G.; Barroca, M.J.; Anjos, O. The link between the consumer and the innovations in food product development. *Foods* **2020**, *9*, 1317. [CrossRef]
29. Triola, M.F.; Flores, V.R.L.F. *Instrodução ÀEstatística*, 12th ed.; LTC: Rio de Janeiro, Brasil, 2017.
30. Levin, J.; Fox, J.A. *Estatística Para Ciências Humanas*, 9th ed.; Pearson: Rio de Janeiro, Brasil, 2004.
31. Fundação Francisco Manuel dos Santos: PORDATA-Base de Dados Portugal Contemporâneo. Available online: https://www.pordata.pt/Home (accessed on 10 December 2020).

32. Cochran, W.G. *Sampling Techniques*, 3rd ed.; John Wiley & Sons: New York, NY, USA, 1977.
33. Levine, D.M.; Stephan, D.F.; Krehbiel, T.C.; Berenson, M.L. *Estatistica Teoria E Aplicacoes Usando O Microsoft Excel Em Portugues*, 5th ed.; LTC: Rio de Janeiro, Brasil, 2008.
34. Florença, S.; Correia, P.; Costa, C.; Guiné, R. Edible insects: Preliminary study about perceptions, attitudes, and knowledge on a sample of Portuguese citizens. *Foods* **2021**, *10*, 709. [CrossRef] [PubMed]
35. Witten, R.; Witte, J. *Statistics*, 9th ed.; Wiley: Hoboken, NJ, USA, 2009.
36. Guiné, R.; Bartkiene, E.; Florença, S.; Djekić, I.; Bizjak, M.; Tarcea, M.; Leal, M.; Ferreira, V.; Rumbak, I.; Orfanos, P.; et al. Environmental issues as drivers for food choice: Study from a multinational framework. *Sustainability* **2021**, *13*, 2869. [CrossRef]
37. Gkargkavouzi, A.; Paraskevopoulos, S.; Matsiori, S. Public perceptions of the marine environment and behavioral intentions to preserve it: The case of three coastal cities in Greece. *Mar. Policy* **2019**, *111*, 103727. [CrossRef]
38. Forleo, M.; Romagnoli, L. Marine plastic litter: Public perceptions and opinions in Italy. *Mar. Pollut. Bull.* **2021**, *165*, 112160. [CrossRef]
39. Putri, N.K. Local concern on plastic bag charge in indonesia: Do we really care? In Proceedings of the 1st International Conference Postgraduate School Universitas Airlangga: "Implementation of Climate Change Agreement to Meet Sustainable Development Goals" (ICPSUAS 2017), Surabaya, Indonesia, 1–2 August 2017; Atlantis Press: Almere, The Netherlands, 2017; pp. 326–329.
40. Conselho Ministros Portuguese Law. Decreto-Lei n.o 102-D/2020-Aprova o regime geral Da gestão de resíduos, o regime jurídico Da deposição de resíduos em aterro e altera o regime Da gestão de fluxos específicos de resíduos, transpondo as diretivas (UE) 2018/849, 2018/850, 2018/851 e 2018/852. *Diário Rep.* **2020**, *25*, 269.
41. The European Parliament; The Council of the European Union. EU Directive (EU) 2019/904 of the European Parliament and of the Council of 5 June 2019 on the Reduction of the Impact of Certain Plastic Products on the Environment [EN]. *Off. J. Eur. Union* **2019**, *155*, 1–19.
42. Presidência do Conselho de Ministros. Decreto-Lei n.° 22-A/2021 de 17 de Março—Prorroga Prazos e Estabelece Medidas Excecionais e Temporárias No Âmbito Da Pandemia Da Doença COVID-19. *Diário Repúb.* **2021**, *53*, 2–8.
43. Assembleia da República. Lei n.° 76/2019 de 2 de Setembro—Determina a Não Utilização e Não Disponibilização de Louça de Plástico de Utilização Única Nas Atividades Do Setor de Restauração e/Ou Bebidas e No Comércio a Retalho. *Diário Repúb.* **2019**, *167*, 31–34.
44. *SPV Embalagens de Plástico Cumprem Metas de Reciclagem*; Sociedade Ponto Verde: Lisboa, Portugal, 2020.
45. Lim, C.; Yusoff, S.; Ng, C.; Lim, P.; Ching, Y. Bioplastic made from seaweed polysaccharides with green production methods. *J. Environ. Chem. Eng.* **2021**, *9*, 105895. [CrossRef]
46. García-Depraect, O.; Bordel, S.; Lebrero, R.; Santos-Beneit, F.; Börner, R.A.; Börner, T.; Muñoz, R. Inspired by nature: Microbial production, degradation and valorization of biodegradable bioplastics for life-cycle-engineered products. *Biotechnol. Adv.* **2021**, 107772, in press. [CrossRef] [PubMed]
47. Aversa, C.; Barletta, M.; Gisario, A.; Pizzi, E.; Prati, R.; Vesco, S. Design, manufacturing and preliminary assessment of the suitability of bioplastic bottles for wine packaging. *Polym. Test.* **2021**, *100*, 107227. [CrossRef]
48. Nigam, S.; Das, A.K.; Patidar, M.K. Valorization of Parthenium hysterophorus weed for cellulose extraction and its application for bioplastic preparation. *J. Environ. Chem. Eng.* **2021**, *9*, 105424. [CrossRef]
49. Schwark, F. Influence factors for scenario analysis for new environmental technologies–The case for biopolymer technology. *J. Clean. Prod.* **2009**, *17*, 644–652. [CrossRef]
50. Peelman, N.; Ragaert, P.; De Meulenaer, B.; Adons, D.; Peeters, R.; Cardon, L.; Van Impe, F.; Devlieghere, F. Application of bioplastics for food packaging. *Trends Food Sci. Technol.* **2013**, *32*, 128–141. [CrossRef]
51. Jõgi, K.; Bhat, R. Valorization of food processing wastes and by-products for bioplastic production. *Sustain. Chem. Pharm.* **2020**, *18*, 100326. [CrossRef]

 *sustainability*

Article

Functional Bakery Snacks for the Post-COVID-19 Market, Fortified with Omega-3 Fatty Acids

Haralabos C. Karantonis [1,*], Constantina Nasopoulou [1] and Dimitris Skalkos [2]

[1] Laboratory of Food Chemistry, Biochemistry and Technology, Department of Food Science and Nutrition, School of Environment, University of Aegean, Metropolitan Ioakeim 2, 81400 Mytilene, Greece; knasopoulou@aegean.gr

[2] Laboratory of Food Chemistry, Department of Chemistry, University of Ioannina, 45110 Ioannina, Greece; dskalkos@uoi.gr

* Correspondence: chkarantonis@aegean.gr; Tel.: +30-225-408-3111

Abstract: Flaxseed is a natural ingredient with health benefits because of its rich contents of omega-3 fatty acids and fiber. In this study, whole-meal sliced bread, chocolate cookies, and breadsticks, which were enriched with flaxseed (*Linum usitatissimu*) were produced as a natural enrichment source in order to provide functional baked goods. The three innovative products were tested as sources of omega-3 fatty acids in terms of α-linolenic acid according to EU 1924/2006 as well as for their in vitro antithrombotic/anti-inflammatory effect. The results showed that omega-3 fatty acids had high concentrations (>0.6 g per 100 g of product) in all products even after the heating treatment with constant stability during the time of consumption. All the enriched products exerted higher, but in different grade, in vitro antithrombotic/anti-inflammatory activity compared to the conventional products. The products were evaluated positively by a panel of potential consumers without significant differences compared to conventional corresponding products. Enriched bakery products with omega-3 fatty acids may represent a novel opportunity for the development of functional foods that can be locally consumed, thereby contributing to public health prevention measures that the post-COVID-19 era demands.

Keywords: omega-3 fatty acids; bakery snacks; sensory evaluation; in vitro nutritional functionality; antithrombotic; anti-inflammatory

Citation: Karantonis, H.C.; Nasopoulou, C.; Skalkos, D. Functional Bakery Snacks for the Post-COVID-19 Market, Fortified with Omega-3 Fatty Acids. *Sustainability* **2022**, *14*, 4816. https://doi.org/10.3390/su14084816

Academic Editor: Dario Donno

Received: 18 March 2022
Accepted: 14 April 2022
Published: 17 April 2022

Publisher's Note: MDPI stays neutral with regard to jurisdictional claims in published maps and institutional affiliations.

1. Introduction

In the new post-COVID-19 period, the consumer is searching for quality, healthy foods with natural ingredients in order to protect themself from diseases, protect the environment, and provide sustainability to local economies [1–3].

Bakery snacks are an important part of the human diet and will continue to be one of the main food choices in the new era. Bread is said to be the world's earliest functional food. Functional foods are modified, enhanced, or improved foods that supply important nutrients to the human body when consumed as part of a diverse and balanced diet [4,5].

The addition of probiotics and omega-3 fatty acids to bakery products has become increasingly popular in recent years [6]. The global production of omega-3 products is estimated to be 3.3 million metric tons and worth $9.1 billion in 2018. Plant omega-3 production values are expected to grow twice as fast as marine production values during the next few years—accounting for 52% of the production value when compared to 48% for marine. In parallel, over the last few years, there have been 255 bread launches containing omega-3, which represent 7.3% of total bread launches [7,8]. Most of these have come from North America (51%), followed by Europe (21%), Asia Pacific (14%), and Latin America (14%).

At the same time linseed is a raw material that is rich in omega-3 fatty acids and fibers, making it an excellent ingredient for fortifying bakery products [9]. A legal list of authorized nutritional claims [10] on food categorization as "source" or "high content" in omega-3 fatty acids has been established and a health claim related to essential fatty acids has been approved and placed in Commission rules [11], stating that "essential fatty acids are needed for the proper development of children".

The majority of research focuses on the experimental production of foodstuffs fortified with omega-3 fatty acids, the study of their sensory quality [12–16], and the bioavailability levels of omega-3 fatty acids in fortified foods [17–19]. Moreover, marine- or plant-originated omega-3 polyunsaturated fatty acids have been proposed as health promoting constituents for cardiovascular health.

The mechanisms though, through which those fatty acids exert their beneficial activities, are not clear [20–23]. Nevertheless, platelet activating factor (PAF) [24] has been recognized as one of the most potent lipid inflammatory and thrombotic mediators that activates various cells through its specific receptor, such as platelets [25]. Activated platelets are important contributors to thrombosis and inflammation and represent an important linkage between inflammation, thrombosis, and atherogenesis [25,26]. The in vitro inhibition of PAF-induced platelet activation from food components have been used as a research tool to investigate the nutritional functionality of those foods and their possible preventive effect against chronic disease development when consumed as part of a balanced diet [27]. Interestingly, fish oil-derived omega-3 fatty acids have been shown to suppress in vitro a fundamental process in many acute and chronic inflammatory diseases—monocyte-endothelium interaction—by inhibiting PAF activity and production [28].

The objective of the present study was to manufacture functional bakery snacks that are enriched with omega-3 fatty acids using linseed as a natural source and investigate their quality in terms of sensory evaluation, omega-3 fatty acids content, and the antithrombotic/anti-inflammatory activity through a nutritive index, as well as forecast the five-year gross sales of the products produced by local companies.

2. Materials and Methods

2.1. Manufacture of Bakery Products

2.1.1. Design of the Products

The first step of the product design was to convene the HACCP food safety team of local enterprises. In this meeting, there was a full description of the products, including the necessary ingredients and raw materials, the manufacturing process of the product, the packaging and distribution, and the final characteristics of the food, as well as the documentation procedure for how to document the allergens in the nutritional details [29]. According to the legislation [30], products are not suitable for consumption for certain sensitive groups, i.e., allergic persons, therefore allergic ingredients are to be referred to on the label of the product.

2.1.2. Design of Flowchart, HACCP Plan—Recipe and Manufacture of the Products

In the design of the flowchart and the confirmation of the HACCP plant in practice, three CCP (Critical Control Points) were confirmed—one for each product in the same stage of thermal processing. The critical limits were: 83 °C/1 min for sliced bread, 78 °C/40 s for cookies, and 80 °C/30 s for breadsticks [29]. During the tests of the recipes, it was decided that 50% of the linseed added to the product should be milled to reduce the feeling of the whole seed and, furthermore, to increase the stability of the fatty acids in the whole grain.

2.1.3. Product Production

The creation of the three functional bakery snacks is depicted in a flow chart in Figure 1, while the ingredients for the recipes are listed in Table 1.

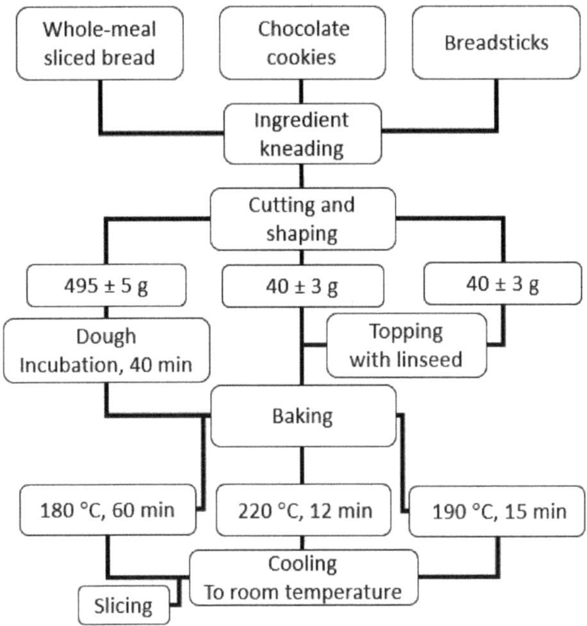

Figure 1. Flow chart presenting the production of the three functional bakery products.

Table 1. Ingredient recipes for whole-meal sliced bread, chocolate cookies, and breadsticks.

Constituents	Whole-Meal Sliced Bread [1]	Chocolate Cookies [1]	Breadsticks [1]
Hard wheat flour (g)	20.00		
Gluten (g)	2.50		
Yeast (g)	0.85		1.00
White wheat flour (g)	35.00	45.00	40.00
Flaxseed (g)	16.00	8.00	12.00
Sugar (g)	0.03	9.00	5.00
Salt (g)	0.05	0.01	1.00
Margarine (g)	3.00	9.00	7.00
Water (g)	22.60	1.00	34.00
Cacao (g)		5.00	
Chocolate drops (g)		5.00	
Egg (g)		10.00	
Milk (g)		5.00	
Honey (g)		3.00	

[1]: Functional bakery products enriched with flaxseed as source of omega-3 fatty acids.

2.2. Sensory Evaluation

Omega-3 fatty acid-enriched bakery products, namely whole-meal sliced bread, chocolate cookies, and breadsticks, were evaluated by 16 trained panelists. Each sample was evaluated on six main attributes: appearance, odour, texture, taste, aftertaste, and overall acceptance. Rating of the sensory attributes was carried out using a 9-point hedonic scale where 1 = nonexistent and 9 = too intense [11,12]. The final sensory profile of the products was determined by the average and the acceptable point (5 at the above scale) in duplicate experiments.

2.3. Chemical Analysis

The protein content was determined using the AOAC Kjeldahl method [31], the moisture was determined by an air oven [31], the fiber (crude) was determined by digesting the sample in a 1.25% (*v*/*w*) H_2SO_4 followed by 1.25% (*v*/*w*) NaOH solution [32]. Furthermore, the total fat was determined after acid hydrolysis and diethyl ether/petroleum ether extraction and fatty acids determination was performed with gas chromatography after methylation in methanol with boron trifluoride (BF3) as a catalyst [31]. The fatty acid methyl ester profiles were measured by gas–liquid chromatography on a Shimadzu 2010 chromatograph (Shimadzu Corporation, Tokyo, Japan), which was fitted with an automatic sampler AOC-20 and flame ionization detector. A fused-silica capillary column was used for the FAME analysis; DB-23, 60 m × 0.251 mm i.d., 0.25 µm (J&W, Agilent Technologies, Palo Alto, CA, USA). The oven temperature value sequence was initially 120 °C for 5 min, was raised to 180 °C at a rate of 10 °C per min, then to 220 °C at a rate of 20 °C per min, and, finally, was isothermal at 220 °C for 30 min. The injector and detector temperatures were maintained at 220 and 225 °C, respectively. The carrier gas was a high purity helium with a linear flow rate of 1 mL per min and a split ratio of 1:50. The individual FAMEs were identified by comparison with the relative retention time of the FAMEs peaks from the samples with the standard mixtures 37 Component FAME Mix (47885-U Supelco, Bellefonte, PA, USA) and Qualmix Fish S (89-5550 Larodan Fine Chemicals AB, Malmö, Sweden). The carbohydrates and energy were calculated from proximate analysis values.

All products were subjected to the determination of omega-3 fatty acids before and after baking in order to assure their availability after baking.

2.4. Shelf-Life Determination

The chemical and sensorial shelf life of the enriched bakery products was determined by means of sensory scoring and omega-3 content to meet the legislative limits for nutritional claims for products kept at room temperature for 24 days. The measurements were performed on days 0, 4, 7, 18, and 24. The overall market shelf life of the products was determined as the combination of the chemical and sensory values.

2.5. Anti-Thrombotic and Anti-Inflammatory Activity

All chemical reagents and solvents were of analytical grade and were supplied by Merck (Darmstadt, Germany). Platelet-activating factor (β-Acetyl-γ-*O*-hexadecyl-*L*-α-phosphatidylcholine hydrate) and bovine serum albumin (BSA) were obtained from Sigma (St. Louis, MO, USA).

2.5.1. Lipid Extraction

An amount of 15.0 mL of methanol was mixed with 3.0 g of flour sample and the mixture was agitated at 200 rpm on a GFL 3017 orbital shaker (GFL, Burgwedel, Germany) for 15 min. After that, 7.5 mL of chloroform was added to the mixture, which was followed by agitation at 200 rpm for 15 min. In the next step, after taking into consideration the moisture content of the samples, 6.0 mL of either distilled water (first version of lipid extraction) or 1 m aqueous sodium chloride solution 0.5% in acetic acid (second version of lipid extraction) was placed in the mixture in order to achieve a ratio for the solvents of methanol/chloroform/water that was equal to 1/2/0.8 (*v*/*v*/*v*), and then an extra agitation was performed at 200 rpm for 15 min. Then, depending on the version of the method, 7.5 mL of methanol were added along with either 7.5 mL of distilled water (first version) or 7.5 mL of 1 m aqueous sodium chloride solution 0.5% in acetic acid (second version), thereby achieving a ratio for the solvents of methanol/chloroform/water that was equal to 1/1/0.9 (*v*/*v*/*v*). After a final agitation at 200 rpm for 15 min, the samples were stored overnight at 4 °C. The samples were then centrifuged for 5 min at 2000× *g* in a Hermle Z 383 centrifuge (Hermle Labortechnik, Wehingen, Germany) and the lower phase of the biphasic solvent system was collected into a pre-weighed glass flask along with a 5.0 mL chloroform rinse of the upper phase. The samples were dried using a Lab Tech

EV 311 Rotary evaporator (Lab Tech, Milan, Italy), which was weighed on a KERN ABJ analytical balance (Kern and Sohn GmbH, Balingen, Germany), suspended in 2.0 mL of chloroform/methanol: 1/1 (v/v), and stored at −40 °C until further study.

2.5.2. In Vitro Anti-Thrombotic and Anti-Inflammatory Activity

The in vitro antithrombotic and anti-inflammatory activity of the lipid extracts that were enriched in omega-3 fatty acids or conventional food products were evaluated on a Chrono-Log 500-Ca aggregometer (Chrono-Log Co., Havertown, PA, USA) that was connected to a computer (Aggro/Link software; Chrono-Log, Hawertown, PA, USA) according to their ability to inhibit the thrombotic and inflammatory lipid mediator of PAF towards platelet rich plasma (PRP) [33]. Aliquots of dissolved lipid extracts that were enriched with omega-3 fatty acids or conventional food products and PAF solution were evaporated under a stream of nitrogen and reconstituted in BSA (2.5 mg/mL saline). The platelet response induced by PAF (10^{-7} M, final concentration) was measured in PRP before (considered as 0% inhibition) and after the addition of various concentrations of the examined sample. Consequently, the plot of the percentage inhibition (ranging from 20 to 80%) versus the different concentrations of the sample is linear. From this curve, the amount of lipids required for a 50% inhibition (inhibitory amount for 50% inhibition; IA50) against PAF was calculated and expressed in μg for the lipid extracts.

2.6. Marketing Plan: Development

The 5-year marketing plan in this research was investigated in order to target the objectives for the gross sales of the products under study by a local bakery and cheese company that were involved. The plan was based on sales: (a) in the first two years they would be at the local markets, where the company is active and well-known; (b) in the third- and fourth-year there would be expansion to the national market; and (c) in the fifth year there would be promotion to the European market, targeting two or three specific countries such as Germany, France, and England. Data analysis was based on the last three years of the company's performance gross sales for its conventional products that were successfully promoted in the local and national Greek market. In addition, the overall company's performance and its plans for growth and expansion in new areas of activities—such as healthy food snacks—were considered when formulating the specific marketing plan. Market trends were also considered at the global level by evaluating existing marketing data on omega-3 functional foods in general and on omega-3 bakery products specifically. The formulated plan considered conventional activities such as advertising, exhibitions, events, etc., as well as modern means such as social media, internet usage, etc. for the promotion of the products and their health claims. Based on the proposed activities each year, and combined with the Greek and global market trends, the feasible six-year projected sales were forecasted and presented in the section below.

2.7. Statistics

A comparison of the results was performed using the statistical package SPSS v.21 using the average value comparison assay for independent samples (Student's t-test for independent samples) at a significance level equal to 0.05.

3. Results

3.1. Manufacture of Bakery Products

A representative photo of the three products is given in Figure 2.

3.2. Sensory Evaluation of the New Products

The organoleptic properties and acceptance of the new products are presented in Figure 3. The overall acceptability is above the set limit for acceptance (5.00) and high enough (7.29 and 7.86 for whole-meal sliced bread and chocolate cookies, followed by 6.36

for breadsticks) to indicate that their fate in the market could be successful from a sensorial point of view.

Figure 2. Representative photos for the three functional bakery snacks.

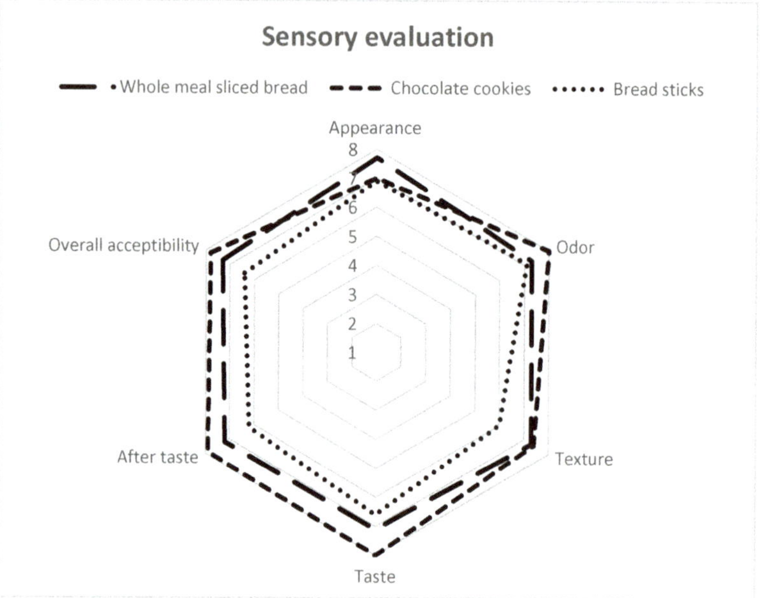

Figure 3. Sensory evaluation of whole-meal sliced bread, chocolate cookies, and breadsticks. Results are expressed as a mean of the responses from 16 trained panelists.

3.3. Chemical Analysis of the Products

The nutritional composition of the new products is depicted in Table 2. According to Regulation 1924/2006 on nutritional claims [10], whole-meal sliced bread is a "source of fiber" since it contains 5.4 g fibers/100 g of product, while the limit for this claim is 3 g fibers/100 g of product and, furthermore, it is a "source of protein" since 16.7 ± 100% of the energy of the product comes from proteins, while the respective limit for this claim is 12% according to the formula: % energy from protein s = [(g proteins/100 g × 4 kcal/g)/(total energy/100 g)] × 100. Moreover, chocolate cookies (7.03 ± 0.30 g/100 g product) and breadsticks (7.08 ± 0.12 g/100 g product) are both a "source of fiber".

Table 2. Nutritional composition of whole-meal sliced bread, chocolate cookies, and breadsticks.

Constituents	Whole-Meal Sliced Bread [1]	Chocolate Cookies [1]	Breadsticks [1]
Fat (g/100 g)	10.0 ± 0.83	17.52 ± 1.39	15.33 ± 0.78
Humidity (g/100 g)	29.80 ± 1.12	10.29 ± 0.49	8.41 ± 0.39
Proteins (g/100 g)	12.30 ± 0.59	10.03 ± 0.42	6.77 ± 0.33
Carbohydrates (g/100 g)	39.40 ± 1.69	50.90 ± 0.15	57.98 ± 1.38
Ash (g/100 g)	3.80 ± 0.13	4.20 ± 0.10	4.62 ± 0.11
Fibers (g/100 g)	5.40 ± 0.21	7.03 ± 0.30	7.08 ± 0.12
Energy Kcal/100 g	294.74 ± 2.36	401.50 ± 2.09	396.97 ± 2.63

[1]: Each result is the mean value of triplicate experiment.

The results of the effect of baking on the content of omega-3 fatty acids are shown in Table 3. According to the results, baking does not influence the omega-3 fatty acids content of the final products.

Table 3. Effect of baking on omega-3 fatty acids concentration.

Constituents	Whole-meal Sliced Bread [1]		Chocolate Cookies [1]		Breadsticks [1]	
	Before	After	Before	After	Before	After
Total fat (g/100 g)	10.90 ± 0.99	10.01 ± 0.83	18.12 ± 1.34	17.74 ± 1.39	15.99 ± 0.96	15.33 ± 0.78
Fatty acids (g/100 g fat)						
Myristic acid C14:0	0.50 ± 0.09	0.60 ± 0.11	1.00 ± 0.08	0.90 ± 0.07	0.20 ± 0.01	0.20 ± 0.01
Palmitic acid C16:0	18.90 ± 2.91	22.70 ± 2.99	25.30 ± 0.95	25.00 ± 1.11	8.30 ± 0.40	7.90 ± 0.03
Stearic acid C18:0	7.80 ± 0.30	9.20 ± 0.41	6.90 ± 0.31	6.80 ± 0.31	2.90 ± 0.19	2.90 ± 0.15
Oleic acid C18:1n9c	24.30 ± 1.51	25.40 ± 1.67	33.00 ± 2.11	31.10 ± 2.43	25.30 ± 1.21	24.40 ± 1.69
Vaccenic acid C18:1n7c	0.70 ± 0.03	0.70 ± 0.04	0.90 ± 0.051	0.90 ± 0.05	0.80 ± 0.07	0.76 ± 0.06
Linoleic acid C18:2n6c	24.60 ± 1.98	23.10 ± 0192	22.50 ± 1.01	22.70 ± 0.99	41.70 ± 2.56	38.10 ± 2.68
α-Linolenic acid C18:3n3c	20.40 ± 1.49	19.60 ± 1.79	9.10 ± 0.72	9.50 ± 0.61	18.30 ± 1.08	19.00 ± 1.23

[1]: Results from experiments, which were run in triplicate, are expressed as mean value ± standard deviation.

Based on the results and Regulation 1924/2006 on nutritional claims [10], whole-meal sliced bread has a "high content" of omega-3 fatty acids since their concentration is 1.96 ± 0.18 g/100 g of the product, while the respective limit for this nutritional claim is 0.6 g/100 g of the product. Moreover, chocolate cookies and breadsticks also have a "high content" of omega-3 fatty acids (0.95 ± 0.06 and 1.90 ± 0.12 g/100 g product, respectively).

3.4. Shelf-Life Assessment

The shelf life, both sensorial and chemical, of the new products is presented in Table 4. The shelf life, concerning the minimum concentration of omega-3 fatty acids for the nutritional claim, is 18 days at room temperature for whole-meal sliced bread and more than 24 days for chocolate cookies and breadsticks. The sensorial shelf life is 7 days at room temperature for whole-meal sliced bread and 24 days for chocolate cookies and breadsticks. Given this, the combined shelf life of the new products at room temperature is 7 days for whole-meal sliced bread and 24 days for chocolate cookies and breadsticks, which is acceptable for the food market.

3.5. Lipid Extraction and In Vitro Antithrombotic and Anti-Inflammatory Activity

In order to optimize lipid extraction, two versions of the Bligh and Dyer method [34] for lipid extraction were evaluated for their capacity in enriched flour samples in triplicate experiments. The results showed that the second version that used 1 m aqueous sodium chloride solution 0.5% in acetic acid instead of distilled water for the first version extracted lipids more efficiently. More specifically, 123.3 ± 8.0 mg of lipids per g of flour were extracted by performing the second version of the extraction versus 106.7 ± 7.2 mg of lipids per g of flour, which were extracted by performing the first version of the extraction ($p < 0.05$). The procedure of the second version was further applied to the prepared foods to extract their total lipids in order to evaluate their in vitro antithrombotic and anti-inflammatory activities.

Table 4. Monitoring of sensorial attributes and omega-3 fatty acid concentration.

Attribute	Product	Day 0	Day 4	Day 7	Day 18	Day 24
Appearance [1]	Whole-meal sliced bread	7.7 ± 0.3	8.1 ± 0.3	7.8 ± 0.3	Na [2]	NA
	Chocolate Cookies	7.1 ± 0.4	7.8 ± 0.3	7.0 ± 0.3	6.5 ± 0.3	5.3 ± 0.2
	Breadsticks	6.9 ± 0.3	6.9 ± 0.3	6.8 ± 0.3	6.6 ± 0.2	5.8 ± 0.2
Odor [1]	Whole-meal sliced bread	7.3 ± 0.4	8.0 ± 0.3	7.7 ± 0.3	NA	NA
	Chocolate Cookies	8.1 ± 0.3	8.0 ± 0.3	7.5 ± 0.3	6.8 ± 0.2	6.0 ± 0.3
	Breadsticks	7.1 ± 0.4	7.3 ± 0.4	6.8 ± 0.3	6.1 ± 0.3	5.4 ± 0.3
Texture [1]	Whole-meal sliced bread	7.3 ± 0.3	7.3 ± 0.4	7.4 ± 0.4	NA	NA
	Chocolate Cookies	7.4 ± 0.3	7.9 ± 0.4	6.5 ± 0.3	6.0 ± 0.2	5.1 ± 0.2
	Breadsticks	6.0 ± 0.3	7.4 ± 0.4	7.3 ± 0.4	6.6 ± 0.3	5.3 ± 0.2
Taste [1]	Whole-meal sliced bread	7.1 ± 0.3	6.9 ± 0.3	7.1 ± 0.4	NA	NA
	Chocolate Cookies	8.0 ± 0.4	7.9 ± 0.4	7.5 ± 0.4	6.8 ± 0.3	5.4 ± 0.3
	Breadsticks	6.6 ± 0.3	7.6 ± 0.3	7.2 ± 0.3	5.9 ± 0.2	5.2 ± 0.2
After taste [1]	Whole-meal sliced bread	7.2 ± 0.2	7.2 ± 0.2	7.3 ± 0.3	NA	NA
	Chocolate Cookies	7.9 ± 0.2	7.8 ± 0.2	7.2 ± 0.3	6.6 ± 0.3	5.2 ± 0.1
	Breadsticks	6.2 ± 0.3	7.7 ± 0.2	7.0 ± 0.3	5.8 ± 0.3	5.0 ± 0.3
Overall acceptability [1]	Whole-meal sliced bread	7.3 ± 0.4	7.3 ± 0.3	7.3 ± 0.3	NA	NA
	Chocolate Cookies	7.9 ± 0.3	7.9 ± 0.3	7.2 ± 0.3	6.5 ± 0.2	5.2 ± 0.3
	Breadsticks	6.4 ± 0.3	7.3 ± 0.3	7.1 ± 0.4	5.9 ± 0.3	5.1 ± 0.3
α-Linolenic acid C18:3n3c (g/100 g product) [3]	Whole-meal sliced bread	1.3 ± 0.2	1.0 ± 0.1	1.0 ± 0.1	0.59 ± 0.1	NA
	Chocolate Cookies	1.3 ± 0.2	0.9 ± 0.1	1.1 ± 0.1	0.9 ± 0.1	1.2 ± 0.1
	Breadsticks	2.9 ± 0.1	6.9 ± 0.2	7.2 ± 0.3	6.8 ± 0.4	6.2 ± 0.3

[1] Results are expressed as mean ± standard deviation of the responses from 16 trained panelists; [2] NA: not analysed because mold has been grown on the product; [3] Results are expressed as mean ± standard deviation from triplicate experiment.

The results from the in vitro inhibition of PAF-induced platelet activation are presented in Table 5. Flour enriched in omega-3 fatty acids, whole-meal sliced bread, chocolate cookies, and breadsticks showed higher antithrombotic and anti-inflammatory activities compared to the corresponding conventional samples. More specifically, the enriched flour, sliced bread, chocolate cookies, and breadsticks showed ×4.7, ×4.6, ×1.1, and ×1.6 higher activities compared to their respective conventional samples. The results show that the enrichment of omega 3 fatty acids may result in food products with increased nutritional value, but this enrichment may be affected by the processing that is followed to prepare each product.

Table 5. In vitro antithrombotic and anti-inflammatory activity.

Food Product	[1] IA_{50}	
	Conventional	Enriched
Flour	7.60 ± 0.08	1.61 ± 0.07 [2]
Whole-meal sliced bread	86.13 ± 1.01	18.73 ± 0.40 [2]
Chocolate cookies	55.72 ± 0.63	49.72 ± 0.60 [2]
Breadsticks	73.92 ± 0.83	46.72 ± 0.73 [2]

[1]: Inhibitory amount for 50% inhibition of PAF activity toward washed platelets expressed in μg of lipid extracts. Values from experiments in triplicate are expressed as mean ± standard deviation. [2]: Statistically significant difference of enriched versus conventional products at significance level equal to 0.05.

Lipid mediators are a heterogeneous group of molecules that mediate several physiological cellular functions that, when they are dysfunctional, may lead to the development of many chronic diseases. Food-derived bioactive compounds can beneficially influence metabolism, thereby offering an attractive way to prevent the establishment of chronic diseases.

An increased intake of omega-3 fatty acids from foodstuffs is related to a favorable clinical profile of various chronic diseases. Platelet activating factor (PAF) is one of the most potent inflammatory and thrombotic lipid mediators, playing a crucial role in the initiation and propagation of atherosclerosis. Therefore, the omega-3 fatty acid-induced PAF inhibition is very important in terms of their nutritional value.

3.6. Marketing Plan

The results of the marketing plan developed in terms of forecasted productions and sales over the first 5 years are shown in Table 6. The first two years, when the products will be promoted to the local market, project that the sales will have a 102.7% annual increase, reaching 75,000€. A 120% sales increase is expected in the third year due to the promotion of the products in the overall Greek market, which will be followed by a 45% medium increase the coming year within the same market. Finally in the last year, with the promotion of the products to international markets, an initial 20% increase in sales is expected, which will increase steadily by 20–30% annually in the years to come based on more promotion internationally. A similar annual increase is expected for the production capacity from year to year as well: 2nd from 1st year (103.3%), 3rd from 2nd year (131.7%), 4th from 3rd year (42.1%), and 5th from 4th year (20.9%). The expected differences in production and sales between the three products over the years are based on the company's production and sales for the corresponding conventional products, which are already on the market.

Table 6. Expected productions/sales of the new omega-3 bakery products.

		Whole-Meal Sliced Bread	Chocolate Cookies	Breadsticks	TOTAL
1st year [1]	Quantity(Kg)	6600	2500	3000	12,100
	Sales (€)	10,000	12,000	15,000	37,000
2nd year [1]	Quantity(Kg)	13,300	53,000	6000	24,600
	Sales (€)	20,000	25,000	30,000	75,000
3rd year [2]	Quantity(Kg)	33,300	11,700	12,000	57,000
	Sales (€)	50,000	55,000	60,000	165,000
4th year [2]	Quantity(Kg)	46,000	17,000	18,000	81,000
	Sales (€)	70,000	80,000	90,000	240,000
5th year [3]	Quantity(Kg)	56,000	20,000	22,000	98,000
	Sales (€)	84,000	96,000	108,000	288,000

[1]: Sales at the local market only; [2]: Sales at the Greek market as well; [3]: Sales in targeted European countries as well.

The proposed marketing plan predicts 288,000€ expected annual sales in the 5th year from the three selected products. This amount, for a company of 5,000,000€ total sales with hundreds of products produced, is a satisfactory sales target from three products alone (5.7%), and a good driving force for the company to invest in for the implementation of the five-year proposed marketing plan.

4. Discussion

The relevance and innovation of the present study lies in the incorporation of omega-3 fatty acids as an ingredient for manufacturing healthy bakery snacks that offer new opportunities for the local Greek snack market. The three bakery snacks manufactured are rich in omega-3 fatty acids since their content is higher than 0.6 g of omega-3 fatty acids per 100 g of baked product, and they are also a source of dietary fiber since their content is higher than 3 g of fiber per 100 g of final product. Moreover, whole-meal sliced bread is, in addition, a source of protein since more than 12% of the % energy comes from it.

The produced whole-meal sliced bread, chocolate cookies, and breadsticks have 2.0 g, 1.7 g, and 2.9 g of -linolenic acid, respectively, and daily bread consumption of 160 g [35,36] is higher than that of the other two goods that are taken on a more irregular basis.

Taking into account the frequency and quantity of an adult's average consumption of manufactured products—given that they need 2.22 g of a-linolenic acid per day [37]—

whole-meal sliced bread looks to practically meet the daily requirements for a-linolenic acid, being a better source of omega-3 fatty acids than chocolate cookies or breadsticks.

Omega-3 fatty acids incorporation also enhanced the antithrombotic and anti-inflammatory bioactivity in terms of the in vitro inhibition of platelet activating factor against platelet rich plasma, thereby indicating health protective properties compared to the conventional formulations.

Omega-3 fatty acid-enriched bakery products were well accepted by panelists for their sensorial attributes.

The findings of this study demonstrated that omega-3 fatty acids might represent a valuable ingredient to improve the nutritional and health-protective properties of bakery snacks. Indeed, linseed (*Linum usitatissimum* L.)—generally known as flaxseed—has been demonstrated to boost immunity against viral infections as well as control cytokine storms and inflammatory mediators [38–43]. Moreover, due to the high number of both soluble and insoluble dietary fibers, flaxseed is beneficial for gut health, thereby boosting the immune system [44,45].

5. Conclusions

In conclusion, the design and preparation of enriched bakery snacks with flaxseed as a source of omega-3 fatty acids may represent a novel opportunity for the development of functional foods that are sustainable in the food market and could also contribute to the prevention of public health issues when consumed.

Author Contributions: Conceptualization, supervision, methodology H.C.K. and D.S.; investigation, writing—original draft preparation and writing—review and editing H.C.K., C.N. and D.S.; software, data curation and resources H.C.K. All authors have read and agreed to the published version of the manuscript.

Funding: This research received no external funding.

Institutional Review Board Statement: Not applicable.

Informed Consent Statement: Not applicable.

Data Availability Statement: Not applicable.

Conflicts of Interest: The authors declare no conflict of interest.

References

1. Ceniti, C.; Tilocca, B.; Britti, D.; Santoro, A.; Costanzo, N. Food Safety Concerns in "COVID-19 Era". *Microbiol. Res.* **2021**, *12*, 53–68. [CrossRef]
2. Olaimat, A.N.; Shahbaz, H.M.; Fatima, N.; Munir, S.; Holley, R.A. Food Safety during and after the Era of Covid-19 Pandemic. *Front. Microbiol.* **2020**, *11*, 1854. [CrossRef] [PubMed]
3. Galanakis, C.M. The Food Systems in the Era of the Coronavirus (CoVID-19) Pandemic Crisis. *Foods* **2020**, *9*, 523. [CrossRef]
4. Hasler, C.M.; Bloch, A.S.; Thomson, C.A. Position of the American Dietetic Association: Functional Foods. *J. Am. Diet. Assoc.* **2004**, *104*, 814–826. [CrossRef]
5. Kaur, R.; Sood, A.; Kanotra, M.; Arora, S.; Subramaniyan, V.; Bhatia, S.; Al-Harrasi, A.; Aleya, L.; Behl, T. Pertinence of Nutriments for a Stalwart Body. *Environ. Sci. Pollut. Res.* **2021**, *28*, 54531–54550. [CrossRef]
6. Ruiz, J.C.R.; Vazquez, E.L.L.O.; Campos, M.R.S. Encapsulation of Vegetable Oils as Source of Omega-3 Fatty Acids for Enriched Functional Foods. *Crit. Rev. Food Sci. Nutr.* **2017**, *57*, 1423–1434. [CrossRef]
7. Schoefield, L. The Omega 3 Market Essentially Innovative. Available online: https://www.nutraceuticalsworld.com/issues/2013-09/view_features/the-Omega-3-market-essentially-innovative/ (accessed on 9 September 2013).
8. Nieburg, O. Omega-3 Bread: Health, Claims, Sources and Dosage. Specially Edition "Riding the Health Claims". Nutra Ingredients.Com. Available online: https://www.nutraingredients.com/article/2013/04/18/Omega-3-bread-and-health-claims (accessed on 17 April 2021).
9. Friedberg, J. Next-Generation Fats and Oils for Snack and Bakery R&D. Available online: https://www.preparedfoods.com/articles/126620-next-generation-fats-and-oils-for-snack-and-bakery-r-and-d (accessed on 17 February 2022).
10. *Regulation (EC) No 1924/2006 of the European Parliament and of the Council of 20 December 2006 on Nutrition and Health Claims Made on Foods OJ L 404*; European Parliament, Council of the European Union: Brussels, Belgium, 2006; pp. 9–25.

11. *Commission Regulation (EC) No 983/2009 of 21 October 2009 on the Authorisation and Refusal of Authorisation of Certain Health Claims Made on Food and Referring to the Reduction of Disease Risk and to Children's Development and Health*; European Commission: Brussels, Belgium, 2009.
12. Ayerza, R.; Coates, W.; Lauria, M. Chia Seed (*Salvia hispanica* L.) as an ω-3 Fatty Acid Source for Broilers: Influence on Fatty Acid Composition, Cholesterol and Fat Content of White and Dark Meats, Growth Performance, and Sensory Characteristics. *Poult. Sci.* **2002**, *81*, 826–837. [CrossRef]
13. Umesha, S.S.; Manohar, R.S.; Indiramma, A.R.; Akshitha, S.; Naidu, K.A. Enrichment of Biscuits with Microencapsulated Omega-3 Fatty Acid (Alpha-Linolenic Acid) Rich Garden Cress (Lepidium Sativum) Seed Oil: Physical, Sensory and Storage Quality Characteristics of Biscuits. *LWT-Food Sci. Technol.* **2015**, *62*, 654–661. [CrossRef]
14. Kumar, N.A.; Rao, U.J.S.P.; Jeyarani, T.; Indrani, D. Effect of Ingredients on Rheological, Physico-Sensory, and Nutritional Characteristics of Omega-3-Fatty Acid Enriched Eggless Cake. *J. Texture Stud.* **2017**, *48*, 439–449. [CrossRef]
15. Alejandre, M.; Astiasarán, I.; Ansorena, D. Omega-3 Fatty Acids and Plant Sterols as Cardioprotective Ingredients in Beef Patties Composition and Relevance of Nutritional Information on Sensory Characterization. *Food Funct.* **2019**, *10*, 7883–7891. [CrossRef]
16. Faccinetto-Beltrán, P.; Gómez-Fernández, A.R.; Orozco-Sánchez, N.E.; Pérez-Carrillo, E.; Marín-Obispo, L.M.; Hernández-Brenes, C.; Santacruz, A.; Jacobo-Velázquez, D.A. Physicochemical Properties and Sensory Acceptability of a Next-Generation Functional Chocolate Added with Omega-3 Polyunsaturated Fatty Acids and Probiotics. *Foods* **2021**, *10*, 333. [CrossRef] [PubMed]
17. Puranik, S.S. *Emulsions of Omega-3 Fatty Acids for Better Bioavailability and Beneficial Health Effects*; Springer Publishing: New York, NY, USA, 2016; ISBN 9783319404585.
18. Köhler, A.; Heinrich, J.; Von Schacky, C. Bioavailability of Dietary Omega-3 Fatty Acids Added to a Variety of Sausages in Healthy Individuals. *Nutrients* **2017**, *9*, 629. [CrossRef] [PubMed]
19. Punia, S.; Sandhu, K.S.; Siroha, A.K.; Dhull, S.B. Omega 3-Metabolism, Absorption, Bioavailability and Health Benefits–A Review. *PharmaNutrition* **2019**, *10*, 100162. [CrossRef]
20. Calviello, G.; Su, H.-M.; Weylandt, K.H.; Fasano, E.; Serini, S.; Cittadini, A. Experimental Evidence of ω-3 Polyunsaturated Fatty Acid Modulation of Inflammatory Cytokines and Bioactive Lipid Mediators: Their Potential Role in Inflammatory, Neurodegenerative, and Neoplastic Diseases. *BioMed Res. Int.* **2013**, *2013*, 743171. [CrossRef] [PubMed]
21. Barrett, S.J. The Role of Omega-3 Polyunsaturated Fatty Acids in Cardiovascular Health. *Altern. Ther. Health Med.* **2013**, *19* (Suppl. 1), 26–30. [PubMed]
22. Bu, J.; Dou, Y.; Tian, X.; Wang, Z.; Chen, G. The Role of Omega-3 Polyunsaturated Fatty Acids in Stroke. *Oxid. Med. Cell. Longev.* **2016**, *2016*. [CrossRef] [PubMed]
23. Podzolkov, V.I.; Pisarev, M.V. Role of Omega-3 Polyunsaturated Fatty Acids in Cardiovascular Risk Management. *Cardiovasc. Ther. Prev.* **2020**, *19*, 86–94. [CrossRef]
24. Demopoulos, C.A.; Pinckard, R.N.; Hanahan, D.J. Platelet-Activating Factor. Evidence for 1-0-Alkyl-2-Acetyl-Sn-Glyceryl-3-Phosphorylcholine as the Active Component (a New Class of Lipid Chemical Mediators). *J. Biol. Chem.* **1979**, *254*, 9355–9358. [CrossRef]
25. Demopoulos, C.A.; Karantonis, H.C.; Antonopoulou, S. Platelet Activating Factor-A Molecular Link between Atherosclerosis Theories. *Eur. J. Lipid Sci. Technol.* **2003**, *105*, 705–716. [CrossRef]
26. Badimon, L.; Suades, R.; Fuentes, E.; Palomo, I.; Padró, T. Role of Platelet-Derived Microvesicles as Crosstalk Mediators in Atherothrombosis and Future Pharmacology Targets: A Link between Inflammation, Atherosclerosis, and Thrombosis. *Front. Pharmacol.* **2016**, *7*, 293. [CrossRef]
27. Antonopoulou, S.; Fragopoulou, E.; Karantonis, H.C.; Mitsou, E.; Sitara, M.; Rementzis, J.; Mourelatos, A.; Ginis, A.; Phenekos, C. Effect of Traditional Greek Mediterranean Meals on Platelet Aggregation in Normal Subjects and in Patients with Type 2 Diabetes Mellitus. *J. Med. Food* **2006**, *9*, 356–362. [CrossRef] [PubMed]
28. Mayer, K.; Merfels, M.; Muhly-Reinholz, M.; Gokorsch, S.; Rosseau, S.; Lohmeyer, J.; Schwarzer, N.; Krüll, M.; Suttorp, N.; Grimminger, F.; et al. ω-3 Fatty Acids Suppress Monocyte Adhesion to Human Endothelial Cells: Role of Endothelial PAF Generation. *Am. J. Physiol.-Heart Circ. Physiol.* **2002**, *283*, H811–H818. [CrossRef] [PubMed]
29. *ELOT EN ISO 22000: 2005*; Food Safety Management Systems-Requirements for Any Organization in the Food Chain, European Standard. Hellenic Organization for Standardization: Peristeri, Greece, 2005; pp. 11–27.
30. *Regulation (EU) No 1169/2011 of the European Parliament and of the Council of 25 October 2011 on the Provision of Food In-formation to Consumers, OJ L 304*; European Parliament, Council of the European Union: Brussels, Belgium, 2011; pp. 18–63.
31. *AOAC Official Methods of Analysis*, 17th ed.; Association of Official Analytical Chemists: Washington, DC, USA, 2000.
32. *AOAC Official Methods of Analysis*, 14th ed.; Association of Official Analytical Chemists: Washington, DC, USA, 1985.
33. Thanou, K.; Kapsi, A.; Petsas, A.S.; Dimou, C.; Koutelidakis, A.; Nasopoulou, C.; Skalkos, D.; Karantonis, H.C. Ultrasound-Assisted Extraction of Texas Variety Almond Oil and in Vitro Evaluation of Its Health Beneficial Bioactivities. *J. Food Process. Preserv.* **2021**, e16144. [CrossRef]
34. Bligh, E.G.; Dyer, W.J. A Rapid Method of Total Lipid Extraction and Purification. *Can. J. Biochem. Physiol.* **1959**, *37*, 911–917. [CrossRef]
35. AIBI. *AIBI Bread Market Report 2012*; AIBI: Brussels, Belguim, 2013.
36. United States Department of Agriculture, E.R.S.W. Role in the U.S.D. Available online: https://www.ers.usda.gov/topics/crops/wheat/wheats-role-in-the-us-diet (accessed on 28 April 2017).

37. Simopoulos, A.P. Importance of the Ratio of Omega-6/Omega-3 Essential Fatty Acids: Evolutionary Aspects. *World Rev. Nutr. Diet.* **2003**, *92*, 1–22.
38. Thibault, R.; Seguin, P.; Tamion, F.; Pichard, C.; Singer, P. Nutrition of the COVID-19 Patient in the Intensive Care Unit (ICU): A Practical Guidance. *Crit. Care* **2020**, *24*, 447. [CrossRef]
39. López-Gómez, J.J.; Lastra-González, P.; Gómez-Hoyos, E.; Ortolá-Buigues, A.; Jiménez-Sahagún, R.; Cuadrado-Clemente, L.; Benito-Sendín-Plaar, K.; Cuenca-Becerril, S.; Portugal-Rodríguez, E.; De Luis Román, D.A. Evolution of Nutrition Support in Patients with COVID-19 Disease Admitted in the Intensive Care Unit | Evolución Del Soporte Nutricional En El Paciente Con Enfermedad COVID-19 Ingresado En La Unidad de Cuidados Intensivos. *Endocrinol. Diabetes Y Nutr.* 2022; *in press*. [CrossRef]
40. Chang, J.P.-C.; Pariante, C.M.; Su, K.-P. Omega-3 Fatty Acids in the Psychological and Physiological Resilience against COVID-19. *Prostaglandins Leukot. Essent. Fat. Acids* **2020**, *161*, 102177. [CrossRef]
41. Hirayama, D.; Iida, T.; Nakase, H. The Phagocytic Function of Macrophage-Enforcing Innate Immunity and Tissue Homeostasis. *Int. J. Mol. Sci.* **2018**, *19*, 92. [CrossRef]
42. Messina, G.; Polito, R.; Monda, V.; Cipolloni, L.; Di Nunno, N.; Di Mizio, G.; Murabito, P.; Carotenuto, M.; Messina, A.; Pisanelli, D.; et al. Functional Role of Dietary Intervention to Improve the Outcome of COVID-19: A Hypothesis of Work. *Int. J. Mol. Sci.* **2020**, *21*, 3104. [CrossRef]
43. Ren, G.-Y.; Chen, C.-Y.; Chen, G.-C.; Chen, W.-G.; Pan, A.; Pan, C.-W.; Zhang, Y.-H.; Qin, L.-Q.; Chen, L.-H. Effect of Flaxseed Intervention on Inflammatory Marker C-Reactive Protein: A Systematic Review and Meta-Analysis of Randomized Controlled Trials. *Nutrients* **2016**, *8*, 136. [CrossRef] [PubMed]
44. Zhu, L.; Sha, L.; Li, K.; Wang, Z.; Wang, T.; Li, Y.; Liu, P.; Dong, X.; Dong, Y.; Zhang, X.; et al. Dietary Flaxseed Oil Rich in Omega-3 Suppresses Severity of Type 2 Diabetes Mellitus via Anti-Inflammation and Modulating Gut Microbiota in Rats. *Lipids Health Dis.* **2020**, *19*, 20. [CrossRef] [PubMed]
45. Zhang, C.; Zarepoor, L.; Lu, J.T.; Power, K.A. Functional Foods and Gut Health. In *Nutraceuticals and Functional Foods: Natural Remedy*; Brar, S.K., Kaur, S., Dhillon, G.S., Eds.; Nova Science, Inc.: New York, NY, USA, 2014; ISBN 9781629487939.

Article

An Insight into the Level of Information about Sustainability of Edible Insects in a Traditionally Non-Insect-Eating Country: Exploratory Study

Raquel P. F. Guiné [1,2,*], Sofia G. Florença [1,3], Ofélia Anjos [4,5,6], Paula M. R. Correia [1,2], Bruno M. Ferreira [7] and Cristina A. Costa [1,2]

1 Agrarian School of Viseu, Polytechnic Institute of Viseu, 3500-606 Viseu, Portugal; sofiaguine@gmail.com (S.G.F.); paulacorreia@esav.ipv.pt (P.M.R.C.); amarocosta@esav.ipv.pt (C.A.C.)
2 CERNAS Research Centre, Polytechnic Institute of Viseu, 3504-510 Viseu, Portugal
3 Faculty of Food and Nutrition Sciences, University of Porto, 4200-465 Porto, Portugal
4 School of Agriculture, Polytechnic Institute of Castelo Branco, 6001-909 Castelo Branco, Portugal; ofelia@ipcb.pt
5 Forest Research Centre, School of Agriculture, University of Lisbon, 1349-017 Lisbon, Portugal
6 Centro de Biotecnologia de Plantas da Beira Interior, 6001-909 Castelo Branco, Portugal
7 School of Technology and Management of Viseu, Polytechnic Institute of Viseu, 3504-510 Viseu, Portugal; morgado.ferreira@estgv.ipv.pt
* Correspondence: raquelguine@esav.ipv.pt

Abstract: Insects have been reported as a possible alternative solution to help feed the growing world population with less stress on the planet, thus contributing to the preservation of the environment and natural ecosystems. However, the consumption of edible insects (EIs), although culturally accepted for some communities, is not readily accepted for others. Hence this work explores the level of information that people in a traditionally non-insect-eating country have about the sustainability issues related with EIs, and also some possible reasons that could motivate their consumption. The study was based on a questionnaire survey and the results were explored by descriptive statistic tools, tree classification analysis, factor analysis and cluster analysis. The results showed that the level of information is still low in general, with most people not manifesting an opinion. However, some aspects are relatively familiar to the participants (88.9% know that the ecological footprint of insects is smaller than other meats and 86.9% know that they efficiently convert organic matter into protein). Factor and cluster analysis showed three classes: cluster 1—people not informed about the facts disclosed through the true statements and also not able to distinguish the false information; cluster 2—people not informed about the facts disclosed through the true statements but who were able to distinguish the false information; and cluster 3—people well informed about the facts disclosed through the true statements but who were marginally unable to distinguish the false information. It was also found that education, sex and professional area are the most relevant sociodemographic factors associated with the level of information, and the highest motivations to consume EIs are their contribution to preserve the environment and natural resources followed by being a more sustainable option (for 64.7% and 53.4% of participants, respectively). Hence it was concluded that, although some work still needs to be done to better inform people about EIs, there is already some conscientiousness that they constitute a good and more sustainable alternative to other types of meat.

Keywords: edible insects; sustainability; information; questionnaire survey

Citation: Guiné, R.P.F.; Florença, S.G.; Anjos, O.; Correia, P.M.R.; Ferreira, B.M.; Costa, C.A. An Insight into the Level of Information about Sustainability of Edible Insects in a Traditionally Non-Insect-Eating Country: Exploratory Study. *Sustainability* **2021**, *13*, 12014. https://doi.org/10.3390/su132112014

Academic Editor: Piotr Prus

Received: 29 September 2021
Accepted: 28 October 2021
Published: 30 October 2021

1. Introduction

The planet faces in the modern times a most prominent challenge associated with the need to feed the increasing world population, while producing food in sustainable ways, so as to preserve the environment and the biosystems [1]. The sustainability aspects must be sought throughout the entire food chain, from primary production to industrial processing, transport and storage, consumption and final disposal of leftovers and packaging materials [2,3].

Intensive food production is leading to unsustainable practices around the planet with consequences, such as global warming due to increasing greenhouse gas (GHG) emissions, loss of natural habitats and deforestation or animal overexploitation, with these stresses being caused by both vegetable and animal food production [4–7]. It has been reported that nearly 80% of the GHG emissions resulting from the food sector derives from livestock, including the emissions generated from forage growing, cattle rearing, transportation of meat to the processing companies and to the sales points [8].

Insects have emerged as one of the possible solutions to help feed the world population with a lower impact on the environment. One of the reasons for this is associated with the high feed conversion ratio, which allows to obtain animal protein with a considerably lower need for land, feed or water, while at the same time generating less GHG. All these result in a much lower ecological footprint [9–14].

EIs are a good source of nutrients, besides protein and their essential amino acids, they also contain fiber, fat (including polyunsaturated fatty acids), vitamins (particularly those of group B) and dietary minerals (for example calcium, iron and magnesium) [15–17]. However, they may also contain anti-nutrients, such as oxalates and phytic acid [18]. On the other hand, there is some debate about the food safety aspects that may affect not only producers but also consumers of EIs [19]. Some risks are associated with a possible microbial contamination or with chemical hazards, such as toxins or heavy metals. Additionally, while for some people eating insects is safe, for others it may be problematic because sensitive people may suffer from allergic reactions [20].

Although billions of people consume insects in many countries worldwide [21], in other cultures eating insects, especially among individuals from Western societies, is still a taboo and people experience a high degree of neophobia with this practice. This is particularly intense if the insects are presented whole, making the consumers more prone to not start eating foods that contain insects [22–24].

Portugal is a country situated in Europe, more precisely on the Iberian Peninsula, and therefore the dietary habits are Western and traditional diets are typically Mediterranean. The Mediterranean diet was recognized by UNESCO (United Nations Educational, Scientific and Cultural Organization) as an Intangible Cultural Heritage of Humanity in 2010, initially applying to four Mediterranean countries (Greece, Italy, Morocco and Spain), but in 2013 the list of countries was expanded to include Cyprus, Croatia and Portugal [25]. Therefore, this study aimed to explore how people perceive EIs in a country where eating insects is a strange habit because it is not part of the traditional dietary patterns. Additionally, and having in mind the need to have in the near future more sustainable diets in order to feed the world population, this study also intends to investigate whether people are informed about the role of EIs as a possible more sustainable food in the future. In particular, our research questions were as follows: (1) Are people informed about the sustainability aspects that relate to EIs, either associated with their production or consumption? (2) What sociodemographic factors may influence people's level of information? (3) What reasons could influence people to consume EIs?

2. Literature Review

In this section are presented some insights into the scientific literature related to each of the three research questions addressed in this study.

2.1. Information about the Sustainability of EIs

The potential of EIs to constitute a more environmentally friendly alternative to other protein sources, for example beef, has been pointed out as a significant advantage that could potentially influence people towards a better acceptance of entomophagy [1].

Verbeke [2] studied the consumer acceptance of EIs in a sample of 368 consumers in Belgium, by means of an online survey, and one of their explanatory variables was the attention paid to the environmental impact of food choice. They observed significant effects of food choice motivation from the importance people attribute to the environmental impact of the food they consume. They quantified this influence as so: "an increase of one unit in the importance attached to the environmental impact of food choice increased the likelihood of being ready to adopt insects by 71%" [2].

Kostecka et al. [3] have investigated insect-based food acceptance among a sample of Polish consumers under a view of less resource-consuming food systems. They concluded that although Polish consumers are not prone to incorporating insect-based food into their diet, they recognized the importance of the food sector to the preservation of natural resources. In this way, they believe that consumers must be informed about the advantages of production or use of insect biomass originating from natural ecosystems. As a result, increasing the acceptance of alternative sources of protein may contribute to an effective reduction in the pressure generated by the food systems on the environment [3]. Although Kostecka et al. [3] did not evaluate the knowledge of consumers about facts related with sustainability of EIs, it gives some insight into the role of knowledge and information in the food choice process.

The study by Lensvelt and Steenbekkers [4] investigated consumer acceptance of edible insects through an online survey conducted with 134 participants from the Netherlands and 75 from Australia. They concluded that information is one of the key factors to positively influence the participant's willingness towards entomophagy, and therefore, "education is a pivotal key to be addressed". However, this research focused on information about social norms and trust and on information about physiological factors, and therefore the sustainability was not directly studied.

2.2. Sociodemographic Factors That Influence People's Knowledge

It has been widely known that sociodemographic characteristics influence people's food behavior, including food choices. The work by Guiné et al. [5] investigated the influence of environmental issues on consumer's food choice, by an online questionnaire survey using a non-probabilistic sample of 10,067 participants from 13 countries. Their results showed that people attributed importance to the sustainability of their food choices. They also reported significant differences in the motivations for food choice across sociodemographic groups (age, sex, marital status, education, professional area, living environment and country), with country being the more influential variable, followed by age and sex.

In related research, Guiné et al. [6] reported that most consumers admitted to basing their food choices on some environmental issues. Additionally, they conducted a factor analysis that showed two factors: purely environmental concerns and sustainability related to quality concerns. They also conducted a cluster analysis which allowed them to conclude that more than half of participants paid attention to both types of concerns when making their food choices, which is indicative that consumers are becoming more aware of their role in the sustainability of the food chain.

Sarić et al. [7] reported for a sample of 1534 participants from Croatia that sociodemographic factors influenced the food choices for more sustainable options. In their research they reported that older and female participants with higher education level (university degree) and married were more concerned about environmental friendly food choices.

There are other works that address the effect of sociodemographic factors on the way people act and how they attribute importance to sustainability. However, although there is some environmental awareness about EIs, no specific studies were found that focused on the evaluation of the level of knowledge about sustainability of EIs and the way that knowledge varies across sociodemographic groups.

2.3. Motivations to Consume EI

Although insects are a highly appreciated food source in numerous parts of the world, it is also a known fact that for most Western cultures EIs are not considered as an appropriate food source, and therefore negative attitudes continue to be dominant [8].

Some studies carried out in different countries highlight that individuals who pay attention to the environmental impact of the foods they consume, and who are informed about the ecological benefits of EIs, are more open to entomophagy [9].

The study by Tan et al. [10] conducted a cross-cultural qualitative study investigating how cultural exposure and individual experience can shape the perceptions of people who usually consume EIs and those who do not. They used eight focus groups, four in the Netherlands and four in Thailand, being a total of 54 participants. One of the factors that was pointed out by some participants as motivation for consumption was sustainability.

Niva and Vainio [11] investigated consumers' willingness to replace the consumption of beef by alternative sources, including insect-based protein products, in a sample of 1000 Finish consumers. They observed that a quarter of participants intended to increase the consumption of insect-based products, and this was driven by a wish to comply with more sustainable systems.

The work by Orsi et al. [9] addressed the determinants of consumer acceptance of EIs in Germany through an online survey on a sample constituted by 393 participants. Their study revealed a low willingness of Germans to try insects as food. They also were able to identify some obstacles to the consumption related with the prevalence of psychological and personality barriers, including a sense of disgust and food neophobia. Nevertheless, they also found that processed insect products might be a better solution to introduce EIs into the diets of Germans than whole insects. However, this study did not consider the effect of EIs as a possible more sustainable source of protein.

In a study conducted through an online survey with 820 Australian consumers, Wilkinson et al. [12] reported that factors, such as taste, appearance, safety and quality could motivate the willingness to try eating insects. Nevertheless, the consumer's attitudes towards EIs were relying to a great extent on food neophobia. Again, the authors found that the incorporation of insects into familiar products (e.g., biscuits, snacks) or cooked meals could improve the motivation to consume them, but no attempt was made to investigate if sustainability could influence the willingness to consume EIs.

3. Materials and Methods

This research was based on a questionnaire survey undertaken through internet invitation. The instrument used to collect the data was developed under the objectives of project "FZ—Drone Flour", which is under development and aims to investigate the technological possibility to produce innovative drone flour to commercialize in the Portuguese food market. This flour was obtained from the beehives held by the Portuguese beekeepers, as a way to mitigate the harmful effects of the Varroa mite in the beehives, while at the same time having a socioeconomic impact, providing extra income to the farmers.

3.1. Instrument and Data Collection

The questionnaire was developed purposely for this work, and submitted to the ethics committee at the Polytechnic Institute of Viseu, who approved it under ref. no. 06/SUB/2020, dated 11 September 2020. The full questionnaire is shown in Appendix B. Only after approval the questionnaire was deployed into the Google Forms platform, ensuring the anonymity of all answers received. The participation was voluntary and data collection occurred between September and October 2020. All ethical issues were respected when designing the research and collecting the data, and the participants only answered the questionnaire after giving informed consent or declaring that they were 18 years of age or older.

Taking into count the nature of the data collection strategy, the questionnaire was applied to a convenience sample, defined in terms of facility of recruitment and disposition to take part in the research. It is an unquestionable fact that convenience samples have some limitations, namely in what concerns the extrapolation of the conclusions to the whole population. However, they have also been reported as having some advantages, namely being easy to recruit and providing a good tool to undertake exploratory research [26,27].

Although being a convenience sample, some hint of the possible minimum sample size was calculated, as indicated. For this, some assumptions were considered:

- confidence interval = 90%;
- Z score = 1.645;
- power of the test = 5% (minimum acceptable probability of preventing type II error = 0.05) [28,29];
- Portuguese population in 2019 (the latest year available when the data collection started) = 10,283,822 people [30]: assumed that ~7.5 million were adults and the target population was 25% = 1875 thousand.

We targeted only 25% of the population in this research because it was assumed that in a Western country (situated in Europe under influence of Mediterranean diet) where eating insects it not natural or traditional in any way, it might not be expected that more than a quarter of the population might be interested in the near future to shift to this kind of food product. Considering all aforementioned conditions, calculation of the minimum sample size resulted in 203 adults [31,32].

3.2. Data Analysis

The data were analyzed using SPSS software V26 (IBM, Inc., Armonk, NY, USA) and Excel 2016 (Microsoft Corporation, Redmond, WA, USA).

The crosstabs with the chi-square test and Fisher's test were used to investigate the relations between some research variables and the sociodemographic categorical variables. Furthermore, the Cramer's V coefficient was used to quantify the intensity of the significant associations found between variables (considering a level of significance of 5%). The value of V varies between a minimum of zero (corresponding to no association) and a maximum of one (when the association is perfect). Indicative values were considered as the following [15,33]: $V \approx 0.1$—weak association; $V \approx 0.3$—moderate association; and $V \approx 0.5$ or higher—strong association.

The different items used to assess the level of information about sustainability issues related with EIs were submitted to a Factor Analysis (FA) for possible reduction of constructs. FA was completed using extraction by Principal Component Analysis (PCA) method, with quartimax rotation and using the scree plot to determine the number of factors. The percentage of variance explained by the factors extracted was based on the communalities [34]. Factor loadings with absolute below 0.5 were excluded, meaning that variables which had at least 25% of variance explained were only considered in the analysis. Internal consistency of the factors was evaluated through the Cronbach's alpha (α) [34,35].

Cluster Analysis (CA) started by applying five hierarchical methods based on the variables that resulted from the FA (scores saved as variables): (1) average linkage—between groups, (2) average linkage—within groups, (3) complete linkage—furthest neighbor, (4) centroid and (5) Ward. In all cases, it considered the measure for interval by the squared Euclidean distance. Based on the agglomeration schedule, it was possible to identify the most adequate number of clusters. Then, those five solutions were compared for similarity using contingency tables, which allows inferring about possible stability.

After establishing the number of clusters as three, the partitive method K-means was used, as it is commonly recommended for cluster analysis, due to its robustness [36]. The five initial solutions all converged to the same final solution, confirming the stability of the solution [36] and consequent confidence in the results.

Figure 1 presents a schematic representation of the procedures followed in the FA and CA.

Additionally, the items used to assess the level of information about sustainability of EIs were used to calculate an average score accounting for the general level of information, and this variable was submitted to a tree classification analysis against all the sociodemographic variables in the study, to investigate their relative importance to the level of information. For this, the CRT (classification and regression trees) algorithm was used with cross-validation [37], considering a minimum change in improvement equal to 0.001 and establishing the minimum number of cases equal to 10 for parent nodes and 5 for child nodes.

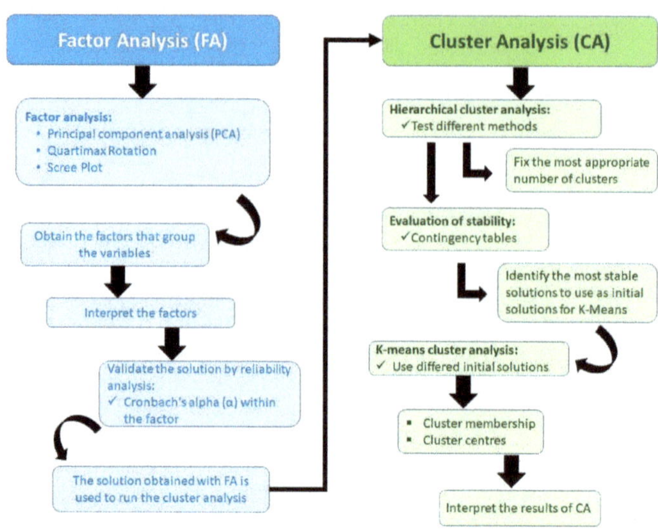

Figure 1. Methodology used to perform FA and CA.

4. Results

4.1. Sample Characterization

The sample consisted of 213 respondents, of whom most were female (79%), and a lower percentage were men (21%) (Table 1). The minimum age of the participants was 18 and the maximum was 80 years old. The participants were classified into groups according to the age, as follows: young adults (18–30 years), representing 24.4%, intermediate adults (31–55 years) accounting for 57.7% and senior adults (56 years or over) representing the remaining 17.8%. The majority of the participants, 78.4%, had a high education level (university graduate or post-graduate). Concerning the area of residence, most lived in urban environments (62.9%). Regarding marital status, most respondents (59.6%) had a life partner, i.e., were married or living together as a couple. Finally, the professional area of the participants was also investigated for its possible association with some variables of interest in the research. For this, it was specifically addressed if the participants were from areas related with food or nutrition (25.4%), agriculture, environment or biology (16.0%) or with other areas (58.7%).

Table 1. Sociodemographic characteristics of the sample used in the research.

Variable	Groups	N (%)
Sex	Women	168 (78.9)
	Men	45 (21.1)
Age group	Young adults (18–30 years)	52 (24.4)
	Intermediate adults (31–55 years)	123 (57.7)
	Senior adults (≥56 years)	38 (17.8)
Education level	Under university level	46 (21.6)
	University level (graduate or post-graduate)	167 (78.4)
Living environment	Urban	134 (62.9)
	Suburban/Rural	79 (37.1)
Marital status	No life partner (Single/Divorced/Widowed)	86 (40.4)
	With life partner (Living together/Married)	127 (59.6)
Professional area	Food/Nutrition	54 (25.4)
	Agriculture/Environment/Biology	34 (16.0)
	Other areas	125 (58.7)
	Total	**213 (100.0)**

4.2. Information about Sustainability Aspects That Relate to EIs

To answer research question (1) "Are people informed about the sustainability aspects that relate with EI, either associated with their production or consumption?", the results of the answers to a set of questions formulated to measure the participants knowledge/degree of information about sustainability facts related with EIs were used. Table 2 presents such results, highlighting the number of answers given to each question and the fraction corresponding to informed or not informed participants. From the 213 participants, a high number had not manifested an opinion about questions 1 to 7 in Table 2. The question that received most answers was Q1 with 160 responses. Among these, a great majority (82.5%) was correct in their perception, which indicates that the respondents were relatively well informed about the possibility of EIs providing protein to fight hunger in the world. On the other extreme end stands Q4, Q5 and Q6 with around 100 responses, all related with the comparison of the use of resources to produce protein. Nevertheless, most participants were well informed about these statements, except for Q5, which was formulated in the reverse mode, and that might have confused the respondents when answering it.

Table 2. Responses for the questions regarding sustainability issues related to EIs.

Facts about Sustainability of EIs (N [1])	Not Informed N (%)	Informed N (%)
Q1. Insects are a possibility to respond to the growing world demand for protein (N = 160).	28 (17.5)	132 (82.5)
Q2. The production of insects for human consumption emits about 10 times less greenhouse gases (GHG) than the production of steak (N = 115).	16 (13.9)	99 (86.1)
Q3. Insects efficiently convert organic matter into protein (N = 122).	16 (13.1)	106 (86.9)
Q4. To produce 1 kg of insect protein, 5 times less food is spent than to produce 1 kg of cow protein (N = 97).	14 (14.4)	83 (85.6)
Q5. To produce 1 kg of chicken protein, 5 times less water is used than to produce 1 kg of insect protein (N = 92) [2].	45 (48.9)	47 (51.1)
Q6. To produce 1 kg of insect protein requires an area 3 times smaller than to produce 1 kg of pig protein (N = 102).	18 (17.6)	84 (82.4)
Q7. The ecological footprint of insects is comparatively smaller when compared to other sources of protein for human consumption (N = 137).	14 (10.2)	123 (89.8)

[1] N = Number of respondents who expressed their opinion on each of the questions. [2] This is a false statement.

The seven questions accounting for sustainability issues related with EIs were submitted to FA with PCA extraction and quartimax rotation, resulting in two components or factors. The total variance explained by the factors was: F1—80.3% and F2—12.3%, with a high cumulative variance of 92.6% explained. The communalities showed that all variables had high variance explained by the solution, with the lowest being for Q3 which was still high (0.820, corresponding to 82% of variance explained). The rotation algorithm converged in three iterations and the results of FA are shown in Table 3.

Table 3. Results of the FA with extraction by PCA and quartimax rotation (factor loadings under 0.5 were excluded).

Items	Loadings	
	Factor F1	Factor F2
Q1. (True)	0.969	
Q2. (True)	0.986	
Q3. (True)	0.905	
Q4. (True)	0.986	
Q5. (False)		0.913
Q6. (True)	0.943	
Q7. (True)	0.932	
Cronbach's alpha	**0.980**	**(*)**

(*) Not calculated because there was only one variable in the factor.

The structure of the factors in Table 3 allows identifying factor F1 as linked with all the true statements, while factor F2 was associated with the false information. All variables presented very high loadings in the factors, being the lowest for Q3 in F1 (0.905), but still corresponding to a very high correlation. Because all the seven variables had loadings with an absolute value higher than 0.5, the solution was satisfactory by including all the variables [38]. Additionally, this solution resulted in a grouping configuration that can be easily interpreted.

Validation through Cronbach's alpha (α) [34] was only possible for factor F1, since F2 included only one variable. The results showed that the internal consistency within factors was 0.980, which is considered very good [39–41]. Furthermore, the value of alpha did not increase by elimination of any of the items, thus meaning that F1 presented a very strong internal uniformity.

The scores obtained with FA were used for the Cluster Analysis (CA). In a first step, CA was applied by five different hierarchical methods in order to define the number of clusters, which in the present case was found to be three according to Figure A1 in Appendix A.

The compatibility between the solutions obtained with the five hierarchical methods, when the number of clusters was fixed as three, was checked though contingency tables, and these results are presented in Appendix A (Table A1). The values of the percentages indicated that the solutions obtained by all the methods converge to a single solution, i.e., the percentage of the cases allocated to the same clusters is the maximum. These results confirm that the ideal number of clusters is three, and that any of the five solutions tested previously were potentially stable, and therefore can be used as an initial solution to the next step, that is to apply the K-means clustering analysis. Furthermore, the application of the K-means to the five initial solutions confirmed that they all converge to a single solution, thus proving stability. The values of the F statistic in ANOVA are high for both factors ($F_{statistic} = 901.6$ for F1 and $F_{statistic} = 2109.6$ for F2, with $p < 0.0005$ in both cases), thus confirming the similarity between the cases within the groups and the differences between groups. Additionally, because both values of $F_{statistic}$ are of similar order of magnitude, they both equally contribute to the discrimination of the groups. Figure 2 shows the final cluster canters that were confirmed by the results of the K-means CA.

The interpretation of the clusters is as follows:

- Cluster 1: people not informed about the facts disclosed through the true statements, and are also not able to distinguish the false information;
- Cluster 2: people not informed about the facts disclosed through the true statements, but who were able to distinguish the false information;
- Cluster 3: people well informed about the facts disclosed through the true statements, but who were marginally unable to distinguish the false information.

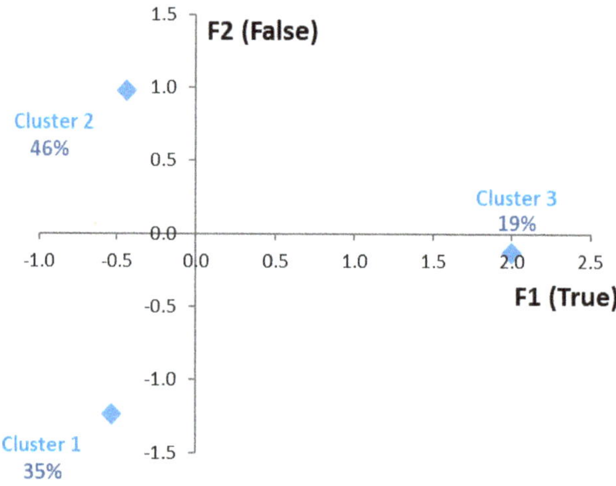

Figure 2. Cluster centers in relation to the factors.

4.3. Influence of Sociodemographic Factors on People's Level of Information

To answer research question (2) "What sociodemographic factors may influence people's level of information?", the responses to the questions Q1 to Q7 were tested against the sociodemographic variables using contingency tables and the chi-square test, being the results presented in Table 4. The results are resented as % in row for each variable and each question because this allows to eliminate the effect of uneven group distribution within each sociodemographic variable. For example, for variable sex, it is % of informed and % of not informed within each group: women or men.

The results showed that for most topics (Q1 to Q7), the sociodemographic variables tested are not associated with the level of information, just with two exceptions:

(1) The information for Q5 (to produce 1 kg of chicken protein, five times less water is used than to produce 1 kg of insect protein), given as a false statement, varies significantly with age ($p = 0.011$) and the association is moderate (V = 0.312);

(2) The information for Q5 varies significantly with marital status ($p = 0.030$) and the association is moderate (V = 0.219).

Table 4. Association between the sociodemographic variables and the responses to questions about EIs and sustainability.

Sociodemographic Variables/Groups	Q1		Q2		Q3		Q4		Q5		Q6		Q7	
	I [1] (%)	NI [2] (%)	I [1] (%)	NI [2] (%)	I [1] (%)	NI [2] (%)	I [1] (%)	NI [2] (%)	I [1] (%)	NI [2] (%)	I [1] (%)	NI [2] (%)	I [1] (%)	NI [2] (%)
Sex														
Women	80.0	20.0	84.4	15.6	85.1	14.9	82.4	17.6	51.4	48.6	80.3	19.7	87.7	12.3
Men	90.0	10.0	92.0	8.0	92.9	7.1	95.7	4.3	50.0	50.0	88.5	11.5	96.8	3.2
p-value [3]	0.112		0.272		0.234		0.102		0.556		0.265		0.127	
V [4]	0.114		0.090		0.097		0.160		0.011		0.094		0.125	
Age														
Young adults	81.6	18.4	84.0	16.0	84.6	15.4	87.5	12.5	29.2	70.8	82.1	17.9	94.4	5.6
Intermediate adults	80.4	19.6	84.7	15.3	86.7	13.3	82.1	17.9	64.2	35.8	80.0	20.0	86.3	13.8
Senior adults	92.0	8.0	94.4	5.6	90.5	9.5	94.1	5.9	40.0	60.0	89.5	10.5	95.2	4.8
p-value [5]	0.391		0.535		0.836		0.447		0.011		0.646		0.270	
V [4]	0.108		0.104		0.054		0.129		0.312		0.093		0.138	
Education														
Under university	78.1	21.9	83.3	16.7	81.8	18.2	81.8	18.2	47.8	52.2	76.9	23.1	86.2	13.8
University	83.6	16.4	86.6	13.4	88.0	12.0	86.7	13.3	52.2	47.8	84.2	15.8	90.7	9.3
p-value [3]	0.311		0.475		0.318		0.395		0.425		0.286		0.339	
V [4]	0.058		0.034		0.070		0.058		0.038		0.083		0.061	
Living Environment														
Urban	80.7	19.3	84.1	15.9	87.5	12.5	86.1	13.9	53.1	46.9	77.8	22.2	92.0	8.0
Suburban/Rural	83.5	16.5	87.3	12.7	86.5	13.5	85.2	14.8	50.0	50.0	84.8	15.2	88.5	11.5
p-value [3]	0.405		0.411		0.550		0.579		0.474		0.263		0.368	
V [4]	0.035		0.045		0.015		0.012		0.030		0.089		0.056	
Marital status														
No life partner	78.8	21.2	84.4	15.6	81.8	18.2	86.1	13.9	37.1	62.9	82.5	17.5	94.4	5.6
With life partner	85.1	14.9	87.1	12.9	89.7	10.3	85.2	14.8	59.6	40.4	82.3	17.7	86.7	13.3
p-value [3]	0.204		0.442		0.167		0.579		0.030		0.597		0.120	
V [4]	0.082		0.038		0.113		0.012		0.219		0.003		0.124	
Professional area														
Food/Nutrition	87.2	12.8	87.9	12.1	90.3	9.7	87.5	12.5	44.0	56.0	82.6	17.4	91.7	8.3
Agric./Env./Biol.	93.3	6.7	90.5	9.5	91.3	8.7	87.5	12.5	50.0	50.0	88.2	11.8	95.5	4.5
Other areas	76.9	23.1	83.6	16.4	83.8	16.2	84.2	15.8	54.9	45.1	80.6	19.4	87.3	12.7
p-value [5]	0.082		0.691		0.529		0.902		0.668		0.767		0.491	
V [4]	0.177		0.080		0.102		0.046		0.094		0.072		0.102	
Total	100	100	100	100	100	100	100	100	100	100	100	100	100	100

[1] I = informed. [2] NI = not informed. [3] Significance of the Fisher's test (level of significance of 5%). [4] V = Cramer's V coefficient. [5] Significance of the chi-square test (level of significance of 5%).

As a complement to the study, a new variable was considered as the average score of all the seven questions, to account for a global level of information for each participant. This new variable was submitted to a tree classification analysis to investigate the relative importance of the sociodemographic variables in the level of information. Figure 3 reveals that the first discriminating sociodemographic variable is education, with the participants with a university level of education being more informed. In level 2 for the people without university education, the next discriminating variable is sex, and then for women the next discriminating variable is professional area. On the other hand, for the ones who had a university degree, the discriminant for level 3 is professional area, in this case the professionals from other areas being less informed than people from Food/Nutrition or Agriculture/Environment/Biology. In level 4, the participants from other areas were separated according to marital status, with the highest information for people with a life partner as compared with those without. Nevertheless, in both these two groups, the last discriminant was living environment.

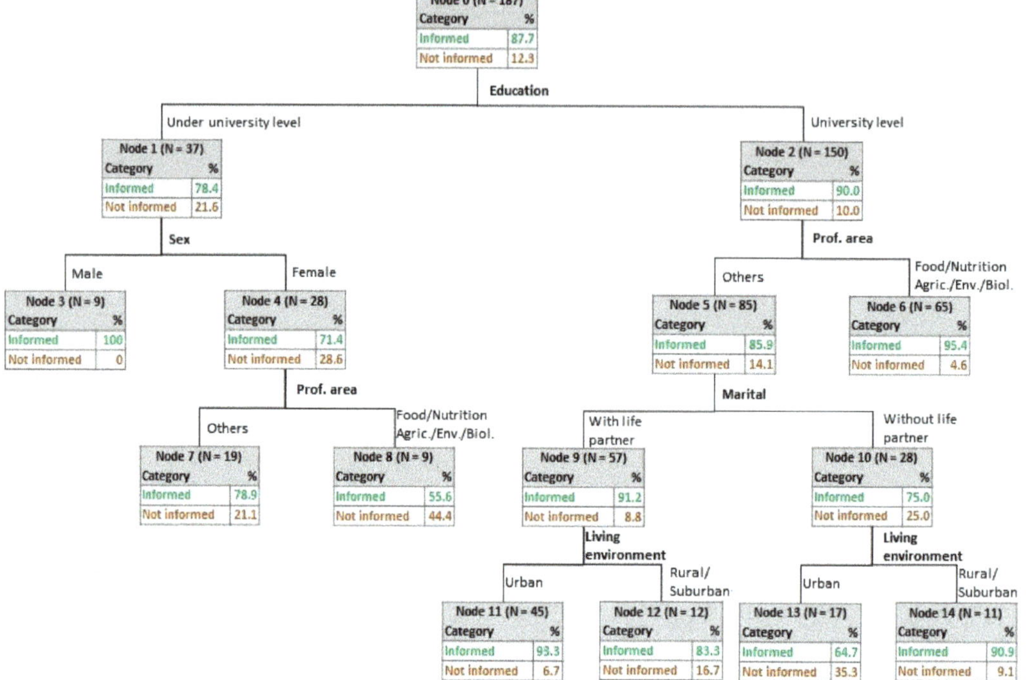

Figure 3. Classification tree for influence of sociodemographic variables on the general level of information.

4.4. Motivations to Consume EIs

To answer the last research question (3) "What reasons could influence people to consume EIs?", a number of possible factors that could motivate people were investigated and the results are presented in Table 5. The reasons that could be more motivating for people to consume EIs included contributing to the preservation of the environment and natural resources (64.7% motivated by this aspect), being a more sustainable option (53.4% motivated by this) or serving to increase the income of the producers' families (50.9% motivated by this). The willingness to follow innovative trends or mimic personalities/influencers is a very weak motivation for the participants (only 8.9% admit this possibility).

Table 5. Responses for the questions regarding motivations to consume EIs.

Possible Motivations to Consume EIs (N [1])	Not Motivated N (%)	Motivated N (%)
Q8. Insects are a more sustainable option (N = 176).	82 (46.6)	94 (53.4)
Q9. Desire to try exotic foods (N = 174).	126 (72.4)	48 (27.6)
Q10. Insects contribute to the preservation of the environment and natural resources (N = 167).	59 (35.3)	108 (64.7)
Q11. Insects contribute to the diversification of food production (N = 169).	90 (53.3)	79 (46.7)
Q12. Insects are a way to increase the income of families that produce them (N = 159).	78 (49.1)	81 (50.9)
Q13. Willing to follow innovative trends or mimic personalities/influencers (N = 192).	175 (91.1)	17 (8.9)
Q14. EIs provide protein foods at cheap prices (N = 164).	92 (56.1)	72 (43.9)

[1] Number of respondents who expressed their opinion.

5. Discussion

Regarding the research question (1) "Are people informed about the sustainability aspects that relate to EIs, either associated with their production or consumption?", the first finding of this study clearly shows that most participants are quite well informed about the facts related to sustainability. These results are in line with the works by Verbeke [2] and Kostecka et al. [3], which concluded that environmental issues are on the rise as factors shaping food choices. Nevertheless, in those studies, the level of knowledge was not assessed as it was in the present study, and therefore, it constitutes a novel approach to quantify the level of knowledge by means of people's accordance with true and false statements provided. To this matter, it was found that half of the participants still accepted false information as true. One explanation for this can be related to the difficulty that people in general experience on a daily basis due to the saturation of information that comes from advertising and online social networks. Fake news, fake videos or biased comments also contribute to this ascertainment. The cluster analysis also confirms this core result, because only one small cluster aggregates the participants that are well informed about the facts which were presented to them in true statements, but who still could not discern the false statements. This result also exposes a lack of skills to distinguish real and fake information [42–45].

The sector of EIs as food or food ingredients is an emerging agricultural sector, with higher potential to grow and a lower associated environmental impact [46]. However, there are some issues that must be identified related to attitudes and knowledge of the consumers towards these products. Additionally, an increase in EI consumption globally could only be possible through insect farming, due to the unfeasibility collecting them in high quantities from natural habitats.

Production of meat foods is responsible for high emissions of GHG, and the results obtained in this work confirm that there is knowledge about this fact, as well as about its association to climate change effects that require urgent dietary changes. In general, a higher percentage of the inquired are well informed about the importance of insects in the food supply chains as a substitute for meat, which is in accordance with an increase in the market for insects or insect protein that may further develop in non-insect-eating countries. However, the cluster analysis evidenced that the possible future consumers need more information to adhere to EIs consumption. Our results showed that practically all of the participants know very well that the footprint of insects is comparatively smaller when compared to other sources of protein for human consumption. Is well known that insects' production emits a considerable lower percentage of methane than cattle, and provide more protein than chicken and beef [9]. This study indicated a good knowledge about these facts. Nevertheless, even though the nutritional value and the positive effects of EIs on the environment are well understood and could be a positive incentive for eating insects, some studies reveal that the sensory aspects and overall experience of eating insects could be an impediment, given the disgust or unfamiliarity issues that can prevent consumers from accepting EIs as food [47]. Additionally, according to Hartmann and Siegrist [48], consumers' disposition to eat EIs is weak.

Regarding the research question (2) "What sociodemographic factors may influence people's level of information?", in a first approach, and considering that women were more participative in this study than men, we could deduce that they would possibly be more interested in this topic. However, it is a fact that the willingness of women to participate in surveys is in general higher than men, regardless of the topic. Therefore, the disposition to participate is not a direct indicator of better acceptability of EIs as foods. To this matter, we must analyze the effect of food neophobia on the participants. Food neophobia could affect the variety of foods in the diet, because some people tend to avoid the consumption of unfamiliar foods [49]. On the other hand, food neophobia could protect the individuals from ingesting possibly toxic or nutritionally inadequate foods, if the consumers are not well informed [50]. Some studies identified men as less food neophobic and less disgusted by insects than women [51,52]. In concordance with this, for the Portuguese sample, the cluster results showed that men are more likely to accept insects as food than woman. This result is also in line with Verbeke [51], which suggests that attitudes towards the use of insects in feed and food in general were significantly more favorable among males than females [51].

The acceptance of EIs as food can be variable according to cultural, geographical, personal and emotional factors [51]. Eating insects is very common in some countries, but can be very disturbing for many people in other countries [53]. In this study, there were not a lot of participants that would be motivated to eat EIs because they desire to try exotic foods. This antipathy could be related to the historical and cross-cultural belief that EIs are disgusting and not edible by humans, except in cases of hunger or malnourishment. In some European countries, such as Belgium [24], Netherlands [54] or Finland [55], a moderate acceptance for EIs has also been shown. The use of EIs as food ingredients might also help increase the adoption of insect-based foods [56–58].

Globally, because responses related to the queries about sustainability of EIs reveal a good knowledge of the positive effects of using EIs as food or food ingredients, we could infer that this could possibly be a good indication that neophobia might not affect the introduction of EIs into the Portuguese consumer's diet. This would be in line with the trend to look for more sustainable foods, so as to preserve the planet resources and defend the natural ecosystems [59]. Portuguese consumers are aware that EIs can help solve some environmental problems by promoting sustainable food choices, in line with other studies that showed consumers have the perception that diets have to adapt to a more sustainable processing and to more environmental friendly food chains [10,60,61].

In regard to research question (3) "What reasons could influence people to consume EI?", among the motivations evaluated, the participants highlighted that aspects, such as being a more sustainable option or contributing to the preservation of the environment and natural resources were stronger motivations. However, the results also show that participants claim that they are not motivated enough to follow innovative trends or mimic personalities or influencers, so these are not aspects valued by possible future consumers of EIs in a non-insect-eating country, such as Portugal. Nevertheless, these results could exhibit cognitive biases when observing the current trends and the time spent by humans on social media. FAO considers insects a sustainable alternative source of animal protein that can respond to population growth. As shown in previous research, the lack of familiarity with EIs can contribute as a barrier in addition to cultural differences, in terms of acceptance of new food [42–45]. For instance, the price dimension combined with the more sustainable option and the fact that it contributes to the diversification of food production should be spotlight by ad campaigns in non-insect-eating countries.

From the point of view of the nutritional value of EIs, the participants also had a good perception that they are rich in protein, if we take into account that Portugal is non-insect-eating country. Due to the presence of proteins, unsaturated fats and fiber, richness in lysine, threonine and tryptophan amino acids [62], as well as in micronutrients, such as iron, zinc, calcium and vitamins [15–19], it is possible to classify EIs as a very good source of nutrients. Some of EIs are also particularly rich in chitin, an insoluble fiber derived from their exoskeleton, which has been found to improve immune responses in humans and decrease allergies [63]. However, EIs can also contain residues of pesticides and heavy metals from the ecosystem and cause human allergic reactions [64], and this point must be well studied in future.

Factors, such as (1) contributing to environmental care; (2) being a sustainable option; (3) economic benefits (increased income); (4) happiness; (5) food security; and (6) long life, can be used to promote consumption of foods from edible insects as a sustainable source of protein [54,65,66]. Some consumer studies disclosed that food choice is primarily motivated by price and health consequences [54,65], which is in line with some of the motivations observed in this study. Furthermore, the economic benefits (higher income) were a motivation highlighted in this work.

6. Conclusions

This study assessed the knowledge regarding information about the sustainability of EIs, either when the questions were related to true or with false information, in a traditionally non-insect-eating country. Additionally, it evaluated the motivations to consume EIs. The results showed that there is a good level of knowledge about sustainability aspects related to EI production and consumption, but also exposed the lack of ability to identify false information as fake. The levels of knowledge seem globally high, but when analyzed in detail, this study finds that the absence of factual knowledge leads to the fact that false information becomes relevant for many participants.

Some limitations can be pointed out, such as the inequality of sociodemographic groups, with more female participants than men, more people with a university degree or more people residing in urban areas as compared to rural environments, which can somewhat bias the study. This heterogeneity results from the fact that we had to use a convenience sample, and woman are more prone to answering questionnaire surveys than men. Furthermore, the contacts used to send the survey included more people with a university degree living in urban areas. Another limitation was related to the method of delivering the survey, through internet, but this limitation was caused by the pandemic situation that the world was facing at the time of the research.

This work brought added value to the identification of the national situation about perceptions of the Portuguese about EIs, and sustainability issues and their implications. It demonstrated that the general public's level of information in traditionally non-insect-eating countries, such as Portugal, needs to be improved and therefore it is imperative to adapt effective strategies to pass the message of sustainability to the wider public. Future educational strategies need to focus on the characteristics of the citizens in non-traditional insect eating countries, and look for ways to shift people's perceptions. As such, producers and brands must educate and inform possible future consumers on this topic. Additional actions could encompass free tastings in shops/supermarkets/restaurants, which in a more direct approach could help overcome some of the barriers for eating IEs. Nevertheless, industries and other actors in the food chain must take into account that some consumers would continue to feel fear, aversion or disgust towards EIs, and therefore not adopt these foods, even knowing about their environmental advantages.

Author Contributions: Conceptualization, R.P.F.G.; methodology, R.P.F.G.; software, R.P.F.G.; validation, R.P.F.G.; formal analysis, R.P.F.G. and S.G.F.; investigation, R.P.F.G., S.G.F. and P.M.R.C.; resources, P.M.R.C. and C.A.C.; data curation, R.P.F.G.; writing—original draft preparation, S.G.F., B.M.F., O.A. and R.P.F.G.; writing—review and editing, R.P.F.G.; visualization, R.P.F.G.; supervision, R.P.F.G.; project administration, C.A.C.; funding acquisition, R.P.F.G., P.M.R.C., B.M.F. and C.A.C. All authors have read and agreed to the published version of the manuscript.

Funding: This research was funded by CI&DETS Research Center (Polytechnic Institute of Viseu, Portugal) in the ambit of the project "FZ—Farinha de zângão: inovar no produto e na proteção da colmeia" from Polytechnic Institute of Viseu, Portugal, with reference PROJ/IPV/ID&I/013. Author Sofia Florença received financial support from FCT—Foundation for Science and Technology through a BII Grant from FCT in the ambit of the program "Verão com Ciência 2020" developed in the Polytechnic Institute of Viseu. The APC was funded by FCT—Foundation for Science and Technology (Portugal) project Ref.ª UIDB/00681/2020.

Institutional Review Board Statement: This research was implemented taking care to ensure all ethical standards and followed the guidelines of the Declaration of Helsinki. The development of the study by questionnaire survey was approved on 11 September 2020 by the ethics committee of Polytechnic Institute of Viseu (Reference No. 06/SUB/2020).

Informed Consent Statement: Informed consent was obtained from all subjects involved in the study.

Data Availability Statement: Data are available from the corresponding author upon reasonable request.

Acknowledgments: This work was supported by the FCT—Foundation for Science and Technology, I.P. Furthermore, we would like to thank the CERNAS Research Center and the Polytechnic Institute of Viseu for their support. This work was prepared in the ambit of the project "FZ—Farinha de zângão: inovar no produto e na proteção da colmeia" from Polytechnic Institute of Viseu, Portugal.

Conflicts of Interest: The authors declare no conflict of interest.

Appendix A

Figure A1 shows the last 20 values of the coefficients of the agglomeration schedule for each hierarchical method tested to evaluate the number of clusters.

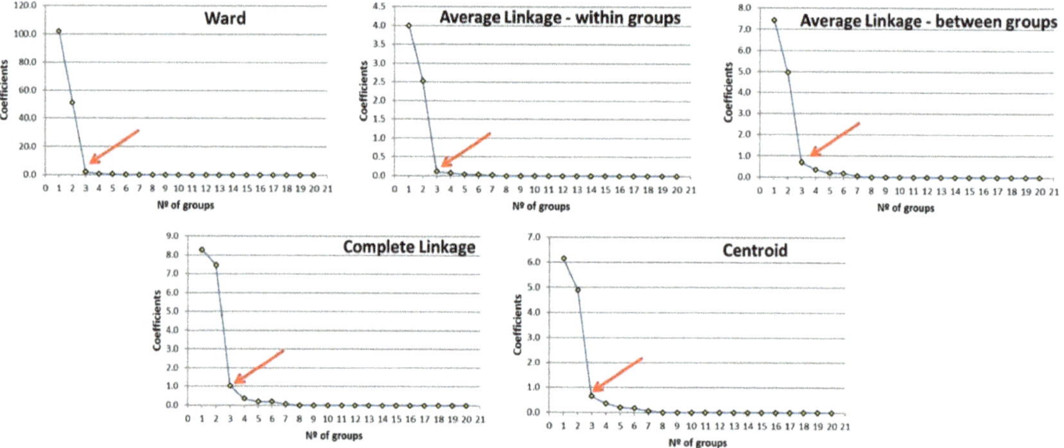

Figure A1. Determination of the number of groups by five hierarchical methods.

Table A1 presents the comparison of the solutions obtained with the five hierarchical methods tested.

Table A1. Comparison of the solutions obtained with the hierarchical methods.

Method [1]	AL-BG	AL-WG	CL-FN	CENT	WARD
AL-BG	—	—	—	—	—
AL-WG	100%	—	—	—	—
CL-FN	100%	100%	—	—	—
CENT	100%	100%	100%	—	—
WARD	100%	100%	100%	100%	—

[1] AL-BG: average linkage—between groups; AL-WG: average linkage—within groups; CL-FN: complete linkage—furthest neighbor; CENT: centroid.

Appendix B

In this appendix the full questionnaire is presented.

I. DEMOGRAPHICS

Age: _____ years

Sex: Female ☐ 1 Male ☐ 2

Highest level of education concluded:

Basic school (9 school years) ☐ 1
Secondary school (12 school years) ☐ 2
University degree ☐ 3
Post-graduate studies (master or PhD) ☐ 4

Living environment:

Rural ☐ 1 Urban ☐ 2 Suburban ☐ 3

Marital status:

Single ☐ 1 Married ☐ 2 Divorced ☐ 3 Widowed ☐ 4

Professional activity/studies related to any of the following areas:

Nutrition ☐ 1 Food ☐ 2 Agriculture ☐ 3 Environment ☐ 4
Biology ☐ 5 Health related activities ☐ 6 None of the previous ☐ 7

II. CHARACTERIZATION OF PARTICIPANT'S HABITS

How often do you eat in restaurants?

Rarely (less than once/month) ☐ 1
Sporadically (between once/week and once/month) ☐ 2
Occasionally (about once/week) ☐ 3
Moderately (2–3 times/week) ☐ 4
Frequently (4 or plus times/week) ☐ 5

When going to restaurants, what kind of establishments do you prefer? (you can choose more than one option)

Portuguese traditional food ☐ 1
Ethnic food (Chinese, Italian, Mexican, Indian, etc . . .) ☐ 2
Gourmet ☐ 3
Convenience food (fast-food) ☐ 4
No preference ☐ 5
Other ☐6 Which: ________________6.a

How often do you travel abroad?

Never ☐ 1
Rarely (about once/year) ☐ 2
Occasionally (about 2–3 times/years) ☐ 3
Frequently (more than 3 times/year) ☐ 4

When traveling abroad, you prefer the type of food you consume?

Typical food from the country visited ☐ 1
Food as similar as possible to Portuguese cuisine ☐ 2
International food (types of food commonly spread around the world) ☐ 3

III. PERCEPTIONS ABOUT EDIBLE INSECTS AND DERIVATIVES

Please indicate, on the scale between Strongly Disagree and Strongly Agree, your opinion on the following information

	Totally Disagree	⟷			Totally Agree	No Opinion
There are more than 2000 species of insects that are consumed by humans in the world	□1	□2	□3	□4	□5	□0
Entomophagy is a dietary practice that consists of the consumption of insects by humans	□1	□2	□3	□4	□5	□0
Some insects can be used to produce animal feed	□1	□2	□3	□4	□5	□0
There are flours for human food produced from insects	□1	□2	□3	□4	□5	□0
There is no consumption of insects in developed countries (INV)	□1	□2	□3	□4	□5	□0
In some European gourmet restaurants it is practice to use edible insects	□1	□2	□3	□4	□5	□0
Insects are part of the gastronomic culture of most countries in the world	□1	□2	□3	□4	□5	□0
Insect consumption is characteristic of less developed countries	□1	□2	□3	□4	□5	□0
Not all insects are edible	□1	□2	□3	□4	□5	□0
In Portugal there are regulations to ensure food safety in the case of edible insects (INV)	□1	□2	□3	□4	□5	□0
Insects are used by some people in traditional medicine	□1	□2	□3	□4	□5	□0

IV. KNOWLEDGE ABOUT EDIBLE INSECTS AND SUSTAINABILITY

Please indicate, on the scale between Strongly Disagree and Strongly Agree, your opinion on the following information

	Totally Disagree	⟷			Totally Agree	No Opinion
Insects are a possibility to respond to the growing world demand for protein	□1	□2	□3	□4	□5	□0
The production of insects for human consumption emits about 10 times less greenhouse gases than the production of beef	□1	□2	□3	□4	□5	□0
Insects efficiently convert organic matter into protein	□1	□2	□3	□4	□5	□0
To produce 1 kg of insect protein, it takes 5 times less food than to produce 1 kg of cow protein	□1	□2	□3	□4	□5	□0
To produce 1 kg of chicken protein, 5 times less water is used than to produce 1 kg of insect protein (INV)	□1	□2	□3	□4	□5	□0
To produce 1 kg of insect protein requires an area 3 times smaller than to produce 1 kg of pig protein	□1	□2	□3	□4	□5	□0
The ecological footprint of insects is comparatively smaller when compared to other sources of protein for human consumption	□1	□2	□3	□4	□5	□0

V. KNOWLEDGE ABOUT NUTRITIVE PROPERTIES OF EDIBLE INSECTS

Please indicate, on the scale between Strongly Disagree and Strongly Agree, your opinion on the following information

	Totally Disagree	⟵		⟶	Totally Agree	No Opinion
Edible insects are a good source of energy	□ 1	□ 2	□ 3	□ 4	□ 5	□ 0
Edible insects are poor in macro and micronutrients (INV)	□ 1	□ 2	□ 3	□ 4	□ 5	□ 0
Edible insects contain group B vitamins	□ 1	□ 2	□ 3	□ 4	□ 5	□ 0
Edible insects are very rich in animal protein	□ 1	□ 2	□ 3	□ 4	□ 5	□ 0
Insect proteins are of poorer quality compared to other animal species (INV)	□ 1	□ 2	□ 3	□ 4	□ 5	□ 0
Edible insects contain minerals of nutritional interest, such as calcium, iron and magnesium	□ 1	□ 2	□ 3	□ 4	□ 5	□ 0
Edible insects contain fat, including polyunsaturated fatty acids.	□ 1	□ 2	□ 3	□ 4	□ 5	□ 0
Edible insects contain bioactive compounds beneficial to human health	□ 1	□ 2	□ 3	□ 4	□ 5	□ 0
Edible insects contain anti-nutrients, such as oxalates and phytic acid	□ 1	□ 2	□ 3	□ 4	□ 5	□ 0
Some edible insects have a proven antioxidant effect	□ 1	□ 2	□ 3	□ 4	□ 5	□ 0
Some edible insects may have anti-inflammatory activity	□ 1	□ 2	□ 3	□ 4	□ 5	□ 0

VI. ATTITUDES REGARDING EDIBLE INSECTS AND DERIVATIVES

Have you ever consumed insects or derived products?

Yes □ 1 No □ 2 I don't know/don't remember □ 3

Under what circumstances did you consume insects or derived products?

a. In Portugal □ 1 Abroad □ 2

b. In a restaurant□ 1 In a hotel □ 2 On the street □ 3

At home □ 4 In the house of friends/family □ 5 In parties □ 6

c. By my own initiative □ 1 Encouraged by friends □ 2

Advised by restaurant professionals □ 3

Other □ 4 Which:____________4.a

Have you ever bought food containing insects?

Yes □ 1 No □ 2 I don't know/don't remember □ 3

If you have already bought food containing insects, where did you get them?

Supermarket □ 1 Internet □ 2 Specialized shop □ 3

Street market □ 4 Other □ 5 Which:__________ __________5.a

What is your acceptability to consume products that contain insect derivatives in their ingredients (e.g., snacks with insect meal)?

Definitely Would Not Eat	⟵		⟶	Definitely Would Eat
□ 1	□ 2	□ 3	□ 4	□ 5

What is your acceptability to consume dishes made with whole insects?				
Definitely Would Not Eat		⟷		**Definitely Would Eat**
☐ 1	☐ 2	☐ 3	☐ 4	☐ 5

What are the motivations that may encourage you to consume foods based on edible insects?

	Very Week Motivation	⟷			**Very Strong Motivation**
Being a more sustainable alternative	☐ 1	☐ 2	☐ 3	☐ 4	☐ 5
Wanting to try exotic foods	☐ 1	☐ 2	☐ 3	☐ 4	☐ 5
Contribute to the preservation of the environment and natural resources	☐ 1	☐ 2	☐ 3	☐ 4	☐ 5
Contribute to the diversification of food production	☐ 1	☐ 2	☐ 3	☐ 4	☐ 5
Contribute to increasing the income of families that can produce them	☐ 1	☐ 2	☐ 3	☐ 4	☐ 5
Follow trends/innovative fashions of personalities/influencers	☐ 1	☐ 2	☐ 3	☐ 4	☐ 5
Possibility of having protein foods at cheap prices	☐ 1	☐ 2	☐ 3	☐ 4	☐ 5

References

1. Ordoñez-Araque, R.; Egas-Montenegro, E. Edible Insects: A Food Alternative for the Sustainable Development of the Planet. *Int. J. Gastron. Food Sci.* **2021**, *23*, 100304. [CrossRef]
2. Krishnan, R.; Yen, P.; Agarwal, R.; Arshinder, K.; Bajada, C. Collaborative Innovation and Sustainability in the Food Supply Chain- Evidence from Farmer Producer Organisations. *Resour. Conserv. Recycl.* **2021**, *168*, 105253. [CrossRef]
3. Thapa Karki, S.; Bennett, A.C.T.; Mishra, J.L. Reducing Food Waste and Food Insecurity in the UK: The Architecture of Surplus Food Distribution Supply Chain in Addressing the Sustainable Development Goals (Goal 2 and Goal 12.3) at a City Level. *Ind. Mark. Manag.* **2021**, *93*, 563–577. [CrossRef]
4. Runyan, C.W.; Stehm, J. Land Use Change, Deforestation and Competition for Land Due to Food Production. In *Encyclopedia of Food Security and Sustainability*; Ferranti, P., Berry, E.M., Anderson, J.R., Eds.; Elsevier: Oxford, UK, 2019; pp. 21–26. ISBN 978-0-12-812688-2.
5. Theurl, M.C.; Lauk, C.; Kalt, G.; Mayer, A.; Kaltenegger, K.; Morais, T.G.; Teixeira, R.F.M.; Domingos, T.; Winiwarter, W.; Erb, K.-H.; et al. Food Systems in a Zero-Deforestation World: Dietary Change Is More Important than Intensification for Climate Targets in 2050. *Sci. Total Environ.* **2020**, *735*, 139353. [CrossRef] [PubMed]
6. Haque, M.M.; Biswas, J.C. Emission Factors and Global Warming Potential as Influenced by Fertilizer Management for the Cultivation of Rice under Varied Growing Seasons. *Environ. Res.* **2021**, *197*, 111156. [CrossRef] [PubMed]
7. Zhuang, M.; Shan, N.; Wang, Y.; Caro, D.; Fleming, R.M.; Wang, L. Different Characteristics of Greenhouse Gases and Ammonia Emissions from Conventional Stored Dairy Cattle and Swine Manure in China. *Sci. Total Environ.* **2020**, *722*, 137693. [CrossRef] [PubMed]
8. McMichael, A.J.; Powles, J.W.; Butler, C.D.; Uauy, R. Food, Livestock Production, Energy, Climate Change, and Health. *Lancet* **2007**, *370*, 1253–1263. [CrossRef]
9. Halloran, A.; Roos, N.; Eilenberg, J.; Cerutti, A.; Bruun, S. Life Cycle Assessment of Edible Insects for Food Protein: A Review. *Agron. Sustain. Dev.* **2016**, *36*, 57. [CrossRef] [PubMed]
10. Huis, A. Potential of Insects as Food and Feed in Assuring Food Security. *Annu. Rev. Entomol.* **2013**, *58*, 563–583. [CrossRef]
11. Huis, A.; Itterbeeck, J.V.; Klunder, H.; Mertens, E.; Halloran, A.; Muir, G.; Vantomme, P. *Edible Insects: Future Prospects for Food and Feed Security*; Food and Agriculture Organization of the United Nations: Rome, Italy, 2013.
12. Nelson, G.C.; Rosegrant, M.W.; Koo, J.; Robertson, R.; Sulser, T.; Zhu, T.; Ringler, C.; Msangi, S.; Palazzo, A.; Batka, M.; et al. *Climate Change: Impact on Agriculture and Costs of Adaptation*; International Food Policy Research Institute: Washington, DC, USA, 2009.
13. Ramos-Elorduy, J. Energy Supplied by Edible Insects from Mexico and Their Nutritional and Ecological Importance. *Ecol. Food Nutr.* **2008**, *47*, 280–297. [CrossRef]
14. Smil, V. Eating Meat: Evolution, Patterns, and Consequences. *Popul. Dev. Rev.* **2002**, *28*, 599–639. [CrossRef]

15. Florença, S.G.; Correia, P.M.R.; Costa, C.A.; Guiné, R.P.F. Edible Insects: Preliminary Study about Perceptions, Attitudes, and Knowledge on a Sample of Portuguese Citizens. *Foods* **2021**, *10*, 709. [CrossRef] [PubMed]
16. Dupont, J.; Fiebelkorn, F. Attitudes and Acceptance of Young People toward the Consumption of Insects and Cultured Meat in Germany. *Food Qual. Prefer.* **2020**, *85*, 103983. [CrossRef]
17. Gahukar, R.T. Edible Insects Collected from Forests for Family Livelihood and Wellness of Rural Communities: A Review. *Glob. Food Secur.* **2020**, *25*, 100348. [CrossRef]
18. Kunatsa, Y.; Chidewe, C.; Zvidzai, C.J. Phytochemical and Anti-Nutrient Composite from Selected Marginalized Zimbabwean Edible Insects and Vegetables. *J. Agric. Food Res.* **2020**, *2*, 100027. [CrossRef]
19. Cappelli, A.; Cini, E.; Lorini, C.; Oliva, N.; Bonaccorsi, G. Insects as Food: A Review on Risks Assessments of Tenebrionidae and Gryllidae in Relation to a First Machines and Plants Development. *Food Control* **2020**, *108*, 106877. [CrossRef]
20. Guiné, R.P.F.; Correia, P.; Coelho, C.; Costa, C.A. The Role of Edible Insects to Mitigate Challenges for Sustainability. *Open Agric.* **2021**, *6*, 24–36. [CrossRef]
21. Baiano, A. Edible Insects: An Overview on Nutritional Characteristics, Safety, Farming, Production Technologies, Regulatory Framework, and Socio-Economic and Ethical Implications. *Trends Food Sci. Technol.* **2020**, *100*, 35–50. [CrossRef]
22. Sidali, K.L.; Pizzo, S.; Garrido-Pérez, E.I.; Schamel, G. Between Food Delicacies and Food Taboos: A Structural Equation Model to Assess Western Students' Acceptance of Amazonian Insect Food. *Food Res. Int.* **2019**, *115*, 83–89. [CrossRef]
23. Orsi, L.; Voege, L.L.; Stranieri, S. Eating Edible Insects as Sustainable Food? Exploring the Determinants of Consumer Acceptance in Germany. *Food Res. Int.* **2019**, *125*, 108573. [CrossRef] [PubMed]
24. Megido, R.C.; Gierts, C.; Blecker, C.; Brostaux, Y.; Haubruge, É.; Alabi, T.; Francis, F. Consumer Acceptance of Insect-Based Alternative Meat Products in Western Countries. *Food Qual. Prefer.* **2016**, *52*, 237–243. [CrossRef]
25. Trichopoulou, A. Mediterranean Diet as Intangible Heritage of Humanity: 10 Years On. *Nutr. Metab. Cardiovasc. Dis.* **2021**. [CrossRef]
26. Guiné, R.P.F.; Florença, S.G.; Villalobos Moya, K.; Anjos, O. Edible Flowers, Old Tradition or New Gastronomic Trend: A First Look at Consumption in Portugal versus Costa Rica. *Foods* **2020**, *9*, 977. [CrossRef] [PubMed]
27. Guiné, R.P.F.; Florença, S.G.; Barroca, M.J.; Anjos, O. The Link between the Consumer and the Innovations in Food Product Development. *Foods* **2020**, *9*, 1317. [CrossRef] [PubMed]
28. Triola, M.F.; Flores, V.R.L.F. *Instrodução ÀEstatística*, 12th ed.; LTC: Rio de Janeiro, Brasil, 2017.
29. Levin, J.; Fox, J.A. *Estatística Para Ciências Humanas*, 9th ed.; Pearson: Rio de Janeiro, Brasil, 2004.
30. Fundação Francisco Manuel dos Santos: PORDATA—Base de Dados Portugal Contemporâneo. Available online: https://www.pordata.pt/Home (accessed on 10 December 2020).
31. Cochran, W.G. *Sampling Techniques*, 3rd ed.; John Wiley & Sons: New York, NY, USA, 1977.
32. Levine, D.M.; Stephan, D.F.; Krehbiel, T.C.; Berenson, M.L. *Estatistica Teoria e Aplicacoes Usando o Microsoft Excel em Portugues*, 5th ed.; LTC: Rio de Janeiro, Brasil, 2008; ISBN 978-85-216-1634-4.
33. Witten, R.; Witte, J. *Statistics*, 9th ed.; Wiley: New Jersey, NJ, USA, 2009.
34. Broen, M.P.G.; Moonen, A.J.H.; Kuijf, M.L.; Dujardin, K.; Marsh, L.; Richard, I.H.; Starkstein, S.E.; Martinez-Martin, P.; Leentjens, A.F.G. Factor Analysis of the Hamilton Depression Rating Scale in Parkinson's Disease. *Parkinsonism Relat. Disord.* **2015**, *21*, 142–146. [CrossRef] [PubMed]
35. Tanaka, K.; Akechi, T.; Okuyama, T.; Nishiwaki, Y.; Uchitomi, Y. Development and Validation of the Cancer Dyspnoea Scale: A Multidimensional, Brief, Self-Rating Scale. *Br. J. Cancer* **2000**, *82*, 800–805. [CrossRef] [PubMed]
36. Dolnicar, S. A Review of Data-Driven Market Segmentation in Tourism. *Fac. Commer.-Pap.* **2002**, *12*, 1–22. [CrossRef]
37. Guiné, R.P.F.; Florença, S.G.; Ferrão, A.C.; Bizjak, M.Č.; Vombergar, B.; Simoni, N.; Vieira, V. Factors Affecting Eating Habits and Knowledge of Edible Flowers in Different Countries. *Open Agric.* **2021**, *6*, 67–81. [CrossRef]
38. Stevens, J.P. *Applied Multivariate Statistics for the Social Sciences*, 5th ed.; Routledge: New York, NY, USA, 2009; ISBN 978-0-8058-5903-4.
39. Hair, J.F.H.; Black, W.C.; Babin, B.J.; Anderson, R.E. *Multivariate Data Analysis*, 7th ed.; Prentice Hall: Hoboken, NJ, USA, 2009; ISBN 978-0-13-813263-7.
40. Maroco, J.; Garcia-Marques, T. Qual a fiabilidade do alfa de Cronbach? Questões antigas e soluções modernas? *Lab. Psicol.* **2006**, *4*, 65–90. [CrossRef]
41. Davis, F.B. *Educational Measurements Their Interpretation*; Wadsworth Pub. Co.: Belmont, CA, USA, 1964.
42. Cicatiello, C.; De Rosa, B.; Franco, S.; Lacetera, N. Consumer Approach to Insects as Food: Barriers and Potential for Consumption in Italy. *Br. Food J.* **2016**, *118*, 2271–2286. [CrossRef]
43. Schösler, H.; de Boer, J.; Boersema, J.J. Can We Cut out the Meat of the Dish? Constructing Consumer-Oriented Pathways towards Meat Substitution. *Appetite* **2012**, *58*, 39–47. [CrossRef]
44. House, J. Consumer Acceptance of Insect-Based Foods in the Netherlands: Academic and Commercial Implications. *Appetite* **2016**, *107*, 47–58. [CrossRef]
45. Gallen, C.; Pantin-Sohier, G.; Peyrat-Guillard, D. Familiarisation et diffusion de l'entomophagie en France. *Innovations* **2021**, *64*, 153–182. [CrossRef]
46. Huis, A. Insects as Food and Feed, a New Emerging Agricultural Sector: A Review. *J. Insects Food Feed* **2020**, *6*, 27–44. [CrossRef]

47. La Barbera, F.; Verneau, F.; Amato, M.; Grunert, K. Understanding Westerners' Disgust for the Eating of Insects: The Role of Food Neophobia and Implicit Associations. *Food Qual. Prefer.* **2018**, *64*, 120–125. [CrossRef]
48. Hartmann, C.; Siegrist, M. Consumer Perception and Behaviour Regarding Sustainable Protein Consumption: A Systematic Review. *Trends Food Sci. Technol.* **2017**, *61*, 11–25. [CrossRef]
49. Ritchey, P.N.; Frank, R.A.; Hursti, U.-K.; Tuorila, H. Validation and Cross-National Comparison of the Food Neophobia Scale (FNS) Using Confirmatory Factor Analysis. *Appetite* **2003**, *40*, 163–173. [CrossRef]
50. Martins, Y.; Pliner, P. Human Food Choices: An Examination of the Factors Underlying Acceptance/Rejection of Novel and Familiar Animal and Nonanimal Foods. *Appetite* **2005**, *45*, 214–224. [CrossRef]
51. Verbeke, W. Profiling Consumers Who Are Ready to Adopt Insects as a Meat Substitute in a Western Society. *Food Qual. Prefer.* **2015**, *39*, 147–155. [CrossRef]
52. Gere, A.; Székely, G.; Kovács, S.; Kókai, Z.; Sipos, L. Readiness to Adopt Insects in Hungary: A Case Study. *Food Qual. Prefer.* **2017**, *59*, 81–86. [CrossRef]
53. Srivastava, S.; Babu, N.; Pandey, H. Traditional Insect Bioprospecting—As Human Food and Medicine. *Indian J. Tradit. Knowl.* **2009**, *8*, 485–494.
54. Lensvelt, E.J.S.; Steenbekkers, L.P.A. Exploring Consumer Acceptance of Entomophagy: A Survey and Experiment in Australia and the Netherlands. *Ecol. Food Nutr.* **2014**, *53*, 543–561. [CrossRef] [PubMed]
55. Elorinne, A.-L.; Niva, M.; Vartiainen, O.; Väisänen, P. Insect Consumption Attitudes among Vegans, Non-Vegan Vegetarians, and Omnivores. *Nutrients* **2019**, *11*, 292. [CrossRef]
56. Cunha, L.M.; Ribeiro, J.C. Sensory and Consumer Perspectives on Edible Insects. In *Edible Insects in the Food Sector: Methods, Current Applications and Perspectives*; Sogari, G., Mora, C., Menozzi, D., Eds.; Springer International Publishing: Verlag, Germany, 2019; pp. 57–71. ISBN 978-3-030-22522-3.
57. Cunha, L.M.; Moura, A.P.; Costa-Lima, R. Consumers' associations with insects in the context of food consumption: Comparisons from acceptors to disgusted. In *Book of Abstracts of the 1st International Conference: Insects to Feed the World (14–17 May 2014)*; Wageningen University: Wageningen, The Netherlands, 2014; p. 108.
58. Cunha, L.M.; Gonçalves, A.T.S.; Varela, P.; Hersleth, M.; Neto, E.M.; Grabowski, N.T.; House, J.; Santos, P.; Moura, A.P. Adoption of insects as a source for food and feed production: A cross-cultural study on determinants of acceptance. In *Book of Abstracts of the 11th Pangborn Sensory Science Symposium (23–27 August 2015)*; European Sensory Science Society: Gothenburg, Sweden, 2015; p. [O.10.06]: 1-1.
59. Guiné, R.P.F.; Bartkiene, E.; Florença, S.G.; Djekić, I.; Bizjak, M.Č.; Tarcea, M.; Leal, M.; Ferreira, V.; Rumbak, I.; Orfanos, P.; et al. Environmental Issues as Drivers for Food Choice: Study from a Multinational Framework. *Sustainability* **2021**, *13*, 2869. [CrossRef]
60. Rodriguez-Oliveros, M.G.; Bisogni, C.A.; Frongillo, E.A. Knowledge about Food Classification Systems and Value Attributes Provides Insight for Understanding Complementary Food Choices in Mexican Working Mothers. *Appetite* **2014**, *83*, 144–152. [CrossRef]
61. Machovina, B.; Feeley, K.J. Livestock: Limit Red Meat Consumption. *Nature* **2014**, *508*, 186. [CrossRef] [PubMed]
62. Bukkens, S.G.F. Insects in the human diet: Nutritional aspects. In *Ecological Implications of Minilivestock; Role of Rodents, Frogs, Snails, and Insects for Sustainable Development*; Science Publishers—CRC Press Group: Boca Raton, FL, USA, 2005; pp. 545–577.
63. Lee, K.P.; Simpson, S.J.; Wilson, K. Dietary Protein-Quality Influences Melanization and Immune Function in an Insect. *Funct. Ecol.* **2008**, *22*, 1052–1061. [CrossRef]
64. Kouřimská, L.; Adámková, A. Nutritional and Sensory Quality of Edible Insects. *NFS J.* **2016**, *4*, 22–26. [CrossRef]
65. Roininen, H.; Ohgushi, T.; Zinovjev, A.; Virtanen, R.; Vikberg, V.; Matsushita, K.; Nakamura, M.; Price, P.; Veteli, T. Latitudinal and Altitudinal Patterns in Species Richness and Mortality Factors of the Galling Sawflies on Salix Species in Japan. In *Galling Arthropods and Their Associates: Ecology and Evolution*; Springer: Tokyo, Japan, 2006; pp. 3–19.
66. Arsil, P.; Li, E.; Bruwer, J.; Lyons, G. Exploring Consumer Motivations towards Buying Local Fresh Food Products: A Means-End Chain Approach. *Br. Food J.* **2014**, *116*, 1533–1549. [CrossRef]

MDPI

Article

The Effect of Whey Protein Films with Ginger and Rosemary Essential Oils on Microbiological Quality and Physicochemical Properties of Minced Lamb Meat

Maria Tsironi, Ioanna S. Kosma and Anastasia V. Badeka *

Laboratory of Food Chemistry, Department of Chemistry, University of Ioannina, 45110 Ioannina, Greece; m.tsironi@outlook.com.gr (M.T.); i.kosma@uoi.gr (I.S.K.)
* Correspondence: abadeka@uoi.gr; Tel.: +30-2651-008705

Abstract: Consumers' constant search for high-quality and safe products, with the least possible preservatives and additives, as well as extended shelf life, has led industries to research and develop alternative forms of food preservation and packaging. The purpose of this research was the study of the effect of natural antimicrobials and, in particular, the essential oils of ginger (*Zingiber Officinale Roscoe*) and rosemary (*Rosmarinus officinalis* L.) on strengthening whey protein films' properties. Whey protein isolate (WPI) films, alone and with incorporated essential oils (WPI + EO) at different concentrations were prepared and then examined for their possible effect on delaying the deterioration of minced lamb meat. Microbiological and physicochemical measurements were carried out to examine the meat's shelf life. Results showed that films with 1% EO significantly improved the microbiological quality of meat. On day 11, total viable counts, *Pseudomonas* spp., *Br. thermosphacta*, lactic acid bacteria, *Enterobacteriaceae*, and yeasts remained low for films with 1% concentration of essential oil compared with 0.5%. Regarding, physicochemical properties the same pattern was observed for pH while oxidation degree was significantly reduced. Finally, color attributes measurements recorded fluctuations between samples, but overall, no considerable discoloration was observed.

Keywords: edible films; whey protein isolate; essential oils; rosemary; ginger; lamb minced meat; mechanical properties; microbiology

Citation: Tsironi, M.; Kosma, I.S.; Badeka, A.V. The Effect of Whey Protein Films with Ginger and Rosemary Essential Oils on Microbiological Quality and Physicochemical Properties of Minced Lamb Meat. *Sustainability* **2022**, *14*, 3434. https://doi.org/10.3390/su14063434

Academic Editor: Filippo Giarratana

Received: 30 December 2021
Accepted: 10 March 2022
Published: 15 March 2022

1. Introduction

One of the important issues of the food industry for the maintenance, storage, handling, and promotion of safe and high-quality products is the design and selection of packaging materials with the appropriate specifications. The use of new technologies and new methods of food processing and preservation has led to a new packaging, which not only provides passive protection for packaged food but also plays an active role in preserving it by providing high-quality food and longer shelf life, compared with the classic packaging. Bioplastics have become a potentially environmentally friendly replacement for conventional petrochemical plastics.

The development of edible coatings for food packaging has increasingly gained the research interest for preserving quality, extending the product's shelf life, and being environmentally friendly [1]. Edible films and coatings can function as barriers to moisture, gases, etc., they contribute to the protection of lipids, prevent the loss of moisture and aroma from food, and are ideal substitutes for petroleum-derived polymers [1]. Furthermore, edible films and coatings can also function as carriers for antimicrobial and antioxidant agents, to control the diffusion rate of preservatives to the food interior, and as a part of a multilayer food packaging along with non-edible films [2]. Edible films and coatings can be applied to many different products, such as fruits, vegetables, meat products, and others [3]. In cheeses, for example, the edible packaging is primarily used to control microbiological deterioration on the surface of the cheese, to minimize the risk of contamination with pathogenic microorganisms, to prolong the quality of the cheese, as well as to manage the taste, color, and nutritional value [4]. Cerqueira et al. [5] applied membranes from mixtures of chitosan, galactomannan, and corn oil to semi-hard cheeses. This prevented mold growth and reduced water evaporation. The handling of water and water activity (a_w) of food determines its microbiological-physicochemical stability and its organoleptic characteristics. Meat and meat products must also avoid the loss of moisture when packaging fresh or frozen meat, reduce the rate of oxidation, retain freshly cut meat juices, and reduce the loss of volatile aromatic compounds, and the uptake of unwanted odors [6]. A film with low oxygen, moisture, and gas permeability can be used to extend the shelf life of meat and meat products. Edible films and coatings do not, in any way, replace the need to package food with non-edible packaging materials; they help them improve the quality of the product, and extend its life. The new packaging coatings consist mainly of milk proteins, and are considered 500 times more effective in keeping oxygen away from food. Furthermore, protein can be easily broken down and even consumed. For the additional strengthening of the membranes, it is necessary to use additives, such as antioxidants and antimicrobial agents, vitamins, probiotics, and minerals. This way, the packaging will have nutritional value on its own.

Whey protein is a material that can be used in the production of biodegradable and edible food packaging. The positive environmental footprint of such food packaging has led the scientific community to research into the production of alternative and environmentally friendly biologically-based materials. In addition, the development of such active bioplastic and edible packaging not only effectively extends the shelf life of products but is also an effective solution to reduce food waste. Their enhanced functions through the incorporation of antioxidants and antimicrobials, along with the good film-forming capacity, safety, and fast biocompatibility and biodegradability rates are an important development in the field of biodegradable and/or edible packaging films [3,7,8].

Plant extracts are rich sources of active compounds with strong antioxidant and antimicrobial activity. Essential oils, as natural compounds, can be used to produce active packaging that exhibits antimicrobial activity against a variety of microorganisms, including Gram-positive and Gram-negative bacteria, yeasts, and molds [8–10]. There is a growing interest in incorporating essential oils into membranes to improve shelf life and microbiological food safety [11]. Among other things, oregano, rosemary, thyme, and sage essential oils are the ones that show the highest effectiveness against microorganisms. Although many of them are considered safe for consumption, their use as food preservatives is often limited as in some cases, to exhibit antimicrobial activity, they must be present in high concentrations, and as a result, it exceeds levels accepted by consumers [12], while due to their high variability, they can be lost during storage, reducing their antibacterial effectiveness [11].

In this perspective, the aim of the present study (conducted between April and May 2021) was to investigate the effectiveness of whey protein films (alone and with incorporated ginger and rosemary essential oils at different concentrations) for the package of lamb minced meat. The prepared films were applied on burger size samples and were tested during their storage time for microbiological and physicochemical properties. Additionally, the prepared WPI films were tested for their mechanical properties.

2. Materials and Methods

2.1. Preparation of Films

Whey protein isolate (WPI), 90% (Arla Foods Ingredients, Greece) was dissolved in distilled water at room temperature in a final concentration of 8% (*w/w*), stirring constantly, until the solution was homogeneous. The solution was then placed in a water bath at 90 °C for 30 min under constant stirring to denature the proteins and immediately afterward in a water bath with ice water to prevent further denaturation. Glycerol (50%) was added [glycerol/(WPI + glycerol)] on a dry basis as a plasticizer, to overcome the fragility of the membranes and to achieve easier handling for various measurements. To enhance the antimicrobial properties of films, essential oils were added to the solution in appropriate amounts [0.5% and 1% essential oils of ginger (*Zingiber Officinale Roscoe*) and rosemary (*Rosmarinus officinalis* L.) (Vögele Ingredients, Germany)], followed by refrigeration for 24 h to remove the bubbles. The above amounts of essential oils are the results of preliminary tests based on sensory evaluation (data not shown). Finally, they were poured into glass molds (internal dimensions 40 cm × 20 cm) and were let at room temperature under a hood to dry. Five types of coatings were prepared (Table 1) including whey protein films (without the addition of essential oils) which were used as the control (WPI), whey protein films with ginger (WPI + GEO), and rosemary essential oil (WPI + REO).

Table 1. Types and number of films prepared.

Type of Coating	Essential Oil (%)	Abbreviation	Number of Films
Whey protein film	-	WPI	16
Whey protein film + ginger essential oil	0.5	WPI + 0.5%GEO	16
Whey protein film + ginger essential oil	1	WPI + 1%GEO	16
Whey protein film + rosemary essential oil	0.5	WPI + 0.5%REO	16
Whey protein film + rosemary essential oil	1	WPI + 1%REO	16

2.2. Film Characterization

2.2.1. Determination of Film Thickness

The film thickness was determined with a portable digital micrometer (IS 13,109 INSIZE CO., LTD, Japan). The film was measured at six different, random points on their surface. The measurements are provided as mean values ±standard deviations.

2.2.2. Mechanical Tests

The determination of the mechanical properties of the test specimens was performed using a Model 4411 Instron Dynamometer (Instron Engineering Corp., Canton, MA, USA). The tests were performed according to method D882 of the American Society for Testing and Materials (ASTM) [13]. The film samples were prepared in the form of rectangular dimensions (1.5 cm × 10 cm). The tests were performed at a temperature of 25 °C, with a transverse head velocity of 50 mm/min. From the measurements and the stress-strain diagrams, information was collected about the properties of the materials, such as the modulus of elasticity E, the leakage limit σy, the maximum stress σmax, and the percentage deformation at the break-off.

2.2.3. FT-IR Analysis

Infrared spectra of films were collected using attenuated total reflectance Fourier Transform Infrared (ATR-FTIR) spectroscopy Cary 630 (Agilent, Santa Clara, CA, USA). Each film was subjected to 16 scans at 4 cm^{-1} resolution from 4000 to 400 cm^{-1} at room temperature.

2.3. Samples Preparation

Fresh lamb of Greek origin was obtained immediately after grinding by a local butcher shop and transported to the laboratory in polystyrene boxes within 30 min. The minced meat was divided into portions of approximately 100 g, in the shape of a burger (8.5 cm diameter and 1.5 cm width), and after being wrapped with the WPI and WP + EO films, were placed in polystyrene trays and wrapped in a transparent oxygen-permeable household polyethylene film. The samples were then stored in 4 °C (±0.5 °C) refrigerators until spoiled. The main goal of the above process was to simulate the packaged portions of minced meat with the corresponding ones available in retail stores. Sampling was performed on 0, 2, 5, 8, and 11 days.

2.4. Microbiological Analyzes

The microbiological analysis of the samples was performed based on official analysis methods [14]. The following groups of microorganisms were studied: total viable counts (TVC), *Pseudomonas* spp., *Enterobacteriaceae*, lactic acid bacteria (LAB), *Brochothrix thermosphacta*, and yeasts. The TVC was determined using a non-selective tryptic glucose yeast agar substrate [(TGYA) Biolife, Italiana S.r.l., Milano, Italy] which was incubated at 30 °C for 2 to 3 days. Accordingly, *Pseudomonas* spp.: on the selective pseudomonas agar base substrate (Oxoid, Basingstoke, UK), with the addition of the antibiotics cetrimide-fucidin-cephaloridine (C.F.C., Oxoid, Basingstoke, UK) was incubated at 25 °C for 2 to 3 days. *Brochothrix thermosphacta*: on the selective substrate streptomycin thallous acetate-actidione agar base (OXOID, Basingstoke, UK) with the addition of antibiotic (SR0151, OXOID, Basingstoke, UK) was incubated at 25 °C for 2 to 3 days. *Enterobacteriaceae*: on the selective violet red bile glucose agar substrate (Biolife, Italiana S.r.l., Milano, Italy) was incubated at 37 °C for 18 to 24 h. Lactic acid bacteria (LAB): on the selective substrate de Man–Rogosa–Sharpe agar (MRS, Biolife, Italiana S.r.l., Milano, Italy) was incubated at 25 °C for 3 to 5 days. Yeasts: on the selective substrate rose bengal chloramphenicol agar base (RBC, Biolife, Italiana S.r.l., Milano, Italy) was incubated at 25 °C for 5 days.

2.5. Physicochemical Analyses

Measurement of pH, Color Attributes, and Lipid Oxidation/2-thiobarbituric Acid Reactive Substances (TBARS) Assay

The pH was measured using a pH-meter model HD 3456.2 (Delta OHM Srl, Selvazzano Dentro, Italy) as follows: meat samples (20 g) were completely homogenized with 10 mL of distilled water, followed by immersion of the electrode and determination of pH.

Color attributes were measured to assess the color changes during the shelf-life of the samples. For that purpose, a Hunter Lab colorimeter model DP-9000 (Reston, VA, USA) was used. Approximately 70 g of minced the meat sample was placed on a glass plate and the parameters L* (brightness), a* (redness), and b* (yellowness) were measured. For each value, the plate was rotated approximately 60° to determine the color on all sides of the meat mass. The ΔE was calculated by the following equation:

$$\Delta E = \sqrt{(L_s^* - L_c^*)^2 + (a_s^* - a_c^*)^2 + (b_s^* - b_c^*)^2}$$

where L_s^* is the brightness value for each sample, L_c^* is the brightness value for the respective control sample, a_s^* is the redness value for each sample, a_c^* is the redness value for the respective control sample, b_s^* is the yellowness value for each sample, b_c^* is the yellowness value for the respective control sample [15].

Finally, the TBARS value was measured according to the method described by Karabagias et al. [16].

2.6. Sensory Evaluation

After each sampling, meat samples were frozen (−30 °C) until sensory evaluation. The attributes of cooked minced lamb meat on each sampling day were evaluated by a panel of eleven untrained judges (age range 25–60), graduate students, and faculty of the Laboratory of Food Chemistry, University of Ioannina. Panelists were asked to evaluate sensory attributes of cooked samples (ca. 100 g), which were prepared by steaming for ca. 10 min to an internal temperature of 85 °C. Sensory evaluation was conducted in individual booths under controlled conditions of temperature, light, and humidity. A set of five samples (corresponding to five different treatments) with random code numbers were presented to panelists. Along with the test samples, a freshly thawed and cooked meat sample, stored at −30 °C throughout the experiment, was served to the panelists as the master control sample. Panelists were asked to score odor, taste, and overall perception of minced lamb meat using a 1–5 acceptability scale, with 5 corresponding to the most liked sample and 1 corresponding to the least liked sample. A score of 3 was taken as the lower limit of acceptability.

2.7. Statistical Analysis

Experiments were replicated twice while analyses were run in triplicate for each sampling day per treatment (n = 4 × 3 = 12). All analyses data were expressed as mean values ± standard deviations along with the microbiological counts which were converted to log CFU/g and subjected to analysis of variance (ANOVA) with Tukey's multiple range tests using the MINITAB software package version 18.0 [17]. Differences between means of multiple groups were analyzed by three-way ANOVA with Tukey's multiple range test. The main effects plots were constructed to assess the relative significance of various parameters on the response of the system.

3. Results and Discussion

3.1. Film Characterization

3.1.1. Film Thickness

Whey films with and without added essential oils were generally homogeneous, transparent, and yellowish. Membranes incorporated with higher concentrations of essential oils (1.0%) were visually more elastic than the WPI films. Similar visual characteristics with those found in the present study were recorded by Ramos et al. [18] who studied membranes produced from isolated whey protein or protein concentrate, and by Galus and Lenart [19], who studied whey protein membranes fortified with almond and walnut oils.

The characteristics of WPI films as well as those fortified with essential oils are presented in Table 2. The film thickness ranged from 0.090 ± 0.010 mm in WPI to 0.148 ± 0.020 mm in WPI + 1%REO. In general, the films fortified with 1% of essential oil were found to be thicker. Those differences in thickness between the control films (WPI) and the fortified ones can be caused by the addition of essential oils. Bertan et al. [20] observed that the addition of hydrophobic substances promoted an increase in the thickness of the biofilm, as it was necessary to use different ratios for each composition aimed at controlling the thickness for repeatability of measurements and validity of comparisons between properties. In the present study, it can be assumed that the percentage of hydrophobic substances (e.g., GEO and REO) was too low to cause such a variation, and the addition of essential oils did not show significant differences in films thickness other than that the higher concentration of both essential oils results in higher film thickness.

Table 2. Mechanical properties of WPI films alone and with incorporated EOs.

Treatment	Thickness (mm)	% Elongation at Break	Tensile Strength at Break (MPa)	Young's Modulus (MPa)
WPI	0.090 ± 0.01 ^{a}	243.10 ± 50.50 ^{a}	16.83 ± 2.10 ^{a}	160.8
WPI + 0.5%GEO	0.129 ± 0.01 ^{b}	300.66 ± 43.40 ^{a}	17.06 ± 2.90 ^{a}	60.80
WPI + 1%GEO	0.141 ± 0.00 ^{b}	415.20 ± 29.60 ^{b}	13.11 ± 1.70 ^{a}	45.90
WPI + 0.5%REO	0.131 ± 0.01 ^{b}	311.10 ± 33.60 ^{a}	15.99 ± 2.20 ^{a}	63.97
WPI + 1%REO	0.148 ± 0.02 ^{b}	399.70 ± 14.40 ^{b}	13.69 ± 0.90 ^{a}	44.77

Means with different letters in the same column indicate statistically significant differences ($p < 0.05$, Tukey's test).

3.1.2. Mechanical Properties

In terms of mechanical properties and the uniaxial tensile test, films with increased essential oil content (1% for both EOs) have statistically higher % elongation values compared with other films (WPI and WPI + 0.5% EOs) and the two EOs behaved similarly (Table 2).

The addition of any type and concentration of EOs did not significantly affect the values of tensile strength. However, the addition of 1% EOs slightly decreased the tensile strength at break (13.11 ± 1.70 for WPI + 1%GEO and 13.69 ± 0.90 for WPI + 1%REO).

The meaning of Young Modulus is an indication of films' elasticity and lower values show higher elasticity. The addition of EOs improved the films' elasticity compared with WPI films (160.8 MPa). Specifically, 1% concentration of EOs improved elasticity by 3.6 times and 0.5% 2.6 times.

Ma et al. [21] reported an increase in tensile strength and elasticity modulus at lower olive oil concentrations (5–15%) and a decrease in higher oil addition (20%) for gelatin films. However, Fang et al. [22] reported a decrease in tensile strength for whey protein membranes with increasing soybean oil content. Similar results were obtained for whey membranes containing olive oil [23] and quinoa-chitosan protein membranes incorporated with sunflower oil [24].

3.1.3. FT-IR Analysis

The ATR-FTIR spectra of produced films showed no differences among them (Figure S6). Specifically, approximately bands at 3500–3100 cm^{-1} and 2974–2800 cm^{-1} were attributed to O-H/N-H and C-H stretching vibrations, respectively. Major absorption bands of protein were peptide linkages of amide I and II and located approximately 1620 cm^{-1} and 1540 cm^{-1}, respectively. The amide I region is related to the stretching vibrations of C=O and C-N bonding while amide II to the stretching of the C-N. The strong band peaks at 1100 cm^{-1} and 1032 cm^{-1} attributed to C-O stretching of the C-O-H and C-O-C groups of the glucose ring [3,8,25–27].

3.2. Microbiological Analyses

The microbiological analysis showed that the initial microflora of minced lamb meat consisted of *Pseudomonas* spp., *Br. thermosphacta*, LAB, and yeasts. The dynamics of these microorganisms as well as their contribution to the final microflora was influenced by various factors, including the type of packaging used.

The initial TVC value (Figure 1a) of fresh minced lamb meat was 3.65 log CFU/g leading to an acceptable quality of fresh meat [28]. The maximum acceptable level for TVC (7 log CFU/g) [29] was reached on day 8 for WPI films (7.17 log CFU/g), between the 8th and 11th day for WPI + 0.5%EO, and on day 11 for WPI + 1%EO (ginger and rosemary for both cases). The fact that WPI + 1%EO reached the maximum acceptable level on day 11 indicates the possible antimicrobial effect of the tested films. Literature data are in accordance with the results of the present study [30–32] regarding WPI films incorporated with rosemary EO.

Pseudomonas spp. is an indicator of psychrotrophic bacteria, absolutely aerobic and sensitive to CO_2, and is considered as one of the main microorganisms responsible for meat spoilage [33]. The initial *Pseudomonas* spp., value was 1.69 log CFU/g, lower compared with literature [31,34], and reached the maximum on day 11 for WPI + 0.5%EO (8.39 log CFU/g for GEO and 8.13 log CFU/g for REO). The *Pseudomonas* spp., from day 2 to day 11 ranged between 2.6 and 8.4 log CFU/g, and according to Figure 1b the addition of essential oil did not hinder their development. Compared with the concentration of EOs added, films containing 1% EO appear to be more effective than 0.5%. Specifically, for WPI + 1%EO the samples also reached their maximum on day 11 (7.34 log CFU/g for GEO and 7.85 log CFU/g for REO); however, their values were lower compared with WPI + 0.5%EO samples, indicating that by increasing the concentration of EOs incorporated in the films, their inhibitory effect also increased.

Br. thermosphacta is a Gram-positive facultative anaerobe bacterium, constituting part of the natural microflora of fresh packaged meat, and one of the spoilage microorganisms, especially, in pork and lamb meat, as they combine different chemical and biochemical parameters that favor its growth [35,36]. Initial counts of *Br. thermosphacta* (Figure 1c) were 3.28 log CFU/g and reached the maximum on day 11 for WPI films (8.01 log CFU/g). On

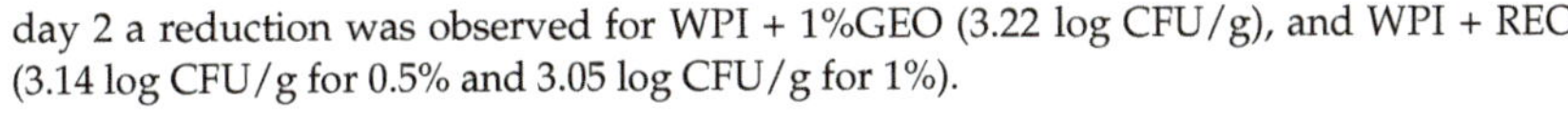
day 2 a reduction was observed for WPI + 1%GEO (3.22 log CFU/g), and WPI + REO (3.14 log CFU/g for 0.5% and 3.05 log CFU/g for 1%).

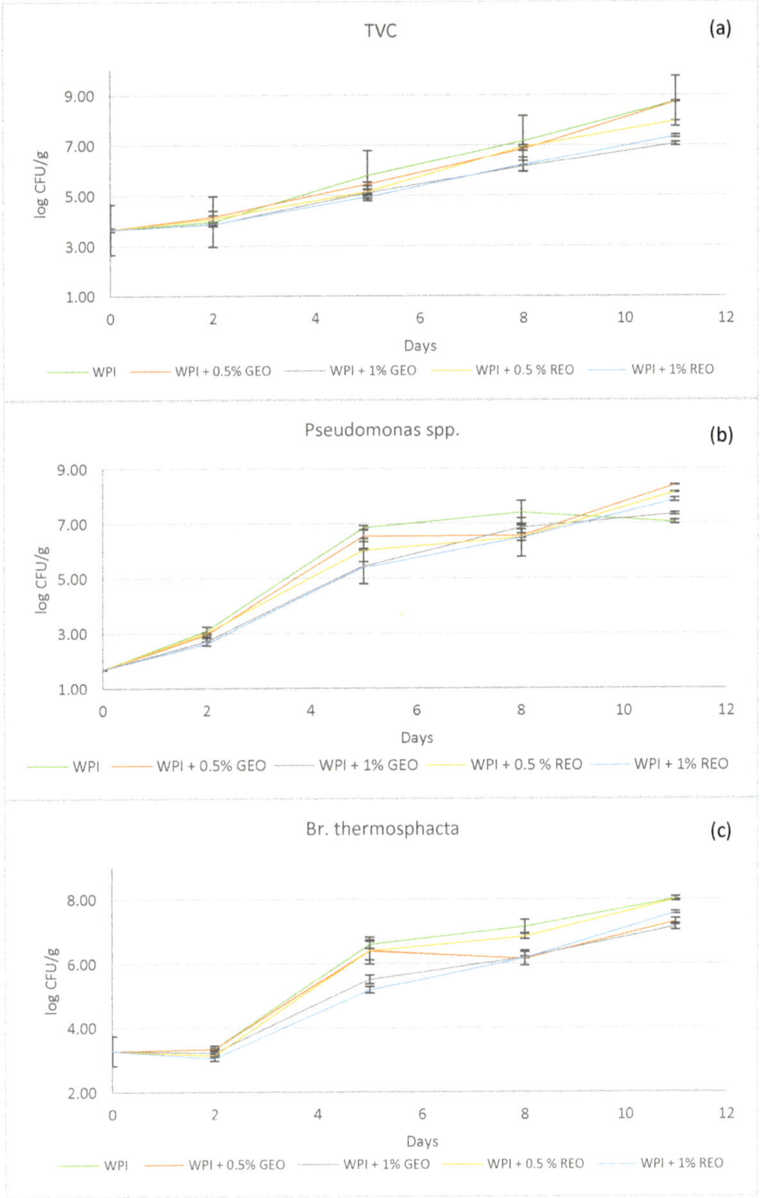

Figure 1. Effect of WPI alone and with incorporated EOs on the growth of (**a**) TVC, (**b**) *Pseudomonas* spp., and (**c**) *Br. thermosphacta*.

LAB are facultative anaerobic bacteria and comprise a significant part of meat microflora, as they can grow under low O_2 concentrations [37]. The growth of LAB during the storage time of the samples ranged from 2.8 to 7.5 log CFU/g. The concentration of the essential oil seems to contribute to the prolongation of the shelf life while the WPI + 1%EO seems to prevent the growth of LAB. Regarding the activity of the two essential oils, rosemary is presented as more active as it has a positive effect on LAB growth. Specifically, the initial population of LAB was 2.81 log CFU/g and reached its maximum on day 11 (7.50 log CFU/g) for WPI, and WPI + 0.5%GEO films. On day 2, samples with films WPI + 1%GEO, and WPI + REO (for both concentrations) recorded a reduction of LAB population over 0.5 log CFU/g (Figure 2a). Over time though, the LAB growth seemed to be suppressed by the incorporated EOs in WPI films as their population remained low, and especially, for WPI + 1%EO where they recorded LAB population under 7 log CFU/g (6.60 log CFU/g for GEO and 6.21 log CFU/g for REO). The results of the present study are in accordance with literature data regarding the reduction of LAB growth by EOs [31,36,38].

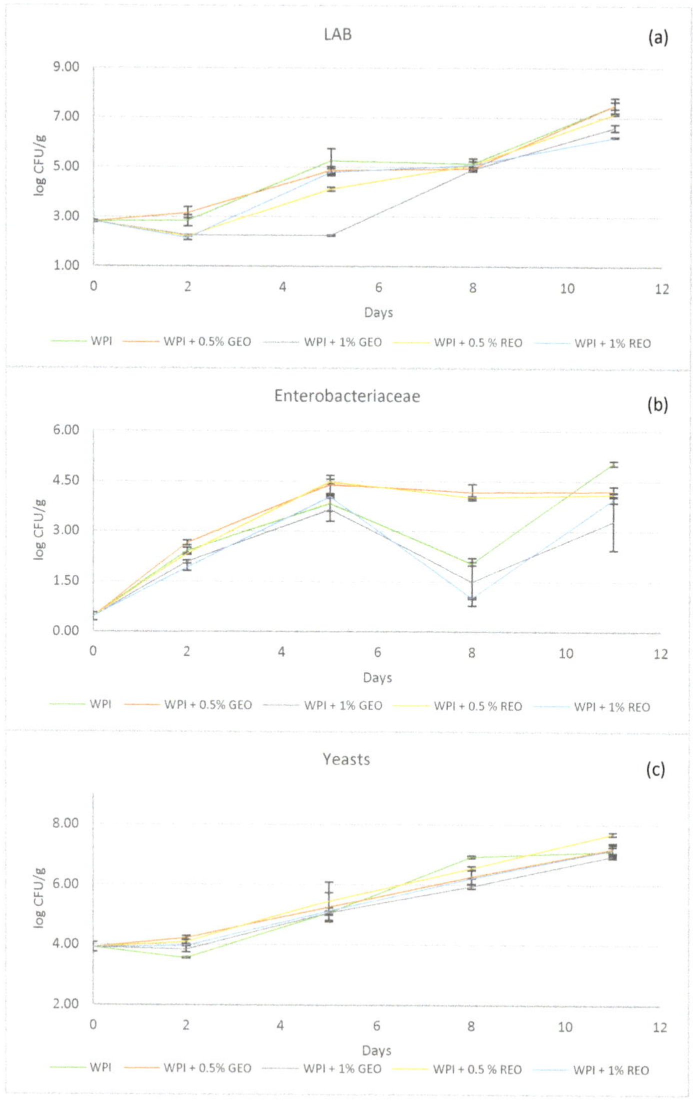

Figure 2. Effect of WPI alone and with incorporated EOs on the growth of (**a**) LAB, (**b**) *Enterobacteriaceae*, and (**c**) yeasts.

Concerning *Enterobacteriaceae*, which is usually considered a hygiene indicator [30] their growth ranged from 0.45 log CFU/g on day 0 to 5.05 log CFU/g on day 11 indicating a good quality of minced lamb meat. Although fluctuations were observed (Figure 2b) during the storage period, WPI + 1%EO samples seem to be more effective against *Enterobacteriaceae* development recording the highest values on day 11 (3.30 log CFU/g for GEO and 3.96 log CFU/g for REO), though lower than those of WPI + 0.5%EO (4.20 log CFU/g for GEO and 4.10 log CFU/g for REO). The final *Enterobacteriaceae* values of the present study are lower than those reported by Soldatou et al. [28] who studied the *Enterobacteriaceae* counts' changes during the storage time of lamb meat products under different package conditions (vacuum and modified atmosphere packaging). Alizadeh Sani et al. [31] also reported higher initial and final values for *Enterobacteriaceae* counts of lamb meat packaged with WPI films, although, as in the present study the inhibitory effect of REO was highlighted.

The yeasts' evolution during minced lamb meat storage is an important factor for its evaluation. The initial and final values (Figure 2c) of the yeasts' counts were higher compared with literature data (3.91 log CFU/g on day 0 and 7.69 log CFU/g on day 11) [30,32]. However, the WPI + 1%EO seemed to be more effective against yeast growth, as their final counts for the respective samples were lower (6.95 log CFU/g for GEO and 7.16 log CFU/g for REO) compared WPI + 0.5%EO (7.20 log CFU/g for GEO and 7.69 log CFU/g for REO).

3.3. Physicochemical Analyses (pH, Color Attributes, and TBARS)

Data of physicochemical analyses are shown in Table 3. The initial pH value of minced lamb meat was 5.72 ± 0.01, which is within the normal range for fresh, raw meat. It seems that the coating films, regardless of the concentration of the essential oils, show an increase in the pH value with the values reaching up to 6.37 ± 0.02 on the last day of storage for WPI + 0.5%REO, while the opposite was observed for the samples with WPI where the values reach marginally at pH = 6. The different types of membranes seem to influence the pH, as the values vary widely both in terms of essential oils and their concentration ($p = 0.034 < 0.05$) as well as in terms of storage time. Fluctuations in pH values during storage are also associated with various changes in the microbial profile of the samples. In general, the main parameters that seem to affect the pH value were found to be the essential oil ($p = 0.014 < 0.05$) and the storage time ($p = 0.000 < 0.05$) (Figure S1).

Table 3. Mean values and SD of physicochemical analyses tested.

Physicochemical Analyses	Days of Storage	Treatment					
		WPI	WPI + 0.5% GEO	WPI + 1% GEO	WPI + 0.5% REO	WPI + 1% REO	
pH	0	5.72 ± 0.01 [a]					
	2	5.73 ± 0.04	5.95 ± 0.04	5.91 ± 0.01	5.82 ± 0.01	5.81 ± 0.00	$p = 0.034 < 0.05$ [b]
	5	5.92 ± 0.02	5.92 ± 0.02	6.02 ± 0.02	6.01 ± 0.01	6.05 ± 0.04	$p = 0.014 < 0.05$ [c]
	8	6.05 ± 0.07	5.80 ± 0.01	5.94 ± 0.05	6.15 ± 0.07	6.05 ± 0.07	$p = 0.000 < 0.05$ [d]
	11	5.99 ± 0.00	6.32 ± 0.02	6.22 ± 0.01	6.37 ± 0.02	6.21 ± 0.01	
TBARS (mg MDA/kg)	0	0.88 ± 0.98 [a]					
	2	0.93 ± 0.00	2.19 ± 0.00	0.37 ± 0.00	1.94 ± 0.35	0.25 ± 0.01	$p = 0.027 < 0.05$ [b]
	5	0.28 ± 0.00	0.52 ± 0.00	0.35 ± 0.00	0.35 ± 0.00	0.24 ± 0.00	$p = 0.875 > 0.05$ [c]
	8	0.54 ± 0.00	0.52 ± 0.00	0.35 ± 0.00	0.45 ± 0.00	0.34 ± 0.00	$p = 0.000 < 0.05$ [d]
	11	0.37 ± 0.00	0.30 ± 0.00	0.45 ± 0.00	0.55 ± 0.00	0.57 ± 0.00	
L*	0	44.40 ± 0.40 [a]					
	2	44.13 ± 0.18	44.25 ± 0.23	44.60 ± 0.02	43.01 ± 0.01	42.84 ± 0.05	$p = 0.033 < 0.05$ [b]
	5	42.48 ± 0.01	42.01 ± 0.72	42.72 ± 0.02	42.91 ± 0.00	41.99 ± 0.02	$p = 0.003 < 0.05$ [c]
	8	44.89 ± 0.02	42.25 ± 0.21	43.50 ± 0.02	42.21 ± 0.03	42.50 ± 0.05	$p = 0.000 < 0.05$ [d]
	11	46.71 ± 0.14	48.36 ± 0.05	44.49 ± 0.02	45.35 ± 0.04	44.17 ± 0.08	

Table 3. *Cont.*

Physicochemical Analyses	Days of Storage	Treatment					
		WPI	WPI + 0.5% GEO	WPI + 1% GEO	WPI + 0.5% REO	WPI + 1% REO	
a*	0	14.73 ± 0.24 [a]					
	2	12.00 ± 0.00	12.65 ± 0.01	13.71 ± 0.02	13.60 ± 0.74	13.65 ± 0.04	$p = 0.109 > 0.05$ [b]
	5	15.34 ± 0.02	15.33 ± 0.04	14.12 ± 0.02	15.51 ± 0.01	16.02 ± 0.02	$p = 0.082 > 0.05$ [c]
	8	15.54 ± 0.04	15.43 ± 0.02	16.21 ± 0.04	14.82 ± 0.04	15.05 ± 0.06	$p = 0.000 < 0.05$ [d]
	11	13.33 ± 0.03	12.32 ± 0.00	14.76 ± 0.06	14.58 ± 0.01	15.17 ± 0.08	
b*	0	13.08 ± 0.24 [a]					
	2	13.33 ± 0.43	13.64 ± 0.00	12.92 ± 0.02	13.41 ± 0.01	13.56 ± 0.01	$p = 0.492 > 0.05$ [b]
	5	12.37 ± 0.01	12.39 ± 0.01	13.10 ± 0.13	13.12 ± 0.01	13.01 ± 0.01	$p = 0.217 > 0.05$ [c]
	8	14.19 ± 0.02	13.73 ± 0.03	14.63 ± 0.04	13.13 ± 0.04	13.83 ± 0.05	$p = 0.000 < 0.05$ [d]
	11	13.93 ± 0.00	13.32 ± 0.01	13.07 ± 0.04	14.89 ± 0.08	14.29 ± 0.04	
ΔE	0	-					
	2	-	1.72 ± 0.09	1.85 ± 0.10	2.02 ± 0.45	2.13 ± 0.16	$p = 0.001 < 0.05$ [b]
	5	-	0.51 ± 0.65	1.45 ± 0.04	0.87 ± 0.03	1.05 ± 0.61	$p = 0.000 < 0.05$ [c]
	8	-	2.69 ± 0.24	1.61 ± 0.03	2.97 ± 0.02	2.47 ± 0.56	$p = 0.000 < 0.05$ [d]
	11	-	2.02 ± 0.07	2.78 ± 0.09	2.08 ± 0.02	3.16 ± 0.02	

[a] Day 0 is the same for all samples, three-way ANOVA results, p-value for each physicochemical analysis between groups of: [b] concentration of essential oil, [c] type of essential oil, and [d] storage time.

The TBARS values are reported in milligrams of malondialdehyde (MDA) per kilogram of the sample (mg MDA/kg). MDA values above 1.5 mg/kg are associated with noticeable and unacceptable organoleptic changes in the meat [39]. The MDA values of meat samples did not show significant antioxidant activity for WPI + EO films compared with the WPI film, as the WPI films appear to have antioxidant activity throughout the storage period of the samples, as the oxidation degree values remained low. Specifically, the values ranged from 0.88 ± 0.98 (1st day) to 2.19 ± 0.00 (2nd day). It is important to note that on the 2nd day an increase in MDA values of WPI + 0.5%GEO and WPI + 0.5%REO was observed, while for the corresponding films with 1% concentration of essential oils the MDA values seem to be more than 50% lower. A similar picture is observed on the 5th day; however, upon comparing the 2nd and 5th days, a decrease in the degree of oxidation is observed mainly for the control films and the WPI + 0.5%EO films (both the GEO and REO). The oxidation degree was found to be statistically significant (Figure S2) as it was affected by the concentration of essential oils ($p = 0.027 < 0.05$), and the storage time ($p = 0.000 < 0.05$) of the samples but did not affect the essential oil ($p = 0.875 > 0.05$). Regarding the TBARS values, the findings of the present study seem to be in contrast with those of Siripatrawan and Noipha [40] who examined the oxidation grade of chitosan films incorporated with green tea extract for 20 days of storage of beefsteaks and reported a reduction in TBARS values for the untreated chitosan film as well as for the film incorporated with the natural antioxidant. A decrease in TBARS values was also observed by Rimini et al. [41] who studied the package conditions for fresh and stored chicken cuts for 12 and 90 days in the presence of a blend of thyme and orange essential oil compared with the control.

Fluctuations were observed in all color attributes' values in all types of films. Specifically, the brightness values (L* parameter), ranged from 44.40 ± 0.40 (1st day) to 48.36 ± 0.05 (11th day). For WPI films, the values increased from the 8th day onwards, while for the WPI + 0.5%EO and WPI + 1%EO films (both the GEO and REO) fluctuations were observed from the 2nd day of sampling onwards, which can be related to the denaturation of proteins in minced lamb meat. Regarding the brightness values, the findings of the present study are in agreement with those of Carvalho et al. [42] who investigated the antioxidant properties of thyme essential oil and whey protein isolate/cellulose nanofiber, nano biopolymers films containing TEO (20%, 30%, and 40% *w/w*) applied on ground beef. The researchers recorded values of the same order of magnitude as those of the present study. In general, the brightness values were affected significantly (Figure S3) by the examined factors [essential oils ($p = 0.003 < 0.05$), concentration ($p = 0.033 < 0.05$), and storage time ($p = 0.000 < 0.05$)].

Regarding the redness values (parameter a*), they were statistically significantly affected only by storage time ($p = 0.000 < 0.05$) [essential oil ($p = 0.082 > 0.05$), and concentration ($p = 0.109 > 0.05$)] (Figure S4). In general, an increase was observed on the 8th day (from 14.82 ± 0.04 for the WPI + 0.5%REO to 16.21 ± 0.04 for the WPI + 1%GEO) followed by a decrease on the 11th day (from 12.32 ± 0.00 for the WPI + 0.5%GEO to 14.58 ± 0.01 for the WPI + 0.5%REO). This phenomenon is evident in all types of films and can be attributed to changes in myoglobin and the accumulation of meta-myoglobin over storage time of the samples. Higher redness values imply the contribution of membranes to the preservation and/or improvement of the meat's red color. However, the redness values of the present study were lower compared with those reported by Carvalho et al. [42] who, in addition, observed a decrease in redness values for all treatments tested.

Regarding the values of yellowness (parameter b*), they ranged from 13.08 ± 0.24 to 14.89 ± 0.08, with fluctuations observed for the WPI and the WPI + GEO films where they recorded their maximum value on day 8. On the other hand, for the WPI + REO films the maximum yellowness values were observed on the 11th day. The results of the present study are in contrast with literature data [42,43] where a decrease in yellowness was observed through storage time. In general, the values of yellowness were statistically significantly affected (Figure S5) by storage time ($p = 0.000 < 0.05$), but not by the concentration ($p = 0.492 > 0.05$) and the type of essential oil ($p = 0.217 > 0.05$).

The ΔE values indicate that there was no significant discoloration between samples and their respective controls (WPI samples) during the days 2 (1.72 ± 0.09 for WPI + 0.5%GEO to 2.13 ± 0.16 for WPI + 1%REO) and 5 (0.51 ± 0.65 for WPI + 0.5%GEO to 1.45 ± 0.04 for WPI + 1%GEO) of storage while this value moderately increased for days 8 (1.61 ± 0.03 for WPI + 1%GEO to 2.97 ± 0.02 for WPI + 0.5%REO) and 11 (2.02 ± 0.07 for WPI + 0.5%GEO to 3.16 ± 0.02 for WPI + 1%REO). The ΔE values between 0 and 1 indicate discoloration not perceptible by the human eye, while the ΔE values between 1 and 2 indicate discoloration perceptible by close observation or only obvious to a trained eye. The ΔE values between 2 and 3.5 that were measured in the sample stored for 11 days indicate discolorations that can be obvious to an untrained eye [44]. These results indicate that most of the prepared films can retain the color of lamb for almost 8 days of storage.

3.4. Sensory Evaluation

The results of sensory (odor, taste, and overall perception) evaluation of cooked minced lamb meat are presented in Figure 3a–c. All three sensory evaluation scores decreased significantly ($p < 0.05$) with storage time. Taste and odor proved to be more sensitive sensory attributes compared with the overall perception. The lower acceptability limit of 3 was reached for taste after day 5 for WPI samples, between day 5 and 8 for WPI + GEO, and after day 8 for WPI + REO samples. A similar pattern was observed for odor scores, the limit of 3 was reached after day 5 for WPI and WPI + GEO samples, and between days 5 and 8 for WPI + REO samples. For both attributes, WPI samples were found unacceptable on day 11 and for that reason, panelists were unable to taste them. The overall perception included the color and the general picture of each sample before consumption. For WPI samples the lower limit of acceptability was reached between days 5 and 8, while for WPI + GEO and WPI + 0.5%REO this limit was reached after day 8, and for WPI + 1%REO on day 11.

The use of EOs in both concentrations retained the sensory properties of lamb meat for almost 5 to 8 days. Specifically, WPI + REO samples reached the limit of acceptability for taste and odor after day 8. At this point, it should be mentioned that REO has a delicate taste compatible with the taste of cooked lamb, while the panelists found it more familiar than GEO.

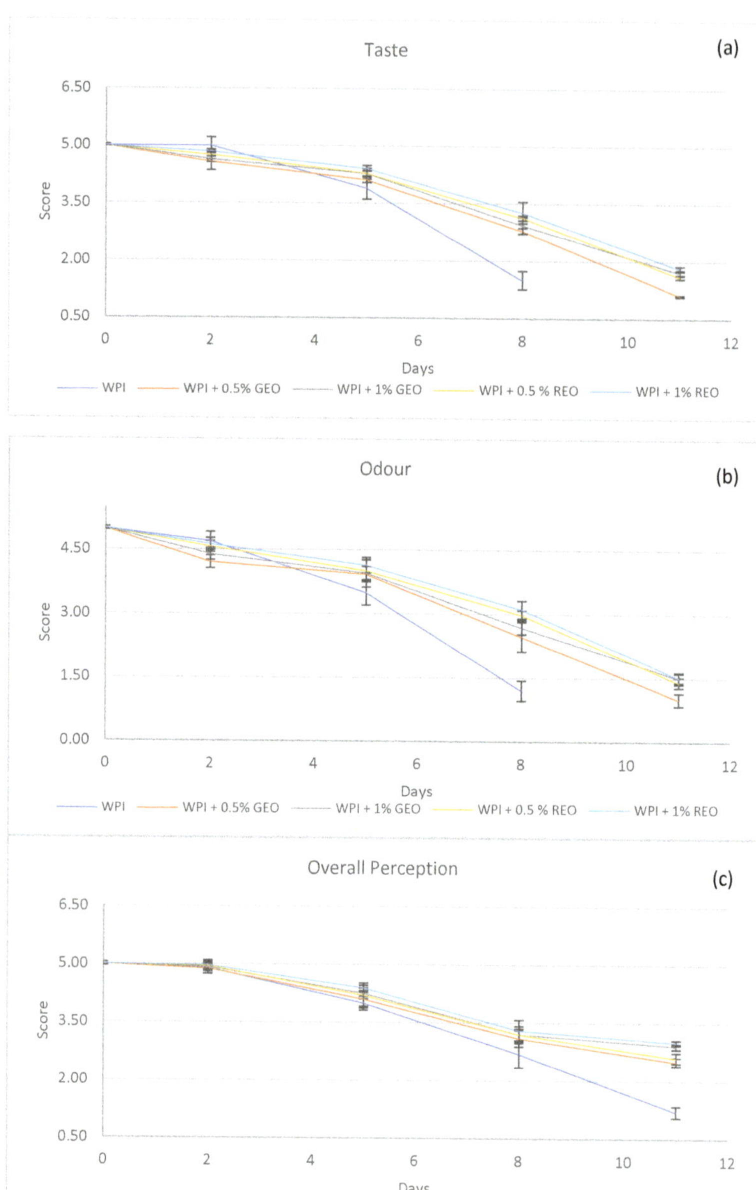

Figure 3. Sensory evaluation scores, taste (**a**), odor (**b**), and overall perception (**c**) of minced lamb meat packaged with WPI films with and without EOs at different concentrations.

Present sensory data were in reasonable agreement with microbiological data (TVC). Differences observed between the two may be attributed to the fact that it is not the total number of microorganisms but rather the number of specific spoilage organisms that are responsible for product deterioration [45]. Alizadeh Sani et al. [31] reported that the use of REO in biodegradable nanocomposite films containing TiO_2 nanoparticles increased significantly the shelf life of lamb meat compared with control samples (for almost 15 days regarding texture, color, and overall acceptability).

4. Conclusions

Edible films/coatings are a great way to diversify the functional food market and a substitute for the packaging and prevailing products. These are promising ways to improve food quality, extend shelf life, ensure safety, maintain functionality, and reduce environmental impact. In addition, these films and coatings can be used as separate bags of homogeneous substances and carriers of the active ingredient. The WPI films prepared in the present study showed a significant delay in microbiological deterioration of minced lamb meat, and especially, the films with 1% incorporated EO (both GEO and REO), while the TBARS values remained low indicating a significant delay in oxidation degree of meat samples. Results showed no significant differences between the GEO and REO 1% films. Furthermore, the color attributes tested as well as the ΔE value showed no significant discoloration of the samples for almost 8 days of storage, while the sensory evaluation test showed that, in terms of taste, and odor, samples packaged with WPI + REO in both concentrations were sensory acceptable for almost 8 days.

Supplementary Materials: The following are available online at https://www.mdpi.com/article/10.3390/su14063434/s1, Figure S1: Main effects plot for essential oil, their concentration and storage time on the pH values of minced lamb meat samples packaged with different types of WPI films. Figure S2: Main effects plot for essential oil, their concentration and storage time on TBARS values of minced lamb meat samples packaged with different types of WPI films. Figure S3: Main effects plot for essential oil, their concentration and storage time on L* parameter (brightness) values of minced lamb meat samples packaged with different types of WPI films. Figure S4: Main effects plot for essential oil, their concentration and storage time on a* parameter (redness) values of minced lamb meat samples packaged with different types of WPI films. Figure S5: Main effects plot for essential oil, their concentration and storage time on b* parameter (yellowness) values of minced lamb meat samples packaged with different types of WPI films. Figure S6: FTIR-ATR spectra of WPI films with and without EOs at different concentrations. Table S1: Compositional analysis (%) of ginger and rosemary essential oils.

Author Contributions: Conceptualization, A.V.B.; methodology, M.T.; software, M.T.; I.S.K.; validation, M.T.; I.S.K.; formal analysis, M.T.; investigation, M.T.; I.S.K.; data curation, M.T.; I.S.K.; writing—original draft preparation, I.S.K.; M.T.; writing—review and editing, I.S.K.; A.V.B.; supervision, A.V.B.; project administration, A.V.B.; All authors have read and agreed to the published version of the manuscript.

Funding: This research was carried out with financial support through project "Development of research infrastructures for the design, production and promotion of the quality and safety characteristics of agri-food and bio-functional products (EV-AGRO-NUTRITION) (MIS 5047235)" which is implemented under the Action "Reinforcement of the Research and Innovation Infrastructure", funded by the Operational Program "Competitiveness, Entrepreneurship and Innovation" (NSRF 2014-2020) and co-financed by Greece and the European Union (European Regional Development Fund).

Institutional Review Board Statement: Ethical review and approval were waived for this study, due to the fact that all materials used are commercially available for food use.

Informed Consent Statement: Informed consent was obtained from all subjects involved in the study.

Data Availability Statement: The data presented in this study are available on request from the corresponding author.

Conflicts of Interest: The authors declare no conflict of interest.

References

1. Fitriani, F.; Aprilia, S.; Arahman, N.; Bilad, M.R.; Suhaimi, H.; Huda, N. Properties of biocomposite film based on whey protein isolate filled with nanocrystalline cellulose from pineapple crown leaf. *Polymers* **2021**, *13*, 4278. [CrossRef] [PubMed]
2. Wittaya, T. Protein-Based Edible Films: Characteristics and Improvement of Properties. In *Structure and Function of Food Engineering*; Eissa, A.A., Ed.; IntechOpen: London, UK, 2012; pp. 43–70.
3. Xu, Y.P.; Wang, Y.; Zhang, T.; Mu, G.Q.; Jiang, S.J.; Zhu, X.M.; Tuo, Y.F.; Qian, F. Evaluation of the properties of whey protein films with modifications. *J. Food Sci.* **2021**, *86*, 923–931. [CrossRef] [PubMed]
4. Kontogianni, V.G.; Kasapidou, E.; Mitlianga, P.; Mataragas, M.; Pappa, E.; Kondyli, E.; Bosnea, L. Production, characteristics and application of whey protein films activated with rosemary and sage extract in preserving soft cheese. *LWT-Food Sci. Technol.* **2022**, *155*, 112996. [CrossRef]
5. Cerqueira, M.A.; Sousa-Gallagher, M.J.; Macedo, I.; Rodriguez-Aguilera, R.; Souza, B.W.S.; Teixeira, J.A.; Vicente, A.A. Use of galactomannan edible coating application and storage temperature for prolonging shelf-life of "Regional" cheese. *J. Food Eng.* **2010**, *97*, 87–94. [CrossRef]
6. Quintavalla, S.; Vicini, L. Antimicrobial food packaging in meat industry. *Meat Sci.* **2002**, *62*, 373–380. [CrossRef]
7. Jariyasakoolroj, P.; Leelaphiwat, P.; Harnkarnsujarit, N. Advances in research and development of bioplastic for food packaging. *J. Sci. Food Agric.* **2020**, *100*, 5032–5045. [CrossRef]
8. Chollakup, R.; Pongburoos, S.; Boonsong, W.; Khanoonkon, N.; Kongsin, K.; Sothornvit, R.; Sukyai, P.; Sukatta, U.; Harnkarnsujarit, N. Antioxidant and antibacterial activities of cassava starch and whey protein blend films containing rambutan peel extract and cinnamon oil for active packaging. *Lwt* **2020**, *130*, 109573. [CrossRef]
9. Sanla-Ead, N.; Jangchud, A.; Chonhenchob, V.; Suppakul, P. Antimicrobial Activity of Cinnamaldehyde and Eugenol and Their Activity after Incorporation into Cellulose-based Packaging Films. In Proceedings of the Packaging and Technology and Science, Berlin, Germany, 16–18 May 2011; Volume 25, pp. 7–17.
10. Balaguer, M.P.; Lopez-Carballo, G.; Catala, R.; Gavara, R.; Hernandez-Munoz, P. Antifungal properties of gliadin films incorporating cinnamaldehyde and application in active food packaging of bread and cheese spread foodstuffs. *Int. J. Food Microbiol.* **2013**, *166*, 369–377. [CrossRef] [PubMed]
11. Zhang, Y.; Ma, Q.; Critzer, F.; Davidson, P.M.; Zhong, Q. Physical and antibacterial properties of alginate films containing cinnamon bark oil and soybean oil. *LWT-Food Sci. Technol.* **2015**, *64*, 423–430. [CrossRef]
12. Alboofetileh, M.; Rezaei, M.; Hosseini, H.; Abdollahi, M. Antimicrobial activity of alginate/clay nanocomposite films enriched with essential oils against three common foodborne pathogens. *Food Control* **2014**, *36*, 1–7. [CrossRef]
13. ASTM D882–12 Standard 486 Test Method for Tensile Properties of Thin Plastic Sheeting. Available online: https://www.astm.org/Standards/D882.htm (accessed on 1 November 2021).
14. APHA. *Compendium of Methods for the Microbiological Examination of Foods*; Salfinger, Y., Tortorello, M.L., Eds.; American Public Health Association: Washington, DC, USA, 2015.
15. International Commission on Illumination. *Commission Internationale de Leclairage Cie 15: Technical Report: Colorimetry*, 3rd ed.; 2004; Volume 552. Available online: https://cie.co.at/publications/colorimetry-3rd-edition (accessed on 30 November 2021).
16. Karabagias, I.; Badeka, A.; Kontominas, M.G. Shelf life extension of lamb meat using thyme or oregano essential oils and modified atmosphere packaging. *Meat Sci.* **2011**, *88*, 109–116. [CrossRef] [PubMed]
17. *Minitab 18 Statistical Software*; Minitab, Inc.: State College, PA, USA, 2017.
18. Ramos, Ó.L.; Fernandes, J.C.; Silva, S.I.; Pintado, M.E.; Malcata, F.X. Edible Films and Coatings from Whey Proteins: A Review on Formulation, and on Mechanical and Bioactive Properties. *Crit. Rev. Food Sci. Nutr.* **2012**, *52*, 533–552. [CrossRef] [PubMed]
19. Galus, S.; Lenart, A. Optical, mechanical, and moisture sorption properties of whey protein edible films. *J. Food Process Eng.* **2019**, *42*, e13245. [CrossRef]
20. Bertan, L.C.; Tanada-Palmu, P.S.; Siani, A.C.; Grosso, C.R.F. Effect of fatty acids and "Brazilian elemi" on composite films based on gelatin. *Food Hydrocoll.* **2005**, *19*, 73–82. [CrossRef]
21. Ma, W.; Tang, C.H.; Yin, S.W.; Yang, X.Q.; Wang, Q.; Liu, F.; Wei, Z.H. Characterization of gelatin-based edible films incorporated with olive oil. *Food Res. Int.* **2012**, *49*, 572–579. [CrossRef]
22. Fang, Y.; Tung, M.A.; Britt, I.J.; Yada, S.; Dalgleish, D.G. Tensile and Barrier Properties of Edible Films Made from Whey Proteins. *J. Food Sci.* **2002**, *67*, 188–193. [CrossRef]
23. Javanmard, M.; Golestan, L. Effect of olive oil and glycerol on physical properties of whey protein concentrate films. *J. Food Process Eng.* **2008**, *31*, 628–639. [CrossRef]
24. Valenzuela, C.; Abugoch, L.; Tapia, C. Quinoa protein-chitosan-sunflower oil edible film: Mechanical, barrier and structural properties. *LWT-Food Sci. Technol.* **2013**, *50*, 531–537. [CrossRef]
25. Meng, Y.; Liang, Z.; Zhang, C.; Hao, S.; Han, H.; Du, P.; Li, A.; Shao, H.; Li, C.; Liu, L. Ultrasonic modification of whey protein isolate: Implications for the structural and functional properties. *LWT* **2021**, *152*, 112272. [CrossRef]
26. Janjarasskul, T.; Tananuwong, K.; Phupoksakul, T.; Thaiphanit, S. Fast dissolving, hermetically sealable, edible whey protein isolate-based films for instant food and/or dry ingredient pouches. *LWT* **2020**, *134*, 110102. [CrossRef]
27. Saxton, R.; McDougal, O.M. Whey protein powder analysis by mid-infrared spectroscopy. *Foods* **2021**, *10*, 1033. [CrossRef] [PubMed]

28. Soldatou, N.; Nerantzaki, A.; Kontominas, M.G.; Savvaidis, I.N. Physicochemical and microbiological changes of "Souvlaki"—A Greek delicacy lamb meat product: Evaluation of shelf-life using microbial, colour and lipid oxidation parameters. *Food Chem.* **2009**, *113*, 36–42. [CrossRef]
29. ICMSF. *Microorganisms in Foods 2. Sampling for Microbiological Analysis: Principles and Scientific Applications*, 2nd ed.; University of Toronto Press: Toronto, ON, Canada, 1986.
30. Uçak, I.; Özogul, Y.; Durmuş, M. The effects of rosemary extract combination with vacuum packing on the quality changes of Atlantic mackerel fish burgers. *Int. J. Food Sci. Technol.* **2011**, *46*, 1157–1163. [CrossRef]
31. Alizadeh Sani, M.; Ehsani, A.; Hashemi, M. Whey protein isolate/cellulose nanofibre/TiO_2 nanoparticle/rosemary essential oil nanocomposite film: Its effect on microbial and sensory quality of lamb meat and growth of common foodborne pathogenic bacteria during refrigeration. *Int. J. Food Microbiol.* **2017**, *251*, 8–14. [CrossRef]
32. Nedorostova, L.; Kloucek, P.; Kokoska, L.; Stolcova, M.; Pulkrabek, J. Antimicrobial properties of selected essential oils in vapour phase against foodborne bacteria. *Food Control* **2009**, *20*, 157–160. [CrossRef]
33. Jay, J.M.; Loessner, M.J.; Golden, D.A. Indicators of Food Microbial Quality and Safety. In *Modern Food Microbiology*; Jay, J.M., Loessner, M.J., Golden, D.A., Eds.; Springer Science & Business Media: New York, NY, USA, 2005; pp. 473–495.
34. Raeisi, M.; Tabaraei, A.; Hashemi, M.; Behnampour, N. Effect of sodium alginate coating incorporated with nisin, Cinnamomum zeylanicum, and rosemary essential oils on microbial quality of chicken meat and fate of Listeria monocytogenes during refrigeration. *Int. J. Food Microbiol.* **2016**, *238*, 139–145. [CrossRef]
35. Labadie, J. Consequences of packaging on bacterial growth. Meat is an ecological niche. *Meat Sci.* **1999**, *52*, 299–305. [CrossRef]
36. Chouliara, E.; Karatapanis, A.; Savvaidis, I.N.; Kontominas, M.G. Combined effect of oregano essential oil and modified atmosphere packaging on shelf-life extension of fresh chicken breast meat, stored at 4 °C. *Food Microbiol.* **2007**, *24*, 607–617. [CrossRef]
37. Lewus, C.B.; Kaiser, A.; Montville, T.J. Inhibition of food-borne bacterial pathogens by bacteriocins from lactic acid bacteria isolated from meat. *Appl. Environ. Microbiol.* **1991**, *57*, 1683–1688. [CrossRef]
38. Zhang, H.; Kong, B.; Xiong, Y.L.; Sun, X. Antimicrobial activities of spice extracts against pathogenic and spoilage bacteria in modified atmosphere packaged fresh pork and vacuum packaged ham slices stored at 4 °C. *Meat Sci.* **2009**, *81*, 686–692. [CrossRef]
39. Lorenzo, J.M.; Batlle, R.; Gómez, M. Extension of the shelf-life of foal meat with two antioxidant active packaging systems. *LWT-Food Sci. Technol.* **2014**, *59*, 181–188. [CrossRef]
40. Siripatrawan, U.; Noipha, S. Active film from chitosan incorporating green tea extract for shelf life extension of pork sausages. *Food Hydrocoll.* **2012**, *27*, 102–108. [CrossRef]
41. Rimini, S.; Petracci, M.; Smith, D.P. The use of thyme and orange essential oils blend to improve quality traits of marinated chicken meat. *Poult. Sci.* **2014**, *93*, 2096–2102. [CrossRef] [PubMed]
42. Carvalho, R.A.; Santos, T.A.; de Oliveira, A.C.S.; de Azevedo, V.M.; Dias, M.V.; Ramos, E.M.; Borges, S.V. Biopolymers of WPI/CNF/TEO in preventing oxidation of ground meat. *J. Food Process. Preserv.* **2019**, *43*, e14269. [CrossRef]
43. Kodal Coşkun, B.; Çalikoğlu, E.; Karagöz Emiroğlu, Z.; Candoğan, K. Antioxidant active packaging with soy edible films and oregano or thyme essential oils for oxidative stability of ground beef patties. *J. Food Qual.* **2014**, *37*, 203–212. [CrossRef]
44. Mokrzycki, W.; Tatol, M. Color difference Delta E—A survey. *Mach. Graph. Vis.* **2011**, *20*, 383–411.
45. Jay, J.; Loessner, M.L.; Golden, D.A. *Modern Food Microbiology*; Springer Science & Business Media: New York, NY, USA, 2008.

Article

Consumers' Trust in Greek Traditional Foods in the Post COVID-19 Era

Dimitris Skalkos [1,*], Ioanna S. Kosma [1], Areti Vasiliou [1] and Raquel P. F. Guine [2]

1 Laboratory of Food Chemistry, Department of Chemistry, University of Ioannina, 45110 Ioannina, Greece; i.kosma@uoi.gr (I.S.K.); pch1162@uoi.gr (A.V.)
2 CERNAS Research Centre, Polytechnic Institute of Viseu, 3504-510 Viseu, Portugal; raquelguine@esav.ipv.pt
* Correspondence: dskalkos@uoi.gr; Tel.: +30-2651008345

Abstract: We are entering a new, unprecedented global economic and social era following the COVID-19 pandemic, in which there will be opportunities and threats for the goods and services provided. Traditional foods (TFs) could have their chances in the new food chain which will be developed, as long as they become the food of choice for the consumers of the future. This paper investigates consumers' trust in Greek TFs, and northwest Greek TFs, in order to assess their potential consumption in the new economy. Trust was tested using the variables of safety, healthiness, sustainability, authenticity and taste, assessing consumers' confidence and satisfaction with the TFs, their raw materials, and the technologies used for their production. A self-response questionnaire survey was carried out in May and June 2021 on a sample of 548 participants through the Google platform. In order to analyze the data, basic descriptive statistical tools were used, combined with crosstabs and chi-square tests. The results show that the participants trust the Greek TFs because they "strongly agree" by an average of 20%, and "agree" by an average of 50% that they are safe, healthy, sustainable, authentic and tasty. A similar pattern was recorded for the regional northwest Greek TFs as well. These results indicate that TFs could be the food of choice because they bear consumers' trust in the coming "new normality", where trust will be a major factor of choice for the purchase of goods and services.

Keywords: traditional foods; consumer trust; confidence and satisfaction; questionnaire survey; post COVID-19 period

Citation: Skalkos, D.; Kosma, I.S.; Vasiliou, A.; Guine, R.P.F. Consumers' Trust in Greek Traditional Foods in the Post COVID-19 Era. *Sustainability* **2021**, *13*, 9975. https://doi.org/10.3390/su13179975

Academic Editors: Richard James Volpe and Mario D'Amico

Received: 17 July 2021
Accepted: 1 September 2021
Published: 6 September 2021

1. Introduction

We are entering a new global economic and social era following the COVID-19 pandemic crisis. Philip Kotler predicts that the slowdown in global economic growth will lead to more unemployment, new consumer behavior, fewer businesses in place, and new measures for accessing the performance of economies [1]. Some countries are now preparing an annual measure of Gross Domestic Happiness (GDH) or Gross Domestic Well-Being (GDW) in order to measure the impact of economic growth in addition to Gross Domestic Product (GDP), which has been used exclusively so far [1]. Surveys conducted in 2020 investigated the food consumption behavior during the pandemic period, attempting to predict the post COVID-19 era as well. A US study in major metropolitan areas showed that patterns for major food groups seem to stay the same, but a large share indicated that they had been snacking more because of the beginning of the pandemic, which was offset by a sharp decline in fast food consumption [2]. A Swiss study revealed that consumers considered having more time to prepare meals themselves as being particularly important to achieving healthier food consumption [3]. An analysis of the datasets of food preparation recipes revealed differences in food consumption patterns in foods such as "Pulses/plants producing pulses", "Pancake/Tortilla/Oatcake", and "Soup/pottage", which increased by 300%, 280% and 100%, respectively, during the pandemic [4]. A unique panel survey of representative households in Addis Ababa implied, at least indirectly, that in the aggregate food value chains have been resilient to the shock associated with the pandemic [5]. Our findings regarding the traditional foods of northwest Greece (TFs) also showed that the pandemic didn't interfere with people's consumption patterns and preferences [6].

Literature Review

TFs are a major economic resource, as the food of choice for many cultures and regions across the world, contributing to their sense of identity, pride and prosperity [7]. They represent key elements of dietary patterns in different countries, and thus they are important for the accurate calculation of the dietary intake by the population [8]. Over the last decade, consumers have shown increased acceptance for TFs, especially in Europe [9,10], which may generate increased growth after the pandemic period [11]. The European Union, since 1992, has had specific rules defining the status under which TFs are designated in three categories [12]: Protected Designation of Origin (PDO), Protected Geographical Indication (PGI), and Traditionally Specific Guaranteed (TSG). These regulations were amended to 509/06 and 510/06, respectively, shortly after the Euro FIR (Food Information Resource) London Congress [13,14]. Currently, EU regulation 1151/12 helps producers of TFs to communicate the products' characteristics and farming attributes to buyers and consumers by establishing voluntary quality schemes [15]. The definition of the term 'traditional' in the above document means proven usage on the domestic market for a period that allows transmission between generations, with this period being at least 30 years.

Greece has incorporated the provisions of the Regulation into the national Legislation with Ministerial Decree (3321/145849) issued by the Hellenic Ministry of Food and Agricultural Development since 2006 [16]. Furthermore, a system of checks at all of the stages of production, processing and distribution of geographical indications and traditional specialties guaranteed was established, and is being implemented by the Hellenic Agricultural Organization, Demetra (AGROCERT). All of the registered traditional Greek foods are shown by their different types in Table 1.

Table 1. Distribution of the Greek recognized foods between the different categories.

Type of Food Products	PDOs	PGIs	TSGs
Wine	33	116	
Olive oil	21		
Meat	2		
Cheese	22	1	
Foods of animal origin	2		
Fish	1		
Fruits & vegetables	29	21	1
Others	6		
Total	116	138	1

Greece has registered 116 PDO products out of a total of 661 in eAmbrosia, the EU Geographical Indication Register [17]. The majority of the Greek PDO foods (33) belong to wines, and 29 belong to the class of fruit and vegetables, fresh or processed. There are also 138 Greek PGI products out of the 881 in the register. In the class of "others" belong the Chios' masticha, Chios' masticha oil, Chios' masticha gum, the safran of Kozani, the Kretan rusk and the melekouni dessert of Rodos. Surprisingly, there is only one Greek TSG food in the register, despite the wide variety of Greek traditional products and food recipes.

The TFs of northwest Greece (namely the region of Epirus) comprise a significant portion of the overall Greek TFs. It is a region with a long history of local traditional food products, such as the traditional green pies used for the feeding of its residents. Livestock—sheep and goats—have been developed as the primary self-employment by the regional farmers for centuries, producing milk, which has been used for cheese. Besides this, the farmers also made other dairy TF products, as well as wine, pasta, honey, oil, herbs or legumes, among others. Selected regional TFs, mainly the PDO cheeses and wines, are exported throughout Europe, thus promoting the regional brand name.

As has been reported for other countries [18], Greek cuisine can be represented by a triangle of influences and connections: food, culture and history. These relationships explain the link between food and the culture of the local community in a region, and this has resulted in the culinary tradition. The longevity associated with the Mediterranean Diet could be partly attributed to Mediterranean traditional foods, which this diet incorporates, including Greek traditional foods [19]. The analysis of several traditional Greek foods indicated that they may contribute to the apparent health benefits of the Greek version of the Mediterranean diet [20–22]. The traditional Greek diet favors plant foods with antioxidant potential, which are considered to provide protection from coronary heart disease and cancer, providing a high antioxidant content to the Greek Mediterranean diet [23–25].

In the post COVID-19 era, a major issue for the customer will be the trust in the products and services he will choose to buy. Food will be included in his daily agenda of preferences and choices. Consumer trust in food has become a major concern in the debate around food policy in recent years [26–28]. Trust was an important predictor of the acceptance of water recycling, both directly and indirectly through the reduction of risk perceptions [29], while trust in the food industry was important in influencing the acceptance of functional foods and foods affected by nanotechnology [30]. Lobb et al. showed significant interactions between trust, risk perceptions, and attitudes in UK consumers' decisions to purchase chicken [31]. Janssen and Hamm showed the importance of trust in the acceptance of certain types of (unfamiliar and more familiar) organic food labels [32]. Jonge et al.'s research showed that consumer trust in the safety of the food supply was mainly related to specific trust dimensions that were different for different products in the food chain [33]. The lack of trust and the ensuing lack of confidence is not only a problem for the food chain actors trying to develop and market food products in the post COVID-19 era but also a barrier for attempts to enact transformations of the food system that are widely believed to be necessary. Recent research illustrated the need for behavioral change by consumers [34,35], most notably towards more sustainable and healthier food choices [36].

The aim of the present work was to assess the factors associated with consumers' trust in Greek TFs in view of the post COVID-19 era in order to predict their future prospect, growth and development in the new rising economy. In order to accomplish this objective, following the existing literature on the parameters of food trust [37–42], the current study examines the following five determinants of consumers' trust in Greek TFs in the post COVID-19 period:

(I) Consumers' trust in the safety of the Greek TFs. This involves characteristics such as hygiene, freshness, traceability, transparency, controlled processes and additives, allergen labels and certified quality.
(II) Consumers' trust in the healthiness of the Greek TFs. This involves characteristics such as being natural, being organic, having fewer chemicals, being less processed, having fewer additives, being good for you, having low sugar/salt, and being vegetarian/vegan.
(III) Consumers' trust in the sustainability of the Greek TFs This involves characteristics such as being local, seasonal and low carbon; fair production; animal welfare; involving less meat and less packaging; being recyclable and organic; and having no chemicals.
(IV) Consumers' trust in the authenticity of the Greek TFs. This involves characteristics such as being genuine, local and nostalgic; having natural ingredients; being unprocessed; having few additives; being non-uniform; and having a certified provenance and traceability.

(V) Consumers' trust in the taste of the Greek TFs. This involves characteristics such as freshness, quality, intense flavor, sensory characteristics, the individual pleasure of eating, and valuing substance over appearance.
(VI) Consumers' trust in the Northwest Greek TFs (Epirus' region). This involves the characteristics of safety, healthiness, sustainability, authenticity, and taste mentioned in I–V above.

The consumers' trust in the technologies and raw materials used for food production was also evaluated in our study. In order to understand better the customers' perception, not only their confidence in each item of trust was evaluated (for the TFs and their technologies) but also their satisfaction with the same items.

2. Materials and Methods

2.1. Data Collection and Sample Characterization

The data collection was based on a questionnaire prepared to investigate the motivations that influence consumers' trust concerning Greek TFs, including the TFs of the Epirus region. The questionnaire was built up in seven parts. Each question was created in such a way that it could provide the best possible information for each section. The parts were built up using a similar previous study [43]. The first part included questions about the social-demographic characteristics of the respondents, specifically gender, age, level of education, civil state job situation, and permanent residency in different parts of Greece. The second part consisted of five questions designed to assess the confidence in the safety of the TFs, their production process and their raw materials, which lead the participants to purchasing in the post COVID era. The third part included five questions focused on the participant's confidence in the healthiness of the TFs, which motivates their purchase. In the fourth part, issues concerning the participants' confidence in the sustainability of TFs were assessed through five questions. The fifth part included five questions that approached the buying behavior of the participants in relation to their confidence in the authenticity of TFs. In the sixth part, using five questions, the participants' preference of the TFs regarding their confidence in the taste of the TFs was assessed. Finally, in the seventh part, using ten questions, the participants were asked to respond and provide information on their trust in the northwest Greek TFs, which can direct their preference to these foods. Issues such as safety, healthiness, sustainability, authenticity and taste were taken into consideration. In order to guarantee the quality of the data obtained through the application of the questionnaire, this was pretested with 30 respondents. This phase was pivotal to ensure that the questions were clear and understandable, such that the respondents could answer them easily. The research was carried out using electronic questionnaires, as it was easier to distribute and collect during the semi-lockdown period. The distribution method chosen was by e-mail, as was similarly performed in papers investigating consumer behaviors [44–46]. A snowball method was used in order to obtain a large number of participants [47]. The sample of the population is very well distributed, because it included a wide range of ages and civil states, etc., and the participants were familiar with the new technologies. A higher rate for female respondents was recorded, at 57.9%; this is similar to the observations of other papers as well [48–51], leading to the conclusion that women respond more willingly to food-related surveys because they are primarily involved in the household organization, consisting of people who are familiar with the concept of TF, and therefore could provide reliable answers (in order to accurately describe their choice to buy these foods). The research questionnaire was created through the Google platform and the Google Forms function due to the ability to directly export the results to an Excel sheet for further processing. The geographical context for the present study was all of the Greek territory, divided into five parts: north–west–central–south and the islands, because the country includes many of them in the Aegean and the Ionian seas. The sample included students, among others, and through them the questionnaire was made available to their families, friends and acquaintances. The respondents received e-mails explaining the purpose of the research and the importance of their participation, while there was an attached link that led to the electronic form of the questionnaire. The responses were anonymous, and no personal information was collected or correlated with any of the responses in order to ensure the protection of the participants.

The survey took place during the period May–June 2021, and consisted of the information shown in Table 2.

Table 2. Sociodemographic characterization of the sample.

Variable	Groups	(%)
Gender	Male	42.1
	Female	57.9
Age	18–25	18.3
	26–35	25.8
	36–45	17.8
	46–55	25.6
	56+	12.5
Level of education	None/Primary school	0.6
	Secondary school	0.4
	High school	14.5
	University	84.6
Civil state	Single	44.3
	Married	49.6
	Divorced	6.1
Job situation	Employed	74.6
	Unemployed	8.2
	Student	17.3
Permanent resident in Greece	NORTH GREECE (regions of Macedonia—Thrace)	23.8
	WEST GREECE (region of Epirus—Etoloakarnania prefecture)	41.3
	CENTRAL GREECE (including Athens)	24.7
	SOUTH GREECE (region of Peloponnese)	2.6
	ISLANDS (Ionian and Aegean)	7.7

Of the 548 participants, 42.1% were male and 57.9% female. Regarding the spacial distribution, 41.3% were permanent residents of west Greece, 24.7% of were residents central Greece (including the capital, Athens), 23.8% were residents of north Greece, 7.7% were residents of the Greek islands, and 2.6% were residents of south Greece, leading to a wide geographic distribution. The majority of the participants were aged 26–35, 46–55, 18–25, and 36–45 years (25.8%, 25.6%, 18.3%, and 17.8% respectively), while the other age group, 56+, was the least represented (12.5%). Regarding the level of education, most of the participants had higher education (university, 84.6%), and only 1% had only completed primary or secondary school, while the employment status category was dominated by employed (74.6%) participants. Regarding the civil state of the participants, 49.6% were married, 44.3% single, and 6.1% were divorced.

2.2. Data Analysis

The exploratory analysis of the data was achieved through basic statistical tools. The survey was prepared in Greek, and was divided into six parts, as detailed above:

Part I. Sociodemographic data (see Table 1).

Part II. Consumers' trust in the safety of the Greek TFs.

Part III. Consumers' trust in the healthiness of the Greek TFs.

Part IV. Consumers' trust in the sustainability of the Greek TFs.

Part V. Consumers' trust in the authenticity of the Greek TFs.

Part VI. Consumers' trust in the taste of the Greek TFs.

Part VII. Consumers' trust in the northwest Greek TFs.

The raw materials and the technologies used for the traditional foods were considered seperately, and therefore they were assessed separately. Furthermore, the consumers' confidence was considered separately from the satisfaction for the trust in the TFs and in the technologies used to produce them.

The sociodemographic characteristics were collected in the first part of the questionnaire. In order to measure the respondents' opinion about a set of statements related to TFs, a 5-point Likert scale, ranging from 1 = strongly disagree to 5 = strongly agree, was used [52].

The statistical processing of the data was performed using IBM SPSS Statistics for Windows (Version 25.0, IBM Corp., Armonk, NY, USA) on the data before proceeding with the other statistical tests. The data obtained from the Likert scale were considered as ordinal values.

Nonparametric tests were used. The nonparametric testing was performed in order to test the distribution of the variables of each group and response based on the hypothesized equal proportions for each variable. The Chi-Square Independence Test was used to determine whether there is an association between the variables. Cramer's V coefficient was used to analyze the strength of the significant relations found between some of the variables in the study. This coefficient ranged from 0 to 1, and can be interpreted as follows: $V \approx 0.1$, the association is considered weak; $V \approx 0.3$, the association is moderate; and $V \approx 0.5$ or over, the association is strong. Sociodemographic characteristics were considered as predictor variables that could affect the other responses of the questionnaire. In all of the tests performed, the level of significance considered was 5% ($p < 0.05$).

3. Results

In the results presented in the tables below, the percentages of strongly disagree (1) and disagree (2) were less than 10% in all of the questions of the study. Answer number 3 corresponds to "neither disagree nor agree" for all of the questions used in the study. Table 3 presents the participants' perceptions of the safety of Greek TFs, their raw materials, and the technologies used to produce them in the post COVID-19 period. The results show that the majority of the participants agree that they will be safe (for the TFs by 51.9% for confidence and 56.9% for satisfaction; for the raw materials by 55.4%; and for the technologies used by 55.1% for confidence and 57.2% for satisfaction). A significant part, more than 20%, strongly agree on the safety of the products (for TFs by 29.1% for confidence and 24.4% for satisfaction, for raw materials by 23%, and for the technologies used by 20% for confidence and 19.3% for satisfaction). …

Table 3. Participants' confidence in the safety of the Greek TFs (Scale from 1 = strongly disagree to 5 = strongly agree).

Questions	Answers According to Scale Points (%)				
	1	2	3	4	5
1. I am confident that the Greek TFs will be **safe** in the post COVID-19 era	0.5	2.4	16.1	51.9	29.1
2. I will be satisfied with the **safety** of the Greek TFs in the post COVID-19 era	0.4	2.6	15.8	56.9	24.4
3. The Greek TFs will be produced with **safe** raw materials in the post COVID-19 era	0.4	2.4	18.8	55.4	23.0
4. I am confident that the food technologies of the Greek TFs will be **safe** in the post COVID-19 era	0.4	3.1	21.4	55.1	20.0
5. I will be satisfied with the **safety** of the food technologies of the Greek TFs in the post COVID-19 era	0.4	2.2	20.9	57.2	19.3

The chi-square test, presented in Table 6, showed that there were significant differences between the perceptions for TFs' safety in terms of:

1. Confidence in the safety of the Greek TFs: only between age (x^2 = 32.714, p = 0.008) and level of education (x^2 = 75.835, p = 0.000).
2. Satisfaction with the safety of the Greek TFs: between age (x^2 = 33.380, p = 0.007), level of education (x^2 = 104.816, p = 0.000), civil state (x^2 = 18.329, p = 0.019) and job situation (x^2 = 16.419, p = 0.037).
3. The safety of the raw materials used: between gender (x^2 = 14.567, p = 0.006), level of education (x^2 = 365.786, p = 0.000) and residency (x^2 = 34.132, p = 0.005)
4. Confidence in the safety of the technologies used: between age (x^2 = 30.135, p = 0.017), level of education (x^2 = 102.641, p = 0.000) and job situation (x^2 = 24.197, p = 0.002).
5. Satisfaction with the safety of the technologies used: between age (x^2 = 27.170, p = 0.040), level of education (x^2 = 106.212, p = 0.000), job situation (x^2 = 26.035, p = 0.001) and residency (x^2 = 29.897, p = 0.019).

Table 4 presents the participants' perceptions of the healthiness of the Greek TFs, their raw materials, and their technologies in the new era after COVID-19. The results show that more than 55% of the participants agree that they are healthy (for the TFs themselves, by 55.4% for confidence, by 57.2% for satisfaction, for the raw materials by 53.8%, and for the technologies used by 56.4% for confidence and 56.3% for satisfaction). A portion around 18% strongly agree that they are healthy (for TFs by 20.8% for confidence and 20.7% for satisfaction, for raw materials by 18.7%, and for the technologies used by 17.1% for confidence, and 16.3% for satisfaction).

Table 4. Participants' confidence in the healthiness of the Greek TFs (Scale from 1 = strongly disagree to 5 = strongly agree).

Questions	Answers According to Scale Points (%)				
	1	2	3	4	5
1. I am confident that the Greek TFs will be **healthy** in the post COVID-19 era	0.2	3.5	20.1	55.4	20.8
2. I will be satisfied with **the healthiness** of the Greek TFs in the post COVID-19 era	0.6	2.4	19.1	57.2	20.7
3. The Greek TFs will be produced with **healthy** raw materials in the post COVID-19 era	0.9	3.1	23.5	53.8	18.7
4. I am confident that the food technologies of the Greek TFs will result in **healthy** food products in the post COVID-19 era	0.4	3.3	22.8	56.4	17.1
5. I will be satisfied with how in the post COVID-19 era the food technologies of the Greek TFs will result in **healthy** food products	0.2	2.9	24.2	56.3	16.3

The results of the chi-square test, presented in Table 6, showed that there were significant differences between the perceptions for TFs' healthiness in terms of the level of education and job situation only:

1. Confidence in the healthiness of the Greek TFs: only between the level of education (x^2 = 184.489, p = 0.000) and job situation (x^2 = 26.619, p = 0.001).
2. Satisfaction with the healthiness of the Greek TFs: only between the level of education (x^2 = 62.647, p = 0.000) and job situation (x^2 = 17.180, p = 0.028).
3. Healthiness of the raw materials used: only between the level of education (x^2 = 47.320, p = 0.000).
4. Confidence in the healthiness of the technologies used: only between the level of education (x^2 = 103.465, p = 0.000).
5. Satisfaction with the healthiness of the technologies used: only between the level of education (x^2 = 22.337, p = 0.034).

Table 5 presents the participants' perceptions of the sustainability of the Greek TFs, their raw materials, and their technologies used after the pandemic. The results show that an average of 50% of the participants agree that they are sustainable (for the TFs by 49.4% for confidence and 49.2% for satisfaction, for the raw materials by 51.9%, and for the technologies used by 49.0% for confidence and 52.0% for satisfaction). A low percentage strongly agree that they are sustainable products (for TFs by 16.9% for confidence and 14.4% for satisfaction, for the raw materials by 13.5%, and for the technologies used by 12.2% for confidence and 12.4% for satisfaction).

Table 5. Participants' confidence in the sustainability of Greek TFs (Scale from 1 = strongly disagree to 5 = strongly agree).

Questions	Answers According to Scale Points (%)				
	1	2	3	4	5
1. I am confident that the Greek TFs will be produced in a **sustainable** way in the post COVID-19 era (i.e., environmentally friendly, resource efficient, ethically responsible)	1.1	7.5	25.1	49.4	16.9
2. I will be satisfied with **the sustainability** of the Greek TFs in the post COVID-19 era (i.e., they will be produced in a way that will be environmentally friendly, resource efficient, ethically responsible)	1.7	6.3	28.5	49.2	14.4
3. The Greek TFs will be produced with raw materials produced in a **sustainable** way in the post COVID-19 era (i.e., environmentally friendly, resource efficient, ethically responsible)	1.8	5.4	27.4	51.9	13.5
4. I am confident that the food technologies of the Greek TFs will be **sustainable** in the post COVID-19 era (i.e., environmentally friendly, resource efficient, ethically responsible)	1.5	7.4	30.0	49.0	12.2
5. I will be satisfied with **the sustainability** of the food technologies of the Greek TFs in the post COVID-19 era (i.e., they will be produced in a way that will be environmentally friendly, resource efficient, ethically responsible)	0.9	7.6	27.1	52.0	12.4

The results of the chi-square test, presented in Table 6, showed that there were significant differences between the perceptions of TFs' sustainability in terms of:

1. Confidence in the sustainability of the Greek TFs: only between civil state ($x^2 = 22.102$, $p = 0.005$).
2. Satisfaction with the sustainability of the Greek TFs: only between the level of education ($x^2 = 24.912$, $p = 0.015$) and job situation ($x^2 = 18.179$, $p = 0.020$).
3. The sustainability of the raw materials used: between level of education ($x^2 = 28.650$, $p = 0.004$), civil state ($x^2 = 19.215$, $p = 0.014$) and job situation ($x^2 = 22.237$, $p = 0.004$).
4. Confidence in the sustainability of the technologies used: only between level of education ($x^2 = 30.864$, $p = 0.002$) and job situation ($x^2 = 28.307$, $p = 0.000$).
5. Satisfaction with the sustainability of the technologies used: between age ($x^2 = 32.331$, $p = 0.009$), civil state ($x^2 = 23.388$ $p = 0.003$) and job situation ($x^2 = 16.995$, $p = 0.030$).

Table 7 presents the participants' perceptions of the authenticity of the Greek TFs, their raw materials, and the technologies used in the post COVID-19 era. The results show that an average of 50% of the participants agree that they are authentic products (for the TFs by 48.7% for confidence and 50.8% for satisfaction, for the raw materials by 48.6%, and for the technologies used by 49.0% for confidence and 51.3% for satisfaction). A low percentage of less than 15% strongly agree that they are authentic products (for TFs by 14.3% for confidence and 14.2% for satisfaction, for the raw materials by 13.0%, and for the technologies used by 12.4% for confidence and 12.4% for satisfaction).

Table 6. Associations between the variables: (A) the safety, (B) the healthiness and (C) the sustainability of Greek TFs, and the sociodemographic variables.

	Gender			Age			Level of Education			Civil State			Job Situation			Residency		
	X^2 *	p **	V ***	X^2	p	V	X^2	p	V	X^2	p	V	X^2	p	V	X^2	p	V
A. Safety of the Greek TFs																		
1. I am confident that the Greek TFs will be **safe** in the post COVID-19 era				32.714	0.008	0.122	75.835	0.000	0.215									
2. I will be satisfied with the **safety** of the Greek TFs in the post COVID-19 era				33.380	0.007	0.124	104.816	0.000	0.254	18.329	0.019	0.130	16.419	0.037	0.124			
3. The Greek TFs will be produced with **safe** raw materials in the post COVID-19 era	14.567	0.006	0.164				365.786	0.000	0.473							34.132	0.005	0.125
4. I am confident that the food technologies of the Greek TFs will be **safe** in the post COVID-19 era				30.135	0.017	0.118	102.641	0.000	0.251				24.197	0.002	0.150			
5. I will be satisfied with the **safety** of the food technologies of the Greek TFs in the post COVID-19 era				27.170	0.040	0.112	106.212	0.000	0.255				26.035	0.001	0.156	29.897	0.019	0.117
B. Healthiness of the Greek TFs																		
1. I am confident that the Greek TFs will be **healthy** in the post COVID-19 era							184.469	0.000	0.336				26.619	0.001	0.157			
2. I will be satisfied with **the healthiness** of the Greek TFs in the post COVID-19 era							62.647	0.000	0.196				17.180	0.028	0.126			
3. The Greek TFs will be produced with **healthy** raw materials in the post COVID-19 era							47.320	0.000	0.170									
4. I am confident that the food technologies of the Greek TFs will result in **healthy** food products in the post COVID-19 era							103.465	0.000	0.252									
5. I will be satisfied with how in the post COVID-19 era the food technologies of the Greek TFs will result in **healthy** food products							22.337	0.034	0.117									
C. Sustainability of the Greek TFs																		
1. I am confident that the Greek TFs will be produced in a **sustainable** way in the post COVID-19 era (i.e., environmentally friendly, resource efficient, ethically responsible)										22.102	0.005	0.143						
2. I will be satisfied with **the sustainability** of the Greek TFs in the post COVID-19 era (i.e., they will be produced in a way that will be environmentally friendly, resource efficient, ethically responsible)							24.912	0.015	0.124				18.179	0.020	0.130			
3. The Greek TFs will be produced with raw materials produced in a **sustainable** way in the post COVID-19 era (i.e., environmentally friendly, resource efficient, ethically responsible)							28.650	0.004	0.133	19.215	0.014	0.134	22.237	0.004	0.144			
4. I am confident that the food technologies of the Greek TFs will be **sustainable** in the post COVID-19 era (i.e., environmentally friendly, resource efficient, ethically responsible)							30.864	0.002	0.138				28.307	0.000	0.162			
5. I will be satisfied with **the sustainability** of the food technologies of the Greek TFs in the post COVID-19 era (i.e., they will be produced in a way that will be environmentally friendly, resource efficient, ethically responsible)				32.331	0.009	0.122				23.388	0.003	0.147	16.995	0.030	0.126			

* Chi-square test, ** Level of significance of 5%: $p < 0.05$, *** Cramer's coefficient.

Table 7. Participants' confidence in the authenticity of the Greek TFs (Scale from 1 = strongly disagree to 5 = strongly agree).

Questions	Answers According to Scale Points (%)				
	1	2	3	4	5
1. I am confident that the Greek TFs will be **authentic** in the post COVID-19 era (real, honest, genuine, not fake or artificial)	0.4	8.2	28.4	48.7	14.3
2. I will be satisfied with **the authenticity** of the Greek TFs in the post COVID-19 era (they will be real, honest, genuine, not fake or artificial)	0.6	6.1	28.4	50.8	14.2
3. The Greek TFs will be produced with **authentic** raw materials in the post COVID-19 era (real, honest, genuine, not fake or artificial)	0.7	6.8	30.8	48.6	13.0
4. I am confident that the food technologies of the Greek TFs will be **authentic** in the post COVID-19 era (real, honest, genuine, not fake or artificial)	0.4	6.7	31.5	49.0	12.4
5. I will be satisfied with the authenticity of the food technologies of the Greek TFs in the post COVID-19 era (they will be real, honest, genuine, not fake or artificial)	0.9	5.7	29.6	51.3	12.4

The results of the chi-square test, presented in Table 10, showed that there were significant differences between perceptions for TFs' authenticity in terms of:

1. Confidence for the authenticity of the Greek TFs: only between the level of education (x^2 = 94.729, p = 0.000) and job situation (x^2 = 19.508, p = 0.012).
2. Satisfaction with the authenticity of the Greek TFs: between gender (x^2 = 9.741, p = 0.045), age (x^2 = 27.304 p = 0.038) and level of education (x^2 = 69.304, p = 0.000).
3. Authenticity of the raw materials used: only between gender (x^2 = 10.551, p = 0.032).
4. Confidence in the authenticity of the technologies used: between gender (x^2 = 10.758, p = 0.029), age (x^2 = 32.726, p = 0.008) and job situation (x^2 = 16.787, p = 0.032).
5. Satisfaction with the authenticity of the technologies used: only between the level of education (x^2 = 57.836, p = 0.000).

Table 8 presents the participants' perception of the taste of the Greek TFs, their raw materials, and their technologies used in the new economy following the pandemic. The results show that a significant percentage—more than 55%—of the participants agree that they are tasty products (for the TFs by 57.4% for confidence and 57.3% for satisfaction, for the raw materials by 57.2%, and for the technologies used by 56.0% for confidence and 55.8% for satisfaction). A relatively high percentage—more than 20%—strongly agree that they are tasty products (for TFs by 25.4% for confidence and 24.8% for satisfaction, for the raw materials by 21.9%, and for the technologies used by 21.1% for confidence and 20.4% for satisfaction).

Table 8. Participants' confidence in the taste of the Greek TFs (scale from 1 = strongly disagree to 5 = strongly agree).

Questions	Answers According to Scale Points (%)				
	1	2	3	4	5
1. I am confident that the Greek TFs will be **tasty** in the post COVID-19 era	0.2	0.9	16.2	57.4	25.4
2. I will be satisfied with **the taste** of the Greek TFs in the post COVID-19 era		0.9	17.0	57.3	24.8
3. The Greek TFs will be produced with **tasty** raw materials in the post COVID-19 era	0.2	1.5	19.3	57.2	21.9
4. I am confident that the food technologies of the Greek TFs will result in **tasty foods** in the post COVID-19 era	0.2	1.5	21.2	56.0	21.2
5. I will be satisfied with how in the post COVID-19 era the food technologies of the Greek TFs will result in **tasty foods**	0.2	1.5	22.1	55.8	20.4

The results of the chi-square test, presented in Table 10, showed that there were significant differences between the perceptions for TFs' taste in terms of:

1. Confidence in the taste of the Greek TFs: only between the level of education ($x^2 = 185.729$, $p = 0.000$)
2. Satisfaction with the taste of the Greek TFs: only between the level of education ($x^2 = 41.300$, $p = 0.000$).
3. Taste of the raw materials used: only between the level of education ($x^2 = 193.799$, $p = 0.000$).
4. Confidence with the technologies used: between gender ($x^2 = 10.439$, $p = 0.034$) and level of education ($x^2 = 182.639$, $p = 0.000$).
5. Satisfaction with the technologies used: between the level of education ($x^2 = 183.483$, $p = 0.000$), job situation ($x^2 = 15.619$, $p = 0.048$) and residency ($x^2 = 35.722$, $p = 0.003$).

Table 9 presents the participants' trust in the northwest (region of Epirus) Greek TFs and their raw materials in the post COVID-19 period. The results show that—by an average of 50%—the participants "agree" that the Epirus' Greek TFs are worthy of being trusted (in terms of safety by 54.2% for the foods, and 53% for their raw materials; in terms of healthiness by 52.3% for the foods and 48.5% for the raw materials; in terms of sustainability by 50.9% for the foods and 49.5% for the raw materials; in terms of authenticity by 50.5% for the foods and 50.8% for the raw materials; and in terms of taste by 52.6% for the foods and 52.5% for the raw materials). A relatively high percentage—more than 20%—"strongly agree" about the safety (22.3% for the foods and 22.2 for the raw materials), the healthiness (21.7% for foods and 22.1% for the raw materials), the sustainability (19.6% for the foods and 19.7% for the raw materials), the authenticity (21.2% for the foods, 20.9% for the raw materials), and the taste (23.9% for the foods, 23.6% for the raw materials) of Epirus' Greek TFs.

Table 9. Participants' trust in the northwest (Epirus' region) Greek TFs (scale from 1 = strongly disagree to 5 = strongly agree).

Questions	Answers According to Scale Points (%)				
	1	2	3	4	5
1. I am confident that the Epirus' Greek TFs will be **safe** in the post COVID-19 era	0.2	1.8	21.4	54.2	22.3
2. The Epirus' Greek TFs will be produced with **safe** raw materials in the post COVID-19 era		2.2	22.6	53.0	22.2
3. I am confident that the Epirus' Greek TFs will be **healthy** in the post COVID-19 era	0.2	1.9	23.9	52.3	21.7
4. The Epirus' Greek TFs will be produced with **healthy** raw materials in the post COVID-19 era	0.2	2.6	26.6	48.5	22.1
5. I am confident that the Epirus' Greek TFs will be produced in a **sustainable** way in the post COVID-19 era (i.e., environmentally friendly, resource efficient, ethically responsible)	0.6	3.7	25.3	50.9	19.6
6. The Epirus' Greek TFs will be produced with raw materials produced in a **sustainable** way in the post COVID-19 era (i.e., environmentally friendly, resource efficient, ethically responsible)	0.2	3.2	27.5	49.5	19.7
7. I am confident that the Epirus' Greek TFs will be **authentic** in the post COVID-19 era (real, honest, genuine, not fake or artificial)	0.0	2.6	25.8	50.5	21.2
8. The Epirus' Greek TFs will be produced with **authentic** raw materials in the post COVID-19 era (real, honest, genuine, not fake or artificial)	0.0	3.7	24.6	50.8	20.9
9. I am confident that the Epirus' Greek TFs will be **tasty** in the post COVID-19 era	0.2	1.3	22.0	52.6	23.9
10. The Epirus' Greek TFs will be produced with **tasty** raw materials in the post COVID-19 era	0.4	1.5	22.1	52.5	23.6

The results of the chi-square test, presented in Table 10, showed that there were significant differences between the perceptions of the TFs' authenticity in terms of:

1. Confidence in the safety of the Epirus' TFs: only between residency ($x^2 = 36.757$, $p = 0.002$).
2. Confidence in the safety of the raw materials for the Epirus' TFs: between gender ($x^2 = 8.235$, $p = 0.045$), level of education ($x^2 = 20.809$, $p = 0.014$) and residency ($x^2 = 28.965$, $p = 0.004$).
3. Confidence in the healthiness of the Epirus' TFs: between the level of education ($x^2 = 185.754$, $p = 0.000$), job situation ($x^2 = 32.460$, $p = 0.000$) and residency ($x^2 = 34.613$, $p = 0.004$).
4. Confidence in the healthiness of the raw materials used for Epirus' TFs: between the level of education ($x^2 = 181.592$, $p = 0.000$), civil state ($x^2 = 17.306$, $p = 0.027$), job situation ($x^2 = 21.989$, $p = 0.005$) and residency ($x^2 = 35.744$, $p = 0.003$)
5. Confidence in the sustainability of Epirus' TFs: between the level of education ($x^2 = 63.737$, $p = 0.000$), job situations ($x^2 = 30.166$, $p = 0.000$) and residency ($x^2 = 29.051$, $p = 0.024$).
6. Confidence in the sustainability of the raw materials used for Epirus' TFs: only between civil state ($x^2 = 16.583$, $p = 0.035$) and residency ($x^2 = 29.856$, $p = 0.019$).
7. Confidence in the authenticity of Epirus' TFs: only between residency ($x^2 = 24.476$, $p = 0.018$).
8. Confidence in the authenticity of the raw materials used for Epirus' TFs: between civil state ($x^2 = 13.852$, $p = 0.031$), job situation ($x^2 = 17.627$ $p = 0.007$) and residency ($x^2 = 29.950$, $p = 0.003$).
9. Confidence in the taste of Epirus' TFs: between gender ($x^2 = 13.235$, $p = 0.010$), level of education ($x^2 = 30.227$, $p = 0.003$) and residency ($x^2 = 38.560$, $p = 0.001$).
10. Confidence in the taste of the raw materials used for Epirus' TFs: only between residency ($x^2 = 31.282$, $p = 0.012$).

Table 10. Association between the variables: (A) the authenticity, (B) the taste of Greek TFs and (C) trust in the northwest (Epirus region) Greek TFs, and the sociodemographic variables.

	Gender			Age			Level of Education			Civil State			Job Situation			Residency		
	χ^2 *	p **	V ***	χ^2	p	V	χ^2	p	V	χ^2	p	V	χ^2	p	V	χ^2	p	V
A. Authenticity of the Greek TFs																		
1. I am confident that the Greek TFs will be **authentic** in the post COVID-19 era (real, honest, genuine, not fake or artificial)							94.729	0.000	0.241				19.508	0.012	0.135			
2. I will be satisfied with **the authenticity** of the Greek TFs in the post COVID-19 era (they will be real, honest, genuine, not fake or artificial)	9.741	0.045	0.134	27.304	0.038	0.112	69.304	0.000	0.206									
3. The Greek TFs will be produced with **authentic** raw materials in the post COVID-19 era (real, honest, genuine, not fake or artificial)	10.551	0.032	0.140															
4. I am confident that the food technologies of the Greek TFs will be **authentic** in the post COVID-19 era (real, honest, genuine, not fake or artificial)	10.758	0.029	0.142	32.726	0.008	0.123							16.787	0.032	0.126			
5. I will be satisfied with the authenticity of the food technologies of the Greek TFs in the post COVID-19 era (they will be real, honest, genuine, not fake or artificial)							57.836	0.000	0.189									
B. Taste of the Greek TFs																		
1. I am confident that the Greek TFs will be **tasty** in the post COVID-19 era							185.729	0.000	0.338									
2. I will be satisfied with **the taste** of the Greek TFs in the post COVID-19 era							41.300	0.000	0.160									
3. The Greek TFs will be produced with **tasty** raw materials in the post COVID-19 era							193.799	0.000	0.347									
4. I am confident that the food technologies of the Greek TFs will result in **tasty foods** in the post COVID-19 era	10.439	0.034	0.139				182.639	0.000	0.335									
5. I will be satisfied with how in the post COVID-19 era the food technologies of the Greek TFs will result in **tasty foods**							183.483	0.000	0.337				15.619	0.048	0.121	35.722	0.003	0.129
C. Trust in the Northwest (Epirus' region) Greek TFs																		
1. I am confident that the Epirus' Greek TFs will be **safe** in the post COVID-19 era																36.757	0.002	0.131
2. The Epirus' Greek TFs will be produced with **safe** raw materials in the post COVID-19 era	8.235	0.041	0.123				20.809	0.014	0.113							28.965	0.004	0.134
3. I am confident that the Epirus' Greek TFs will be **healthy** in the post COVID-19 era							185.754	0.000	0.340				32.460	0.000	0.175	34.613	0.004	0.127
4. The Epirus' Greek TFs will be produced with **healthy** raw materials in the post COVID-19 era							181.592	0.000	0.336	17.306	0.027	0.127	21.989	0.005	0.144	35.744	0.003	0.129
5. I am confident that the Epirus' Greek TFs will be produced in a **sustainable** way in the post COVID-19 era (i.e., environmentally friendly, resource efficient, ethically responsible)							63.737	0.000	0.198				30.166	0.000	0.168	29.051	0.024	0.116
6. The Epirus' Greek TFs will be produced with raw materials produced in a **sustainable** way in the post COVID-19 era (i.e., environmentally friendly, resource efficient, ethically responsible)										16.583	0.035	0.124				29.856	0.019	0.118
7. I am confident that the Epirus' Greek TFs will be **authentic** in the post COVID-19 era (real, honest, genuine, not fake or artificial)																24.476	0.018	0.123
8. The Epirus' Greek TFs will be produced with **authentic** raw materials in the post COVID-19 era (real, honest, genuine, not fake or artificial)										13.852	0.031	0.113	17.627	0.007	0.128	29.950	0.003	0.136
9. I am confident that the Epirus' Greek TFs will be **tasty** in the post COVID-19 era	13.235	0.010	0.157				30.227	0.003	0.137							38.560	0.001	0.134
10. The Epirus' Greek will be produced with **tasty** raw materials in the post COVID-19 era																31.282	0.012	0.120

* chi-square test, ** level of significance of 5%: $p < 0.05$, *** Cramer's coefficient.

4. Discussion

In this research, the consumer's trust regarding TFs, specifically Greek TFs, following the COVID-19 pandemic was studied for the first time. The objective was to predict the future of TFs as the foods of choice in the new global economic and social era, which is already underway. Greece was chosen for the study because it is a Mediterranean, EU country with increased production and use of TFs. The Greek region of Epirus, a mountainous, environmentally intact region with increased TFs, was also used in the study in order to compare the results with the rest of the Greek TFs. The sociodemographic characteristics of the participants of the survey had suitable distribution between the different categories, similar to other recent reports [53]. They were from all different parts of Greece in order to ensure geographical distribution.

The participants in this study showed a high positive perception of all of the parameters studied regarding the safety of TFs, their raw materials and their production technologies used (more than 75%), as shown in Table 3. The results of the chi-square test indicated that there were significant differences regarding safety between: (a) "gender" regarding raw materials, with a weak association (V = 0.164); (b) "age" regarding confidence and satisfaction with TFs and the used technologies used, with a weak association (V = 0.122/0.124/0.118/0.112); (c) "level of education" regarding confidence and satisfaction in/with TFs, the used technologies, and the raw materials used, with a weak to moderate association (V = 0.215/0.254/0.251/0.255/0.473); (d) "civil state" regarding the satisfaction with TFs, with a weak association (V = 0.130); (e) "job situation" regarding the satisfaction with TFs, and confidence and satisfaction with the technologies used, with a weak association (V = 0.124/0.150/0.156); (f) "residency" regarding raw materials and the satisfaction with the used technologies, with a weak association (V = 0.125/0.117). The safety of foods regarding consumers' perceptions has been studied thoroughly in the previous decade as one of the major parameters for the choice of food [33,38]. The results of a recent study, conducted during the pandemic period, indicate that society came to a consensus on trust in the safety of food [54]. The observed differences in outlet and food choices can be explained by income, settlement type, and age, in a pattern similar to our results presented here.

Overall, the participants consider Greek TFs to be healthy foods in the post COVID-19 period. However, the perceived positive result for health, more than 72% (Table 4) was slightly lower than the perceived result for safety. The results of the chi-square test indicated that there were significant differences regarding healthiness between: (a) "level of education" regarding confidence and satisfaction in the TFs, the used technologies, and the raw materials, with a weak to moderate association (V = 0.336/0.196/0.252/0.117/ 0.170); (b) "job situation" regarding confidence and satisfaction with TFs, with a weak association (V = 0.157/0.126). Because of the pandemic, the health parameter as a main reason for food selection was studied thoroughly recently. A study with Romanian participants found that, with aging, there is an increasing concern regarding the practice of a healthy diet [55], with a higher education level being significantly associated with healthier choices. The adoption of healthier food habits for grocery shopping varied significantly with the gender, age and household income of the respondents in another study [56]. Our results, in terms of health issues regarding TFs, agree with the findings of both reports. Other recent studies found environmental factors, together with health, to be the determinants of choices for Italian consumers [57], and that there is a shift towards healthier diets for Russian adults [58]. These findings were also verified by our results in the case of TFs.

The positive results regarding participants' perceptions of the sustainability of the Greek TFs, as shown in Table 5, are slightly lower than the previous two parameters, with an overall positive perception of no more than 65%. The results of the chi-square test indicated that there were significant differences regarding sustainability between: (a) "age" regarding the satisfaction with the used technologies, with a weak association (V = 0.122); (b) "level of education" regarding satisfaction with TFs, and the confidence in the used technologies and the raw materials, with a weak association (V = 0.124/0.138/0.133); (c) "civil state" regarding confidence with the TFs, and satisfaction with the used technologies and the raw materials, with a weak association (V = 0.143/0.147/0.134); (d) "job situation" regarding satisfaction with the TFs, and confidence and satisfaction with the technologies used and the raw materials, with a weak association (V = 0.130/0.162/0.126/0.144). A previous study indicated that attempts at stimulating sustainable consumption might be most effective when differences across consumer segments are taken into account [59]. Motivational imbalance has significant moderating effects, such that consumers who experience motivation imbalance showed consistently weaker intentions to consume sustainable foods than consumers who experience motivation balance [60]. Considering organic food as the mechanism to obtain a more sustainable food production and consumption system, the theoretical implications highlight the importance of the evaluation of more sustainable consumption forms in line with consumer profile particularities [61]. In accordance with these findings, our results indicate that TFs in the perception of the consumer have sustainable characteristics of preference, such as organic foods.

Comparable results with the sustainability parameter were recorded for the authenticity parameter, as shown in Table 7 above, with an overall positive feedback of no more than 64%. The results of the chi-square test indicated that there were significant differences regarding authenticity between: (a) "gender" regarding satisfaction with the TFs, and confidence in the used technologies and the raw materials, with a weak association (V = 0.134/0.142/0.140); (b) "age" regarding satisfaction with the TFs and confidence in the used technologies, with a weak association (V = 0.112/0.123); (c) "level of education" regarding confidence and satisfaction with the TFs, and satisfaction with the used technologies, with a weak to moderate association (V = 0.241/0.206/0.189); (d) "job situation" regarding confidence in the TFs and the technologies used, with a weak association (V = 0.135/0.126). A recent study indicated that organic certificates, traditional and homemade production practices, origin certificates, and information about products' origin country and raw material production region are significant quality cues enabling consumers to judge food authenticity [62]. These are all characteristics that the Greek TFs have as well, which explains our positive results in this parameter in accordance with the exiting literature.

When it comes to the taste of Greek TFs, as shown in Table 8, the positive results recorded were as high as the safety parameter, with a minimum positive perception of 75%. The results of the chi-square test indicated that there were significant differences regarding taste between: (a) "gender" regarding raw materials, with a weak association (V = 0.139); (b) "level of education" regarding confidence and satisfaction with the TFs, and with the used technologies and the raw materials, with a weak to moderate association (V = 0.338/0.160/0.335/0.337/0.347); (c) "job situation" regarding satisfaction with the technologies used, with a weak association (V = 0.121; and (d) "residency" regarding satisfaction with the used technologies, with a weak association (V = 0.129). The literature indicates that even owners of strong food brands cannot trust the ability of their brands to boost a consumer's taste experience if there is no correspondence between his or her central values and the brand symbolism [63]. In another study, parental consumption attitudes were not associated with children's fat, sweet and umami taste preferences [64]. Unfavorable parental consumer attitudes were associated with a lower parental education across Europe. Our findings support the above-mentioned literature for the importance of taste for food selection.

The participants' trust in Epirus' Greek TFs and their raw materials used, as shown in Table 9, compared with the trust in all of the Greek TFs, followed a similar pattern for most of the five parameters tested. The participants' perceptions of the taste of TFs was slightly lower for the Epirus' TFs. The results of the chi-square test indicated that there were significant differences regarding trust for Epirus' TFs between: (a) "gender" regarding the safety of the raw materials and the taste of TFs, with a weak association (V = 0.123/0.157); (b) "level of education" regarding the safety and healthiness of the raw materials, healthiness, sustainability and taste of the TFs, with a weak to moderate association (V = 0.113/0.336/0.340/0.198/0.137); (c) "civil state" regarding the healthiness, sustainability and authenticity of the raw materials, with a weak association (V = 0.127/0.124/0.113); (d) "job situation" regarding the healthiness and sustainability of TFs and healthiness, authenticity of raw materials with a weak association (V = 0.175/0.168/0.144/0.128); (e) "residency" regarding the safety, healthiness, sustainability, authenticity and taste of TFs, and the raw materials used, with a weak association (V = 0.131/0.127/0.116/0.123/0.134/0.134/0.129/0.118/0.136/0.120 respectively). Recent results suggest that COVID-19 psychological pressure was associated with an impulsive approach to buying food [65]. Consequently, food-purchasing behavior is expected to revert to pre-COVID-19 habits when the emergency in over [65]. However, our findings in this, and a recently published study [6], indicate that the increased trust, attitudes and perceptions towards TFs will be long lasting in the new post COVID-19 economy.

5. Conclusions

This research work explored consumers' trust in the Greek TFs at the beginning of an unprecedented and unpredictable social and economic period such as the post COVID-19 era. The present contribution applied the five main parameters of food trust in the TFs in the Greek consumers' mind in order to identify the variables that predicted the preference for the purchase of TFs in this new, unknown period which is changing our lives, our habits and our selections completely. To this purpose, an online survey was used to a sample 548 participants, with gender, age, education, civil state, employment and permanent residency across Greece balanced at the time of the survey conducted recently (May–June 2021). With the drastic change of consumers' behavior for all goods and services, due to the effect of the present pandemic, consumers will change their preference for foods too, in a way which is not clear yet. It is expected that people will spend less for food, in a more selected, personalized way, away from the old massive, unquestionable way. Our results show that the customers of this study appreciate—in order of importance—the safety, the taste, the healthiness, the sustainability, and the authenticity that the TFs, their raw materials, and the used technologies offer in this post COVID-9 era, making them the future foods of choice. Customers are confident and satisfied with the five characteristics associated with TFs. They evaluate the taste equally with the safety, then the healthiness, and last the sustainability and the authenticity of the TFs. They also evaluate, in the same way, the five parameters in all three items, namely the TFs themselves, their raw materials, and the used technologies for their production.

In order to understand whether or not consumers' evaluation of the trust in Greek TFs is driven by local characteristics, a regional TF group of products, namely the northwest Greek (the region of Epirus) TFs were used at the end of the same survey with the same participants. The results showed that customers evaluated in a similar manner Epirus' TFs, raw materials and technologies for the parameters of trust, except the taste, which was considered inferior compared to the taste of the overall Greek TFs.

More women, more people with university degrees, and more people with jobs took part in the survey, and this is a limitation of the study, considering the relatively limited number of responses obtained. Another limitation of the study was the use of the Greek TFs, as compared with the TFs from other countries. However, this study is the first approach to understand the trust in TFs for purchase and consumption in the new period after the pandemic crisis, highlighting which aspects are more relevant for the consumption of these types of products from the consumers' point of view.

The findings of the study are promising for the role of the TFs as the foods of choice, and consequently as the major local economic drivers, in the new post COVID-19 period. However further studies are needed in order to investigate further the parameters of trust in the TFs, the long lasting effects and the adaptation behaviors for the "new normality". The findings will contribute further to the ultimate goal, which is to integrate TFs into the daily consumption of selected consumers in different parts of the world, including Greece. Further studies should expand in two different directions: studying the TFs of other countries, primarily in the EU, either themselves or in comparison, and studying the concept of trust in depth, looking at other parameters as well for Greek TFs, making them more accessible to consumers.

Author Contributions: Conceptualization, supervision, methodology, D.S.; writing—original draft preparation, D.S. and I.S.K.; investigation, A.V.; review and editing, D.S. and R.P.F.G. All authors have read and agreed to the published version of the manuscript.

Funding: This research received no external funding.

Institutional Review Board Statement: Not applicable.

Informed Consent Statement: Not applicable.

Data Availability Statement: Not applicable.

Conflicts of Interest: The authors declare no conflict of interest.

References

1. Kotler, P. The Consumer in the Age of Coronavirus. *J. Creat. Value* **2020**, *6*, 12–15. [CrossRef]
2. Chenarides, L.; Grebitus, C.; Lusk, J.L.; Printezis, I. Food consumption behavior during the COVID-19 pandemic. *Agribusiness* **2021**, *37*, 44–81. [CrossRef] [PubMed]
3. Hansmann, R.; Baur, I.; Binder, C.R. Increasing organic food consumption: An integrating model of drivers and barriers. *J. Clean. Prod.* **2020**, *275*, 123058. [CrossRef]
4. Eftimov, T.; Popovski, G.; Petković, M.; Seljak, B.K.; Kocev, D. COVID-19 pandemic changes the food consumption patterns. *Trends Food Sci. Technol.* **2020**, *104*, 268–272. [CrossRef] [PubMed]
5. Hirvonen, K.; de Brauw, A.; Abate, G.T. Food Consumption and Food Security during the COVID-19 Pandemic in Addis Ababa. *Am. J. Agric. Econ.* **2021**, *103*, 772–789. [CrossRef]
6. Skalkos, D.; Kosma, I.S.; Chasioti, E.; Skendi, A.; Papageorgiou, M.; Guiné, R.P.F. Consumers' Attitude and Perception toward Traditional Foods of Northwest Greece during the COVID-19 Pandemic. *Appl. Sci.* **2021**, *11*, 4080. [CrossRef]
7. Almli, V.L. Consumer Acceptance of Innovations in Traditional Food. Attitudes, Expectations and Perception. Ph.D. Thesis, Norwegian University of Life Sciences, Ås, Norway, 2012.
8. Trichopoulou, A.; Soukara, S.; Vasilopoulou, E. Traditional foods: A science and society perspective. *Trends Food Sci. Technol.* **2007**, *18*, 420–427. [CrossRef]
9. Vanhonacker, F.; Kühne, B.; Gellynck, X.; Guerrero, L.; Hersleth, M.; Verbeke, W. Innovations in traditional foods: Impact on perceived traditional character and consumer acceptance. *Food Res. Int.* **2013**, *54*, 1828–1835. [CrossRef]
10. Kühne, B.; Vanhonacker, F.; Gellynck, X.; Verbeke, W. Innovation in traditional food products in Europe: Do sector innovation activities match consumers' acceptance? *Food Qual. Prefer.* **2010**, *21*, 629–638. [CrossRef]
11. Skalkos, D. Traditional Foods in Europe: Perceptions & Prospects in the New Business Era. *Mod. Concepts Dev. Agron.* **2021**, *8*, 787–789. [CrossRef]
12. Certificates of specific character for agricultural products and food ststus. *Off. J. Eur. Union* **1992**, 9–14. Available online: https://op.europa.eu/en/publication-detail/-/publication/a109261e-f10f-4710-a609-5d4a0a282252/language-en (accessed on 3 September 2021).
13. Agricultural products and food stuffs as traditional specialities quaranteed. *Off. J. Eur. Union* **2006**, 1–11. Available online: https://eur-lex.europa.eu/legal-content/EN/TXT/?uri=celex%3A32006R0509 (accessed on 3 September 2021).
14. Protection of geographical indications and designations of origin for agricultutral products and food stuffs. *Off. J. Eur. Union* **2006**, 12–25. Available online: https://eur-lex.europa.eu/legal-content/EN/TXT/?uri=CELEX%3A32006R0510 (accessed on 3 September 2021).
15. Quality schemes for agricultural products and food stuffs. *Off. J. Eur. Union* **2012**, 31–59. Available online: https://eur-lex.europa.eu/legal-content/EN/TXT/?uri=CELEX%3A32012R1151 (accessed on 3 September 2021).
16. Greek Traditional products (PDO-PGI-TSG). Hellenic Ministry of Agricultural Development and Food. Available online: http://www.minagric.gr/index.php/el/for-farmer-2/2012-02-02-07-52-07 (accessed on 3 September 2021).
17. EU Geographical Indications Register. Available online: https://ec.europa.eu/info/food-farming-fisheries/food-safety-and-quality/certification/quality-labels/geographical-indications-register/ (accessed on 3 September 2021).
18. Palupi, S.; Abdillah, F. Local Cuisine as a Tourism Signature. In *Delivering Tourism Intelligence*; Emerald Publishing Limited: Bingley, UK, 2019; pp. 299–312.
19. Trichopoulou, A.; Vasilopoulou, E.; Georga, K.; Soukara, S.; Dilis, V. Traditional foods: Why and how to sustain them. *Trends Food Sci. Technol.* **2006**, *17*, 498–504. [CrossRef]
20. Trichopoulou, A.; Naska, A.; Orfanos, P.; Trichopoulos, D. Mediterranean diet in relation to body mass index and waist-to-hip ratio: The Greek European Prospective Investigation into Cancer and Nutrition Study. *Am. J. Clin. Nutr.* **2005**, *82*, 935–940. [CrossRef]

21. Trichopoulou, A.; Costacou, T.; Bamia, C.; Trichopoulos, D. Adherence to a Mediterranean Diet and Survival in a Greek Population. *N. Engl. J. Med.* **2003**, *348*, 2599–2608. [CrossRef]
22. Trichopoulou, A.; Bamia, C.; Trichopoulos, D. Mediterranean diet and survival among patients with coronary heart disease in Greece. *Arch. Intern. Med.* **2005**, *165*, 929–935. [CrossRef]
23. Vasilopoulou, E.; Georga, K.; Joergensen, M.B.; Naska, A.; Trichopoulou, A. The antioxidant properties of Greek foods and the flavonoid content of the Mediterranean menu. *Curr. Med. Chem. Immunol. Endocr. Metab. Agents* **2005**, *5*, 33–45. [CrossRef]
24. Trichopoulou, A.; Vasilopoulou, E.; Hollman, P.; Chamalides, C.; Foufa, E.; Kaloudis, T.; Kromhout, D.; Miskaki, P.; Petrochilou, I.; Poulima, E.; et al. Nutritional composition and flavonoid content of edible wild greens and green pies: A potential rich source of antioxidant nutrients in the Mediterranean diet. *Food Chem.* **2000**, *70*, 319–323. [CrossRef]
25. Trichopoulou, A.; Vasilopoulou, E.; Georga, K. Macro- and micronutrients in a traditional Greek menu. *Forum Nutr.* **2005**, *57*, 135–146.
26. Hobbs, J.E.; Goddard, E. Consumers and trust. *Food Policy* **2015**, *52*, 71–74. [CrossRef]
27. Kaiser, M.; Algers, A. Trust in Food and Trust in Science. *Food Ethics* **2017**, *1*, 93–95. [CrossRef]
28. Sapp, S.G.; Arnot, C.; Fallon, J.; Fleck, T.; Soorholtz, D.; Sutton-Vermeulen, M.; Wilson, J.J.H. Consumer Trust in the U.S. Food System: An Examination of the Recreancy Theorem. *Rural Sociol.* **2010**, *74*, 525–545. [CrossRef]
29. Ross, V.L.; Fielding, K.S.; Louis, W.R. Social trust, risk perceptions and public acceptance of recycled water: Testing a social-psychological model. *J. Environ. Manag.* **2014**, *137*, 61–68. [CrossRef] [PubMed]
30. Siegrist, M.; Cousin, M.E.; Kastenholz, H.; Wiek, A. Public acceptance of nanotechnology foods and food packaging: The influence of affect and trust. *Appetite* **2007**, *49*, 459–466. [CrossRef] [PubMed]
31. Lobb, A.E.; Mazzocchi, M.; Traill, W.B. Modelling risk perception and trust in food safety information within the theory of planned behaviour. *Food Qual. Prefer.* **2007**, *18*, 384–395. [CrossRef]
32. Janssen, M.; Hamm, U. Governmental and private certification labels for organic food: Consumer attitudes and preferences in Germany. *Food Policy* **2014**, *49*, 437–448. [CrossRef]
33. De Jonge, J.; van Trijp, H.; Goddard, E.; Frewer, L. Consumer confidence in the safety of food in Canada and the Netherlands: The validation of a generic framework. *Food Qual. Prefer.* **2008**, *19*, 439–451. [CrossRef]
34. Tilman, D.; Clark, M. Global diets link environmental sustainability and human health. *Nature* **2014**, *515*, 518–522. [CrossRef]
35. Willett, W.; Rockström, J.; Loken, B.; Springmann, M.; Lang, T.; Vermeulen, S.; Garnett, T.; Tilman, D.; DeClerck, F.; Wood, A.; et al. Food in the Anthropocene: The EAT–Lancet Commission on healthy diets from sustainable food systems. *Lancet* **2019**, *393*, 447–492. [CrossRef]
36. Afshin, A.; Sur, P.J.; Fay, K.A.; Cornaby, L.; Ferrara, G.; Salama, J.S.; Mullany, E.C.; Abate, K.H.; Abbafati, C.; Abebe, Z.; et al. Health effects of dietary risks in 195 countries, 1990–2017: A systematic analysis for the Global Burden of Disease Study 2017. *Lancet* **2019**, *393*, 1958–1972. [CrossRef]
37. Macready, A.L.; Hieke, S.; Klimczuk-Kochańska, M.; Szumiał, S.; Vranken, L.; Grunert, K.G. Consumer trust in the food value chain and its impact on consumer confidence: A model for assessing consumer trust and evidence from a 5-country study in Europe. *Food Policy* **2020**, *92*, 101880. [CrossRef]
38. De Jonge, J.; Van Trijp, H.; Jan Renes, R.; Frewer, L. Understanding consumer confidence in the safety of food: Its two-dimensional structure and determinants. *Risk Anal.* **2007**, *27*, 729–740. [CrossRef]
39. Gefen, D.; Straub, D.W. Consumer trust in B2C e-Commerce and the importance of social presence: Experiments in e-Products and e-Services. *Omega* **2004**, *32*, 407–424. [CrossRef]
40. Poortinga, W.; Pidgeon, N.F. Exploring the dimensionality of trust in risk regulation. *Risk Anal.* **2003**, *23*, 961–972. [CrossRef]
41. Zhang, A.; Mankad, A.; Ariyawardana, A. Establishing confidence in food safety: Is traceability a solution in consumers' eyes? *J. Verbrauch. Lebensm.* **2020**, *15*, 99–107. [CrossRef]
42. Fernández-Zarza, M.; Amaya-Corchuelo, S.; Belletti, G.; Aguilar-Criado, E. Trust and food quality in the valorisation of geographical indication initiatives. *Sustainability* **2021**, *13*, 3168. [CrossRef]
43. Guiné, R.; Ferrão, A.C.; Ferreira, M.; Correia, P.; Cardoso, A.P.; Duarte, J.; Rumbak, I.; Shehata, A.M.; Vittadini, E.; Papageorgiou, M. The motivations that define eating patterns in some Mediterranean countries. *Nutr. Food Sci.* **2019**, *49*, 1126–1141. [CrossRef]
44. Palmieri, N.; Suardi, A.; Pari, L. Italian consumers' willingness to pay for eucalyptus firewood. *Sustainability* **2020**, *12*, 2629. [CrossRef]
45. Palmieri, N.; Perito, M.A.; Macrì, M.C.; Lupi, C. Exploring consumers' willingness to eat insects in Italy. *Br. Food J.* **2019**, *121*, 2937–2950. [CrossRef]
46. Palmieri, N.; Perito, M.A.; Lupi, C. Consumer acceptance of cultured meat: Some hints from Italy. *Br. Food J.* **2020**, *123*, 109–123. [CrossRef]
47. Palmieri, N.; Perito, M.A. Consumers' willingness to consume sustainable and local wine in Italy. *Ital. J. Food Sci.* **2020**, *32*, 222–233. [CrossRef]
48. De Leeuw, A.; Valois, P.; Ajzen, I.; Schmidt, P. Using the theory of planned behavior to identify key beliefs underlying pro-environmental behavior in high-school students: Implications for educational interventions. *J. Environ. Psychol.* **2015**, *42*, 128–138. [CrossRef]
49. Pappalardo, G.; Lusk, J.L. The role of beliefs in purchasing process of functional foods. *Food Qual. Prefer.* **2016**, *53*, 151–158. [CrossRef]

50. Chinnici, G.; D'Amico, M.; Pecorino, B. A multivariate statistical analysis on the consumers of organic products. *Br. Food J.* **2002**, *104*, 187–199. [CrossRef]
51. Giampietri, E.; Verneau, F.; Del Giudice, T.; Carfora, V.; Finco, A. A Theory of Planned behaviour perspective for investigating the role of trust in consumer purchasing decision related to short food supply chains. *Food Qual. Prefer.* **2018**, *64*, 160–166. [CrossRef]
52. Likert, R. A technique for the measurement of attitudes. *Arch. Psychol.* **1932**, *140*, 44–53.
53. Petrescu-Mag, R.M.; Vermeir, I.; Petrescu, D.C.; Crista, F.L.; Banatean-Dunea, I. Traditional foods at the click of a button: The preference for the online purchase of romanian traditional foods during the COVID-19 pandemic. *Sustainability* **2020**, *12*, 9956. [CrossRef]
54. Skripnuk, D.F.; Davydenko, V.A.; Romashkina, G.F.; Khuziakhmetov, R.R. Consumer Trust in Quality and Safety of Food Products in Western Siberia. *Agronomy* **2021**, *11*, 257. [CrossRef]
55. Bacârea, A.; Bacârea, V.C.; Cînpeanu, C.; Teodorescu, C.; Seni, A.G.; Guiné, R.P.F.; Tarcea, M. Demographic, anthropometric and food behavior data towards healthy eating in romania. *Foods* **2021**, *10*, 487. [CrossRef] [PubMed]
56. Shamim, K.; Ahmad, S.; Alam, M.A. COVID-19 health safety practices: Influence on grocery shopping behavior. *J. Public Aff.* **2021**, e2624. [CrossRef]
57. Wongprawmas, R.; Mora, C.; Pellegrini, N.; Guiné, R.P.F.; Carini, E.; Sogari, G.; Vittadini, E. Food choice determinants and perceptions of a healthy diet among Italian consumers. *Foods* **2021**, *10*, 318. [CrossRef] [PubMed]
58. Ben Hassen, T.; El Bilali, H.; Allahyari, M.S.; Berjan, S.; Fotina, O. Food purchase and eating behavior during the COVID-19 pandemic: A cross-sectional survey of Russian adults. *Appetite* **2021**, *165*, 105309. [CrossRef] [PubMed]
59. Rodríguez-Pérez, C.; Molina-Montes, E.; Verardo, V.; Artacho, R.; García-Villanova, B.; Guerra-Hernández, E.J.; Ruíz-López, M.D. Changes in dietary behaviours during the COVID-19 outbreak confinement in the Spanish COVIDiet study. *Nutrients* **2020**, *12*, 1730. [CrossRef]
60. Elhoushy, S. Consumers' sustainable food choices: Antecedents and motivational imbalance. *Int. J. Hosp. Manag.* **2020**, *89*, 102554. [CrossRef]
61. Feil, A.A.; da Silva Cyrne, C.C.; Sindelar, F.C.W.; Barden, J.E.; Dalmoro, M. Profiles of sustainable food consumption: Consumer behavior toward organic food in southern region of Brazil. *J. Clean. Prod.* **2020**, *258*, 120690. [CrossRef]
62. Chousou, C.; Mattas, K. Assessing consumer attitudes and perceptions towards food authenticity. *Br. Food J.* **2019**, *123*, 1947–1961. [CrossRef]
63. Paasovaara, R.; Luomala, H.T.; Pohjanheimo, T.; Sandell, M. Understanding consumers' brand-induced food taste perception: A comparison of "brand familiarity"—And "consumer value-brand symbolism (in)congruity"—Accounts. *J. Consum. Behav.* **2012**, *11*, 11–20. [CrossRef]
64. Jilani, H.S.; Pohlabeln, H.; Buchecker, K.; Gwozdz, W.; De Henauw, S.; Eiben, G.; Molnar, D.; Moreno, L.A.; Pala, V.; Reisch, L.; et al. Association between parental consumer attitudes with their children's sensory taste preferences as well as their food choice. *PLoS ONE* **2018**, *13*, e0200413. [CrossRef]
65. Russo, C.; Simeone, M.; Demartini, E.; Marescotti, M.E.; Gaviglio, A. Psychological pressure and changes in food consumption: The effect of COVID-19 crisis. *Heliyon* **2021**, *7*, e06607. [CrossRef] [PubMed]

MDPI

Article

Quality Control of Emerging Contaminants in Marine Aquaculture Systems by Spot Sampling-Optimized Solid Phase Extraction and Passive Sampling

Panagiota Martinaiou [1], Panagiota Manoli [1], Vasiliki Boti [1,2,*], Dimitra Hela [1,2,*], Elissavet Makou [1], Triantafyllos Albanis [1,2] and Ioannis Konstantinou [1,2]

[1] Department of Chemistry, University of Ioannina, 45110 Ioannina, Greece; pch1277@uoi.gr (P.M.); p.manoli@uoi.gr (P.M.); pch1248@uoi.gr (E.M.); talbanis@uoi.gr (T.A.); iokonst@uoi.gr (I.K.)

[2] Institute of Environment and Sustainable Development, University Research Center of Ioannina (URCI), University of Ioannina, 45110 Ioannina, Greece

* Correspondence: vboti@uoi.gr (V.B.); dchela@uoi.gr (D.H.)

Abstract: The presence of organic pollutants such as pesticides and pharmaceuticals in the aquatic environment, and especially in regions where fish farms are installed, is a matter of major importance due to their possible risks to ecosystems and public health. The necessity of their detection leads to the development of sensitive, reliable, economical and environmentally friendly analytical methods for controlling their residue in various environmental substrates. In the present work, a solid-phase extraction method was developed, optimized and validated for the analysis of 7 pesticides and 25 pharmaceuticals in seawater using LC-HR-LTQ/Orbitrap-MS. The method was then applied in seawater samples collected from an aquaculture farm located in the Ionian Sea, Greece, in order to evaluate environmental pollution levels. None of the pesticides were detected, while paracetamol was the only pharmaceutical compound that was found (at trace levels). At the same time, passive sampling was conducted as an alternative screening technique, showing the presence of contaminants that were not detected with spot sampling. Among them, irgarol was detected and as far as pharmaceuticals is concerned, trimethoprim and sulfadiazine were found; however, all positive findings were at the very low ppt levels posing no threat to the aquatic environment.

Keywords: passive sampling; marine aquaculture; organic pollutants; solid-phase extraction

Citation: Martinaiou, P.; Manoli, P.; Boti, V.; Hela, D.; Makou, E.; Albanis, T.; Konstantinou, I. Quality Control of Emerging Contaminants in Marine Aquaculture Systems by Spot Sampling-Optimized Solid Phase Extraction and Passive Sampling. *Sustainability* **2022**, *14*, 3452. https://doi.org/10.3390/su14063452

Academic Editor: Dimitris Skalkos

Received: 31 January 2022
Accepted: 9 March 2022
Published: 15 March 2022

1. Introduction

In recent years, the development of the urban environment and industry has led to increasing pollution in the aquatic environment; however, the environmental quality control monitoring in marine aquaculture is one of the main concerns [1]. Aquaculture is among the pressure factors in coastal ecosystems, introducing pollutants such as pharmaceuticals and plant protection compounds. Emerging organic pollutants, including pesticides and pharmaceutical compounds, are a large group of contaminants that can be found in aquatic ecosystems, and therefore various monitoring frameworks have been developed aiming to assess their environmental fate and concentration levels. The EU Marine Strategy Framework Directive (MSFD) (Directive 2008/56/CE), ref. [2] especially, establishes requirements to obtain a good environmental status of the marine environment. Emerging organic pollutants are extensively studied in various aquatic matrices such as wastewater, surface, ground water and drinking water, which are directly affected by them [3–6]; however, studies focusing on the presence of organic pollutants in marine ecosystems are relatively limited [7].

Pesticides are present in the aquatic environment due to their application in agricultural fields [8], and in this way, they can be transported by surface runoff into inland surface waters ending to the sea and by leaching through soil into groundwater [9,10].

Water solubility, as well as the stability of each substance, are major properties affecting the fate of pesticides in the environment [11,12]. In general, pesticides are toxic to living organisms with carcinogenic and mutagenic properties in humans and other animals. Taking into account their potential effects, their persistence and their regular application, these compounds provide a major risk for the environment [13,14]. The concentrations of pesticides that are found in aquatic marine ecosystems are usually at levels of ng L^{-1}. As a result, the use of very reliable, selective and highly sensitive analytical methodologies is very important in order to be detected, involving a pre-concentration step of the target analytes from the seawater sample so that they can be determined at such low levels based on the modern analytical techniques [15].

Pharmaceuticals are widely used for the treatment of diseases in humans and animals, reaching environmental compartments through incomplete removal during decontamination technologies or improper disposal [16]; they are considered as emerging and priority environmental contaminants because of their potential risks both to the environment and human health due to the promotion of microbial resistance to antibiotics [17–19]. The occurrence of pharmaceuticals and their transformation products (TPs) in aquatic systems is highly demonstrated, while they can undergo further transformations into more toxic products in some cases [20–23]. Consequently, the environmental monitoring of these compounds is very important in order to assess their possible environmental risks, and several works have focused on their occurrence in the marine environment [24–27]. The concentration levels of these compounds in fish farm regions are an issue that must be under further investigation while it is directly connected with public health; therefore, these contaminants' presence in marine ecosystems and especially in fish farming areas makes necessary the constant monitoring of pollution levels.

Routine water monitoring mainly relies on spot (grab) sampling at fixed intervals. The analysis of spot samples combined with optimized solid-phase extraction (SPE) procedures and high-resolution chromatographic analysis can detect pollutants at trace levels [8,13], and thus, improvements in extraction and detection methods are highly important; however, this approach provides an instantaneous estimate of the pollutant's concentration at the time and point of sampling and is likely to miss peak inputs in a given aquatic system. Passive sampling (PS) appears as a promising alternative instead of the traditional spot sampling method for environmental contaminants monitoring [28]. This method for detecting such compounds, including pesticides or drugs, offers a great variety of advantages. One significant benefit is that PS, in contrast to instantaneous sampling, allows the determination of the average concentration of pollutants that are present in a sampling area as well as the pre-concentration of pollutants, thus increasing the possibility of their detection in trace concentrations. PS devices are used to enable regular monitoring of chemicals in both spatial and temporal ways in water [29]. In this way, a screening of environmental pollutants could be achieved while during spot sampling some contaminants may not be detected; moreover, the number of required sampling for a reliable analysis is somewhat lower, making this method much more economic as the use of materials and reagents is reduced [30]. Although PS techniques are more challenging due to hydrodynamic regime and calibration requirements, they have been recommended (WFD daughter 2013/39/UE) as complementary methods to improve the level of confidence in surface water monitoring in comparison with grab sampling [31].

In the present study, 7 multiclass pesticides and 25 multiclass pharmaceuticals were selected and studied in seawater samples coming from an aquaculture facility located in the Ionian Sea, Greece. The selection of the target compounds was based on factors such as their extended use in aquaculture facilities, their potential presence in aquatic systems due to the surrounding agriculture activities (pesticides), their multi-purpose use of disease treatment (pharmaceuticals), as well as their detection in surface waters in Greece [1,32–34]. The aim of this work is the development, optimization and validation of an SPE extraction method for conventional spot samples, along with passive sampling screening for seawater quality control in aquatic farm ecosystems, as well as the application of this method in real samples.

2. Materials and Methods

2.1. Chemicals and Reagents

The pesticides selected in the present study were azamethiphos, azoxystrobin, boscalid, irgarol, malathion, pirimiphos-methyl, tebufenozide, and metobromuron (internal standard) were of high purity (>98%) and they were supplied in solid form by Sigma Aldrich (Darmstadt, Germany). The standard pharmaceutical compounds were also of high purity (>98%) and were supplied in solid form (alprazolam, amisulpride, amitriptyline, atenolol, bezafibrate, budesonide, bupropion, carbamazepine, cimetidine, citalopram, diazepam, fluoxetine, haloperidol, ketoprofen, mirtazapine, olanzapine, paracetamol, paroxetine, phenazone, quetiapine, risperidone, sertraline, and venlafaxine). All pharmaceutical compounds were purchased from Sigma Aldrich (Darmstadt, Germany) except olanzapine, amisulpride, amitriptyline, ketoprofen, paroxetine, quetiapine and venlafaxine, which were acquired from Tokyo Chemical Industry, Europe N.V (Oxford, U.K). Mirtazapine and deuterated internal standards D3-olanzapine, D4-haloperidol, D5-fluoxetine, D6-amitriptyline and D10-carbamazepine were obtained from Analytical Standard Solutions, A2S (Saint Jean d'Illac, France). Alprazolam and diazepam are under controlled distribution in Greece with drug control procedures, so they were offered as a donation by the company Adelco (Moschato, Athens, Greece).

Stock standard solutions were prepared for each compound, at concentrations of either 2000 mg L^{-1} or 1000 mg L^{-1}, in methanol. Based on these solutions the mixtures of pesticides and pharmaceutical compounds were prepared at a concentration of 10 mg L^{-1}, in methanol. Both the solutions of individual compounds and the mixtures were stored at −20 °C. The working solutions of the mixtures were prepared in methanol and in concentrations of 50, 100, 250, 500, 1000 and 5000 ng L^{-1}. Both metobromuron and the mixture of deuterated internal standards were prepared in methanol at concentrations of 1000 and 5000 ng L^{-1}.

The solvents methanol and acetonitrile (ACN, acetonitrile) of analytical grade and dichloromethane (DCM, methyl chloride) of purity >99.5% were supplied by Fisher Scientific (Leicestershire, UK). High purity ethanol (ethanol) as well as hexane (n-hexane) were supplied by Lab-Scan (Dublin, Ireland). The purity acetone >99.9% was from Honeywell (Morris Plains, NJ, USA) while the LC–MS purity water was supplied by Fisher Scientific. Formic acid (FA) and ammonium formate (FNH_4) of 98–100% purity was obtained from Merck (Darmstadt, Germany). Oasis HLB extraction cartridges (divinylbenzene/N-vinylpyrrolidone copolymer, 200 mg, 6 mL) were supplied from Waters Corporation (Milford, CT, USA). The samplers POCIS (47 mm i.d. membrane disks) were provided by Exposmeter SA (Tavelsjö, Sweden) with the "generic" configuration for pesticide sampling (pest-POCIS) and for pharmaceutical sampling (pharm-POCIS).

2.2. Sampling

Seawater samples were collected for the development and validation of the extraction method. In addition, a 10-month spot sampling campaign was carried out to estimate the seawater pollution levels. For that purpose, water samples were collected monthly (July 2020 to April 2021) from an aquaculture farm in the Ionian Sea, from two sampling points (one sampling point in the fish farm and one reference point around 1 Km away from the fish farm in the open sea, both at 2 m depth from the surface) (Figure S1). Seawater from the reference point which was previously checked to ensure that it did not contain the selected analytes was also used for the method development and validation. The collection of the samples was carried out in dark glass bottles of 2.5 L. The samples were transferred to the laboratory under refrigeration, filtered through GF/F glass fiber filters (0.7 μm pore size, Whatman International Ltd., Maidstone, UK) and stored at 4 °C until extraction within 24 h. At the same time, passive sampling took place in the same region.

For the passive sampling, both pest-POCIS and pharm-POCIS disks were attached in stainless steel holders and placed in stainless steel canisters as they are provided by Exposmeter SA (Tavelsjö, Sweden). The samplers were mantled in the field before deployment. The three canisters were placed at different sampling sites in the aquaculture region. The first was placed between the coast and the fish farming, the second within fish farming area, while the third was placed away from fish farms in the open sea, all at a depth of 2 to 3 m. The samplers were deployed twice for 3 weeks during the period September 2020–April 2021. At the end of the exposure period, the POCIS were slightly or not biofouled and for this reason, they were rinsed with sea water and ultrapure water to remove any debris and to clean the nuts and bolts, wrapped in aluminum foil, and stored in their original containers, then transported to the laboratory under cooled conditions (~4 °C). The extraction of the target compounds was carried out usually on the same day; otherwise, POCIS were stored, frozen, and the extraction was performed within 24 h. One blank POCIS was exposed to open air during the deployment and retrieval of the POCIS and was transported and analyzed as the deployed POCIS. Procedural blank consisted of POCIS taken through the entire processing and analysis sequence.

2.3. Solid-Phase Extraction (SPE)

Two different analytical procedures for solid-phase extraction were developed, with the first concerning the extraction of pesticides and the second concerning pharmaceutical compounds.

2.3.1. Pesticides

The Oasis HLB extraction columns (200 mg, 6 mL) were placed in a 12-port extraction manifold connected to a vacuum pump and activated by the successive addition of 6 mL of methanol and 6 mL of LC–MS water, which were eluted from the columns with a flow rate ≈1 mL min^{-1}. Immediately after activation and before the adsorbent dries, 250 mL of the aqueous sample percolated through the cartridge by vacuum, with a flow rate ≈2 mL min^{-1}. At the end of the extraction and before the columns dry, the cartridge was washed with 6 mL of LC–MS water and left under a vacuum for 30 min. The elution with 3 mL of dichloromethane, 3 mL of hexane and 3 mL of acetone followed previous reports [35]. This procedure was chosen after testing two extraction protocols in triplicate under different extraction conditions in order to select the optimal conditions for pesticides in water. Oasis HLB extraction columns (200 mg, 6 mL) were used in both cases. The pH of the samples was not adjusted in any protocol as most of the selected compounds do not have chemical moieties that can be ionized; Moreover, the aim of the study was the simultaneous determination of pesticides with different physicochemical properties, and as the pH values of the samples ranged from 6.5 to 7.5, it was finally chosen not to adjust the pH value of the seawater samples. LC–MS grade methanol and water were used as activation solvents in both protocols. The volume of the passing sample in both cases was 250 mL and the flow rate was constant (2 mL min^{-1}). Regarding the rinsing of the cartridges after loading the whole sample and before elution, LC–MS purity water was used in both cases. Testing was performed at the elution stage of the SPE process, where different elution solvents were used among protocols. In "HLB1" 2 × 5 mL MeOH was used while in "HLB2" a combination of solvents was used: 3 mL dichloromethane, 3 mL hexane, 3 mL acetone, successively and in this order; the 9 mL eluate was collected in the same tube. In both cases, the eluate was evaporated to dryness under a gentle stream of nitrogen and redissolved in 500 μL H_2O: MeOH (90:10, v/v) acidified with 0.1% formic acid. Metabromuron was added as an internal standard in the final extracts before chromatographic analysis.

2.3.2. Pharmaceuticals

For the extraction of the pharmaceutical compounds, a different procedure was followed. After the samples were acidified with formic acid (pH = 3–3.5), the mixture of isotopically labeled internal standards (IS) was added. The Oasis HLB extraction columns (200 mg, 6 mL) were placed in an extraction device connected to a vacuum pump and activated by successive addition of 5 mL of methanol and 5 mL of, acidified with formic acid, LC–MS water similarly to the sample, which is eluted from the columns by vacuum application and flow rate ≈1 mL min^{-1}. Immediately after activation, 250 mL of the acidified aqueous sample was extracted through the columns by vacuum application and flow rate ≈2 mL min^{-1}. At the end of the extraction, the cartridges were washed with 5 mL of LC–MS water and were left under vacuum for 30 min, for complete moisture removal. Elution with 2 × 5 mL methanol, in a vacuum, at a flow rate of ≈2 mL min^{-1} followed. The eluents were concentrated under a gentle stream of nitrogen and redissolved in 500 μL of 0.1% formic acid in water/methanol 90/10 (*v*/*v*).

Prior to method validation, three protocols were tested. The pH of the sample was not adjusted before extraction in protocol "HLB3" while in the protocols "HLB4" and "HLB5", the samples were acidified in a final pH value of ≈3. Furthermore, in protocols "HLB3" and "HLB4" 5 mL Na_2EDTA (5% *w*/*v*) were added. Methanol LC–MS and water LC–MS were the activating solvents in all cases. The volume of the passing sample in all 3 cases was 250 mL and the flow rate constant (2 mL min^{-1}). Regarding the rinsing of the cartridges after loading of the whole sample and before elution, in the first 2 cases, LC–MS purity water was used while in the "HLB5" protocol 5 mL of 20% methanol in 2% acetic acid was used. Testing was also performed at the elution stage of the SPE process, where 2 × 5 mL MeOH was used in the first two protocols while 2 × 5 mL methanol was used in "HLB5" with the addition of 2% acetic acid (*v*/*v*). In all 3 cases, the eluate was evaporated to dryness under a gentle stream of nitrogen and redissolved in 500 μL, H_2O: MeOH (90:10, *v*/*v*) acidified with 0.1% formic acid in "HLB4" and H_2O: MeOH (90:10, *v*/*v*) without the addition of formic acid in the other two protocols.

The final extracts from both pesticides and pharmaceutical extraction procedures were then subjected to analysis in LC-HR-LTQ/Orbitrap-MS. The validation of the analytical methods was carried out on fortified samples based on the current instructions 2002/657/EC, [36] and 96/23/EC. After the development, optimization and validation of the SPE methods for the determination of pesticides and pharmaceuticals, the methods were applied in seawater samples. The physicochemical characteristics of the seawater samples were measured using a portable field multimeter sensor (WTW) as shown in Table 1.

Table 1. Physicochemical characteristics of seawater samples.

Parameters	Min	Max	Average	SD
Temperature (°C)	14.9	17.5	15.9	0.82
TDS (mg L^{-1})	77,000	91,000	85,542	4496.3
Conductivity (mS cm^{-1})	45.6	50.5	47.8	1.97
Salinity (‰)	35.3	41.2	38.9	1.98
pH	6.5	7.5	6.8	0.4

2.4. Passive Sampling Procedure

Two configurations of POCIS were used in this study, the generic configuration that contained a mixture of three sorbent materials to sample most pesticides, the pharmaceutical configuration that contained a single sorbent material designed for sampling most pharmaceutical groups. The loaded POCIS were disassembled carefully, and the membranes were detached from the disk. The sorbent was transferred into a glass mortar and left at room temperature until dried. Then, it was weighed and carefully transferred into an empty solid-phase extraction tube (6 mL) and it was packed between two polyethylene (PE) frits (20 μm porosity). The pest-POCIS samplers contained ≈200 mg of a triphasic sorbent admixture: hydroxylated polystyrene-divinylbenzene resin (Isolute ENV+)/carbonaceous sorbent (Ambersorb 572), 80:20 (*w*/*w*), dispersed on styrene-divinylbenzene copolymer (S-X3 Bio Beads) and was enclosed between two hydrophilic polyethersulfone (PES) microporous membranes (130 μm thickness, 0.1 μm pore size). For the extraction of the pesticides, SPE cartridges were disposed on a Visiprep SPE vacuum manifold (Supelco) and were eluted using 30 mL of a mixture of dichloromethane/methanol/toluene (8:1:1, *v:v:v*) [37]. The eluate was reduced to dryness in a gentle stream of nitrogen and the residue was dissolved in 0.5 mL of LC mobile phase.

For the extraction of pharmaceutical compounds, Pharm-POCIS (HLB) sorbent was also transferred into an empty SPE cartridge (6 mL) and packed between two polyethylene frits (6 mL). Elution of the analytes from the sorbent was performed twice with 10 mL of MeOH at 1 mL min^{-1} rate. At the last step, the eluate was evaporated until dryness was almost achieved under a gentle stream of nitrogen, and reconstituted in 1 mL of LC mobile phase. The sampling rate is a parameter that allows the determination of analytes mass in passive sampling devices, as it shows the water volume which is sampled per time units. For the calculation of environmental concentration levels, the following equation for the determination of sampling rates [38] was used:

$$Rs = \frac{Cpocis \times Mpocis}{Cwater \times t}$$

where *Cpocis* (μg g^{-1}) is the concentration of analyte in the sorbent, *Mpocis* (g) is the mass of the sorbent within the POCIS, Cwater (μg L^{-1}) is the mean concentration of the analyte in the water and *t* concerns the sampling period (days). In this study, available *Rs* values were obtained from literature data [23,39], and in this way, the minimum and maximum concentrations of the detected compounds were calculated.

2.5. LC–MS Analysis

An Ultra High Performance Liquid Chromatography (UHPLC) system coupled to a LTQ/Orbitrap FT mass spectrometer was used for the selected pesticides and pharmaceutical compounds determination. The system included an automatic sampler (Accela AS autosampler model 2.1.1), an automatic sample flow pump (Accela quaternary gradient U-HPLC-pump model 1.05.0900) and an LTQ Orbitrap XL 2.5.5 SP1 mass spectrometer from Thermo Fisher Scientific (Bremen, Germany). The selected analytes were separated on a Hypersil GOLD reversed-phase analytical column (50 mm × 2.1 mm, 1.9 μm) from Thermo (Bremen, Germany). The control of the instrument and the processing of the mass spectra was carried out using the Xcalibur v.2.2 software (Thermo Electron, San Jose, CA, USA). Chromatographic analysis in both cases included a gradient elution program. For the pesticides, the mobile phase consisted of (A) H_2O + 5 mM FNH_4 + 0.1% FA and (B) MeOH + 5 mM FNH_4 + 0.1% FA. A 10-min program was used to separate the compounds of interest. The mobile phase gradient started at 90% mobile phase A and was maintained for 0.6 min; then the methanol content (B) increased until it reached 100% at 5.1 min, where it was maintained for 1.2 min. Afterward, the mobile phase was restored to 90% A and maintained over 3 min for re-equilibration. The flow rate was kept constant at 300 μL min^{-1} and the oven temperature was set at 20 °C. For the pharmaceutical compounds, the mobile phase consisted of (A) H_2O + 0.1% FA and (B) MeOH + 0.1% FA with an initial solvent composition of 95% (A) and 5% (B). This was maintained for 1 min. Then, the methanol content increased to 70% in 2 min to reach 100% in 5 min and it was maintained for 2 min, until the system returned to its initial conditions. The flow rate was kept constant at 250 μL min^{-1} and the oven temperature was set at 35 °C. The injection volume was 5 μL in both cases.

All the detected compounds were identified on the basis of their retention time and formation of the protonated molecular ion $[M + H]^+$. The mass range selected for pesticides and pharmaceuticals full scan acquisition was m/z 120–1000 amu. The main instrument parameters were optimized at the instrument tuning sections. Quantification was performed post-acquisition using an isolation window of ±2 amu. The ESI source values and the MS parameters were: spray voltage 3.7 V and 4 V for pesticides and pharmaceuticals, respectively, sheath gas 40, aux gas 15 and sweep gas 0 arbitrary units, capillary temperature

320 °C, capillary voltage 30 V, tube lens 90 V, as well as AGC target 4×10^5 at a resolution of 60,000. Confirmation of the analytes was achieved through their production using Collision Induced Dissociation (CID 35%) and fragmentation process (Data-Dependent mode). The fragment ions produced for the detected compounds were: paracetamol 152.0706 → 110.0597, trimethoprim 291.1452 → 230.1162, sulfadiazine 251.0579 → 156.0114 and irgarol 254.1434 → 198.0811.

3. Results and Discussions

3.1. Optimization and Validation of SPE Method for Pesticides

The optimization of the extraction method was based on different protocols for the multiresidue analysis of pesticides and pharmaceutical compounds. Two extraction protocols (HLB1, HLB2, Table S1) were tested to select the optimal extraction conditions for pesticides from seawater and three extraction protocols (HLB3-HLB5) for pharmaceutical compounds. As shown in Figure 1, the recoveries for most pesticide compounds are in the range of 60–100% in both protocols and the relative standard deviations in the acceptable limits of 0–20%; however, for "HLB1" protocol azamethiphos was not recovered while the relative standard deviations are larger, thus "HLB2" protocol was chosen.

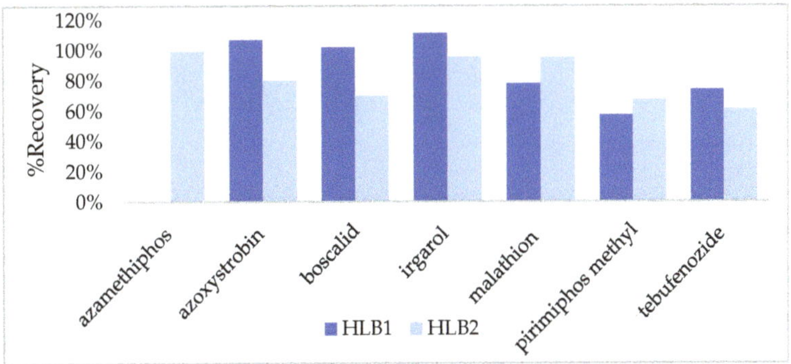

Figure 1. Recoveries (%) of pesticides by applying HLB1 and HLB2 extraction protocols.

The evaluation of the method's trueness was based on the calculation of the recoveries in fortified water samples. The recoveries of pesticides were calculated in three concentration levels, 25 ng L^{-1}, 100 ng L^{-1} and 250 ng L^{-1} that were analyzed in triplicates. The levels selection was based on the concentration levels at which the selected compounds are generally found in the environment. As shown in Table 2, the mean recovery values at the low concentration level ranged from 58.5% (tebufenozide) to 98.9% (pirimiphos-methyl), at the intermediate concentration level from 61, 8% (tebufenozide) to 95% (malathion) and from 58.6% (tebufenozide) to 78.3% (malathion) for the high concentration level. Five samples (n = 5) were fortified and analyzed on the same day to calculate the repeatability of the method (RSDr), for the intermediate concentration level, on five consecutive days to calculate the intermediate precision (RSD_{IP}). The repeatability of the method (RSDr) was always less than 19.5% recorded for the high concentration level. Method intermediate precision (RSD_{IP}) was <11.4%, observed for azoxystrobin.

The use of calibration curves with substrate simulation in conjunction with an internal standard has significantly contributed to minimizing errors in calculations. Slight matrix effect (ME) values (Figure 2) were observed ranging between −20% and 20%. The Limits of Detection (LODs) and Limits of Quantification (LOQs) determined as signal to noise (S/N) ratio 3 and 10 respectively, ranged from 0.2 ng L^{-1} to 7.5 ng L^{-1} and 0.5 ng L^{-1} to 25 ng L^{-1}, respectively (Table 3). Overall, the method presented similar or better performance characteristics to previous SPE-LC–MS/MS methods for the determination of other pesticide compounds in seawater [15].

Table 2. Recovery, repeatability and intermediate precision results of the optimized solid-phase extraction method for the determination of pesticide residues.

Compound	25 ng L^{-1}		100 ng L^{-1}			250 ng L^{-1}	
	(%) R	RSDr (%)	(%) R	RSDr (%)	RSD_{IP} (%)	(%) R	RSDr (%)
Azamethipos	80.5	10.4	79.9	4.0	5.2	70.6	7.9
Azoxystrobin	72.3	13.8	83.6	8.4	11.4	59.7	0.7
Boscalid	68.3	7.8	70.2	5.2	7.0	69.6	19.5
Irgarol	68.2	8.0	69.1	10.6	2.5	74.7	5.4
Malathion	81.4	2.9	95.0	5.4	7.7	78.3	17.6
Pirimiphos-methyl	98.9	4.9	82.0	2.9	9.4	77.2	8.8
Tebufenozide	58.5	1.2	61.9	7.1	8.3	58.6	14.7

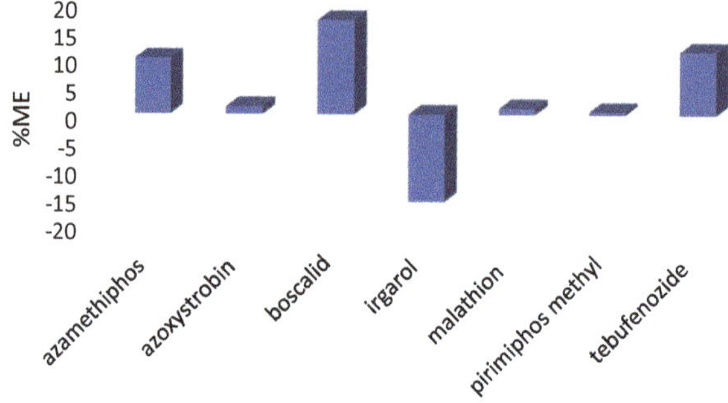

Figure 2. Matrix effect (%ME) for the studied pesticides.

Table 3. Detection and quantification limits, linear range of the method and determination coefficient (R^2) for the studied pesticides.

Compound	LOD (ng L^{-1})	LOQ (ng L^{-1})	Linear Range	R^2
Azamethipos	7.5	25	LOQ-500	0.9994
Azoxystrobin	0.3	1	LOQ-500	0.9992
Boscalid	0.5	1.5	LOQ-500	0.9996
Irgarol	0.2	0.5	LOQ-500	0.9998
Malathion	0.5	1.5	LOQ-500	0.9992
Pirimiphos-methyl	0.5	2	LOQ-750	0.9996
Tebufenozide	3	10	LOQ-500	0.9993

3.2. Optimization and Validation of SPE Method for the Determination of Pharmaceuticals

The optimization of the extraction recovery of pharmaceutical compounds from seawater was based on three different protocols (Table S1). The recoveries ranged from 77 to 143% for "HLB3" protocol, from 62 to 110% for "HLB4" protocol while in the "HLB5" protocol they ranged from 23 to 116% (Figure 3). Therefore, "HLB4" protocol was chosen because all the compounds gave acceptable recovery values as well as the RSD values were within acceptable limits (0–20%). In addition, "HLB4" recoveries are among the higher reported ranges while the whole procedure combines specific conditions (acidification and addition of Na_2EDTA) that improve the extraction performance [40].

For the pharmaceutical compounds, validation procedure was also performed at three concentration levels. Mean recovery values (n = 5) at low level (25 ng L^{-1}) ranged from 52.3% (bupropion) to 128% (diazepam), at intermediate level (100 ng L^{-1} ranged from 65.6% (sertraline) up to 99.6% (sulfadiazine) and at the high level (250 ng L^{-1}) from 47.0% (bupropion) to 128.7% (olanzapine), as shown in Table 4. The repeatability of the method (RSDr) for the low level was less than 14.1% for all compounds, less than 15.8% for the medium level and less than 19.2% for the high level. The intermediate precision (RSD_{IP}) of the method was below 18.2%. It is worth mentioning that the maximum acceptable limit (20%) was not exceeded by any of the pharmaceutical compounds. The linearity of the method was checked by constructing a nine-point curve in fortified samples for a concentration range of LOQ-100LOQ. The determination coefficient (r^2) values were always greater than 0.99, thus indicating excellent linearity for the method. As far as matrix effect, most pharmaceutical compounds showed values between 0 and 20% and only in two cases, i.e., oxolinic acid and bezafibrate presented greater ME but lower than 50% (Figure 4). The washing step with deionized water after the extraction is critical in order to reduced matrix effects of seawater samples in ESI as also denoted previously [41]. The detection and quantification limits for pharmaceutical compounds ranged from 0.5 ng L^{-1} to 10 ng L^{-1} and 2 ng L^{-1} to 30 ng L^{-1}, respectively (Table 5). The low limits of the method indicate that the pre-concentration factor is sufficient to quantify the compounds even in traces.

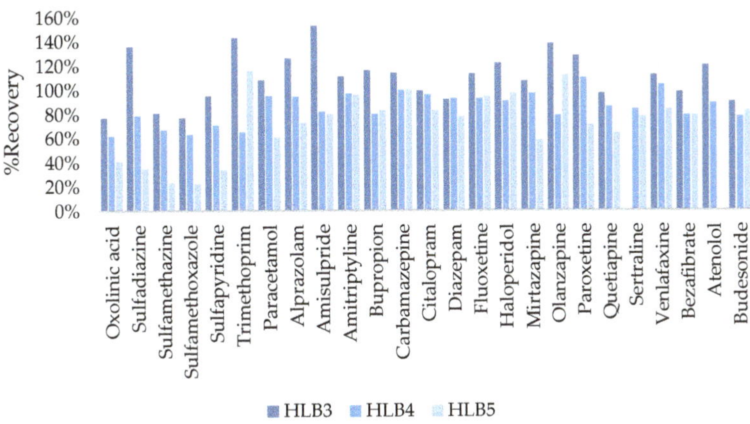

Figure 3. Recoveries (%) of pharmaceutical compounds by applying three extraction protocols.

3.3. *Application to Real Samples*

After the optimization and validation of the SPE method for the determination of pesticides and pharmaceutical compounds, the method was applied in real samples of seawater collected during a ten-month sampling campaign in an aquaculture farm in the Ionian Sea. None of the selected pesticides were detected in the water samples in contrast to pharmaceutical compounds among which paracetamol compound was detected (Figure 5) during the months of December and January in concentrations of 94.0 and 27.4 ng L^{-1}, respectively. Paracetamol is among the compounds previously detected in marine environments.

Passive sampling was conducted twice within the studied period and among drugs, trimethoprim and sulfadiazine were detected at low ppt levels. Passive sampling was not conducted in the months that paracetamol was detected with grab sampling. The concentrations of the detected pharmaceutical compounds were calculated based on literature data for *Rs*. For this reason, the minimum and maximum value of *Rs*, found in literature, were taken into account, as shown in Table 6. As a result, the concentration levels ranged from 0.24–1.14 ng L^{-1} for trimethoprim and 0.91–10.42 ng L^{-1} for sulfadiazine depending on the *Rs* values. Trimethoprim and sulfadiazine were also detected in marine studies, where they were found at concentration levels ranging between 0.02 and 95.8 ng L^{-1} for trimethoprim and 0.207–5.69 ng L^{-1} for sulfadiazine [42,43]. In the previous study a PNEC value of 16 µg L^{-1} was proposed for trimethoprim while the minimum EC50 for sulfadiazine was 0.11 mg L^{-1}, as reported elsewhere [44]. As a result, the detected pharmaceutical concentrations are considered to pose insignificant risk. Among pesticides, Irgarol 1051 was detected at 0.26–0.81 ng L^{-1} only in September (Figure 6). Trimethoprim is a diaminopyrimidine antimicrobial agent used in veterinary medicine. It is commonly used in combination with a sulphonamide (such as sulfadiazine) in a concentration ratio of 1:5. On the other hand, Irgarol–1051 is a booster biocide that has been used to prevent biofouling on submerged surfaces such as boats, navigational buoys, underwater equipment and ships in marine environment. Irgarol was also detected in the study of Muñoz et al., 2010, at 0.36 ng L^{-1} [41] while in Köck-Schulmeyer et al., 2019 study [15], Irgarol was detected in higher levels with a median concentration of 20.2 ng L^{-1}. Annual average and maximum allowable concentration environmental quality standard (EQS) for Irgarol in the marine environment was proposed as 2.5 and 16 ng L^{-1} [45], respectively; thus, the detected concentration levels pose no considerable risk for the aquaculture environment.

Table 4. Validation results of the analytical extraction method for pharmaceuticals.

Compound	25 ng L^{-1}		100 ng L^{-1}			250 ng L^{-1}	
	(%) R	RSDr (%)	(%) R	RSDr (%)	RSD_{IP} (%)	(%) R	RSDr (%)
Oxolinic acid	105.6	12.2	80.8	15.8	7.2	118.3	2.1
Sulfadiazine	–	–	99.6	7.9	15.5	111.2	13.3
Sulfamethazine	–	–	91.9	8.1	10.3	73.2	3.5
Sulfamethoxazole	–	–	90.7	9.2	6.7	113.5	7.1
Sulfapyridine	88.8	8.7	98.1	11.3	18.2	118.8	3.1
Trimethoprim	61.6	9.8	78.4	9.0	9.5	95.4	0.1
Paracetamol	–	–	76.5	3.5	7.3	82.8	0.8
Alprazolam	105.1	10.7	94.2	7.5	6.2	81.6	5.4
Amisulpride	83.7	4.0	89.8	4.7	3.6	81.4	5.4
Amitriptyline	98.1	2.1	94.4	5.2	2.5	99.5	1.3
Bupropion	52.3	14.1	71.5	3.5	9.7	47.0	5.9
Carbamazepine	107.2	2.0	87.0	5.2	4.2	92.1	2.2
Citalopram	94.4	5.2	79.3	6.9	6.6	96.4	6.8
Diazepam	128.0	9.4	93.6	5.6	12.3	94.6	0.2
Fluoxetine	103.2	3.6	89.0	3.9	3.9	95.1	4.1
Haloperidol	96.0	4.2	87.4	4.6	5.6	100.5	5.4
Mirtazapine	113.3	7.5	82.2	6.5	10.1	119.0	3.6
Olanzapine	112.3	9.8	86.1	4.6	4.0	128.7	0.4
Paroxetine	104.0	4.3	79.6	9.6	7.3	118.8	19.2
Quetiapine	109.9	7.4	92.5	7.9	6.6	111.7	12.0
Sertraline	97.3	11.1	65.6	10.3	11.7	103.7	7.9
Venlafaxine	60.0	1.3	76.8	5.8	11.7	59.9	1.3
Bezafibrate	89.6	12.4	79.6	15.7	13.5	122.5	6.3
Atenolol	58.7	5.5	85.4	9.2	8.8	52.1	2.9
Budesonide	118.0	1.7	89.5	7.6	5.3	116.0	3.4

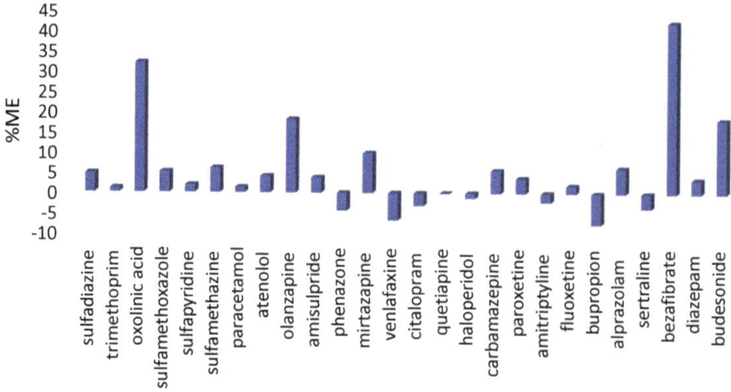

Figure 4. Matrix effect (%ME) for the studied pharmaceuticals.

Table 5. Detection and quantification limits, linear range and determination coefficient (R^2) for the studied pharmaceuticals.

Compound	LOD (ng L^{-1})	LOQ (ng L^{-1})	Linear Range	R^2
Oxolinic acid	1.5	5	LOQ-500	0.9992
Sulfadiazine	10	30	LOQ-750	0.9998
Sulfamethazine	10	30	LOQ-750	0.9990
Sulfamethoxazole	10	30	LOQ-500	0.9986
Sulfapyridine	7.5	25	LOQ-750	0.9997
Trimethoprim	0.5	2	LOQ-500	0.9996
Paracetamol	5	12.5	LOQ-750	0.9999
Alprazolam	1.5	5	LOQ-250	0.9991
Amisulpride	1.5	5	LOQ-250	0.9997
Amitriptyline	0.5	2	LOQ-250	0.9998
Bupropion	3	10	LOQ-250	0.9998
Carbamazepine	0.5	2	LOQ-250	0.9997
Citalopram	1.5	5	LOQ-250	0.9997
Diazepam	0.5	2	LOQ-100	0.9980
Fluoxetine	1.5	5	LOQ-250	0.9994
Haloperidol	5	12.5	LOQ-250	1.000
Mirtazapine	7.5	25	LOQ-250	0.9991
Olanzapine	0.5	2	LOQ-250	0.9993
Paroxetine	7.5	25	LOQ-750	1.000
Quetiapine	0.5	2	LOQ-250	0.9984
Sertraline	1.5	5	LOQ-250	0.9993
Venlafaxine	1.5	5	LOQ-250	0.9998
Bezafibrate	7.5	25	LOQ-750	0.9994
Atenolol	1.5	5	LOQ-500	0.9984
Budesonide	3	10	LOQ-500	0.9992

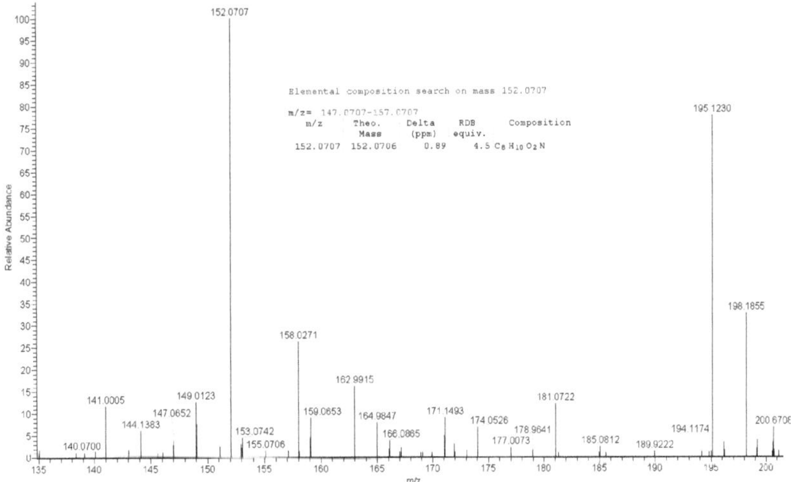

Figure 5. Full scan spectrum of paracetamol.

Table 6. Concentration range of the detected compounds based on passive sampling.

Compound	Rs (L d^{-1}) [a]		C_{water} (ng L^{-1})
	min	max	
Trimethoprim	0.090	0.436	0.24–1.14
Sulfadiazine	0.016	0.184	0.91–10.42
Irgarol	0.041	0.129	0.26–0.81

[a] Range of literature R_S values [23,39].

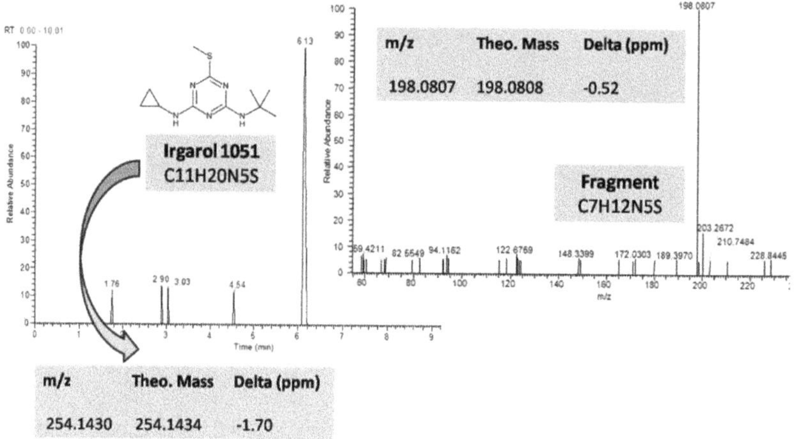

Figure 6. LC-LTQ/Orbitrap MS Extracted Ion Chromatogram (EIC) of seawater passive sampling.

4. Conclusions

In the present study, SPE methods have been developed and validated for the multiresidue determination of pesticides and pharmaceuticals in seawater samples, as selected for their frequent application as well as their occurrence in the environment and in fish farming ecosystems. The suitability of the optimized method for the determination of the selected compounds in seawater was confirmed by the determined performance characteristics (accuracy, repeatability, intermediate precision, linearity LODs-LOQs). The developed analytical methodologies were applied to real samples from aquaculture in the Ionian Sea area with a 10-month monitoring study program (July 2020–April 2021). Regarding pesticide compounds and their detection in seawater, none of them was detected. As far as pharmaceuticals are concerned, only paracetamol was detected twice at concentration levels below 94 ng L^{-1}. At the same time, seawater quality control screening based on passive sampling was carried out. Among pesticides, Irgarol 1051 was detected at 0.26–0.81 only in one case, and among pharmaceuticals, trimethoprim and sulfadiazine were detected at 0.26 to 10.4 ng L^{-1} levels.

Passive sampling can be used successfully for screening purposes in the quality control of emerging contaminants in sea water at low levels due to its integrative nature, although environmental conditions influence the sampling rates and consequently the measured concentrations, indicating that further work is needed in order to improve performance. On the other hand, grab sampling combined with a validated SPE method provides reliable quantitative results, but the time intervals between samplings may lead to misinterpretation of actual environmental concentration levels due to the loss of pollution events detection.

The extensive use of the target analytes and consequently their dispersion in seawater, especially waters hosting aquaculture facilities, illustrates the need for their continuous monitoring in the aquatic environment and relevant organisms. Spot and passive sampling techniques can be used complementarily for the screening and quantitative determination of contaminant levels and potential risks to aquatic environments; however, further improvements such as site-specific sampling rates and frequency are needed in order to maximize their significance in aquatic monitoring.

Supplementary Materials: The following supporting information can be downloaded at: https://www.mdpi.com/article/10.3390/su14063452/s1, Figure S1: Sampling location; Table S1: Solid phase extraction protocols tested for pesticides and pharmaceutical compounds efficient removal from waters; Table S2: Detection parameters for full MS/dd-MS^2 analysis of pesticides; Table S3: Detection parameters for full MS/dd-MS^2 analysis of pharmaceuticals.

Author Contributions: Conceptualization, D.H., V.B. and I.K.; methodology, V.B. and I.K.; validation, P.M. (Panagiota Martinaiou) and V.B.; formal analysis, P.M. (Panagiota Martinaiou), V.B., P.M. (Panagiota Manoli); investigation, P.M. (Panagiota Martinaiou) and V.B.; resources, T.A. and I.K.; data curation, P.M. (Panagiota Martinaiou), V.B. and E.M.; writing—original draft preparation, P.M. (Panagiota Manoli), D.H., I.K.; writing—review and editing, D.H., V.B. and I.K.; visualization, P.M. (Panagiota Manoli) and V.B.; supervision, T.A., D.H.; project administration, V.B.; funding acquisition, T.A. and I.K. All authors have read and agreed to the published version of the manuscript.

Funding: This research was funded by Environmental Aquatic Management (ENVIAQUAMAN) project, grant number HП1AB-00270 which is co-financed by the European Regional Development Fund (ERDF) under the Operational Program «Epirus 2014–2020», NRSF 2014–2020.

Institutional Review Board Statement: Not applicable.

Informed Consent Statement: Not applicable.

Data Availability Statement: The data are contained within the article and the supporting information file.

Acknowledgments: The authors acknowledge the support of this work by the project « Environmental Aquatic Management» (ENVIAQUAMAN (grant number HП1AB-00270)) which is co-financed by the European Regional Development Fund (ERDF) under the Operational Program «Epirus 2014–2020», NRSF 2014–2020. The authors would like to thank the Unit of Environmental, Organic and Biochemical high-resolution analysis–Orbitrap-LC–MS of the University of Ioannina for providing access to the facilities.

Conflicts of Interest: The authors declare no conflict of interest.

References

1. Hernando, M.D.; Martínez-Bueno, M.J.; Fernández-Alba, A.R. Seawater Quality Control of Microcontaminants in Fish Farm Cage Systems: Application of Passive Sampling Devices. *Bol. Inst. Esp. Oceanogr.* **2005**, *21*, 37–46.
2. European Commission. *Directive 2008/56/EC, Official Journal of the European Union, L 164/19, 25.6.2008*. 2008. Available online: https://eur-lex.europa.eu/LexUriServ/LexUriServ.do?uri=OJ:L:2008:164:0019:0040:EN:PDF (accessed on 31 January 2021).
3. Batt, A.L.; Furlong, E.T.; Mash, H.E.; Glassmeyer, S.T.; Kolpin, D.W. The Importance of Quality Control in Validating Concentrations of Contaminants of Emerging Concern in Source and Treated Drinking Water Samples. *Sci. Total Environ.* **2017**, *579*, 1618–1628. [CrossRef] [PubMed]
4. Caldas, S.S.; Bolzan, C.M.; Guilherme, J.R.; Silveira, M.A.K.; Escarrone, A.L.V.; Primel, E.G. Determination of Pharmaceuticals, Personal Care Products, and Pesticides in Surface and Treated Waters: Method Development and Survey. *Environ. Sci. Pollut. Res.* **2013**, *20*, 5855–5863. [CrossRef] [PubMed]
5. Stuart, M.; Lapworth, D.; Crane, E.; Hart, A. Review of Risk from Potential Emerging Contaminants in UK Groundwater. *Sci. Total Environ.* **2012**, *416*, 1–21. [CrossRef]
6. Vergeynst, L.; Haeck, A.; De Wispelaere, P.; Van Langenhove, H.; Demeestere, K. Multi-Residue Analysis of Pharmaceuticals in Wastewater by Liquid Chromatography–Magnetic Sector Mass Spectrometry: Method Quality Assessment and Application in a Belgian Case Study. *Chemosphere* **2015**, *119*, S2–S8. [CrossRef]
7. Vanryckeghem, F.; Huysman, S.; Van Langenhove, H.; Vanhaecke, L.; Demeestere, K. Multi-Residue Quantification and Screening of Emerging Organic Micropollutants in the Belgian Part of the North Sea by Use of Speedisk Extraction and Q-Orbitrap HRMS. *Mar. Pollut. Bull.* **2019**, *142*, 350–360. [CrossRef]
8. Belmonte Vega, A.; Garrido Frenich, A.; Martínez Vidal, J.L. Monitoring of Pesticides in Agricultural Water and Soil Samples from Andalusia by Liquid Chromatography Coupled to Mass Spectrometry. *Anal. Chim. Acta* **2005**, *538*, 117–127. [CrossRef]
9. Larson, S.J.; Capel, P.D.; Goolsby, D.A.; Zaugg, S.D.; Sandstrom, M.W. Relations between Pesticide Use and Riverine Flux in the Mississippi River Basin. *Chemosphere* **1995**, *31*, 3305–3321. [CrossRef]

10. Schulz, R. Rainfall-Induced Sediment and Pesticide Input from Orchards into the Lourens River, Western Cape, South Africa: Importance of a Single Event. *Water Res.* **2001**, *35*, 1869–1876. [CrossRef]
11. Kearney, P.; Wauchope, R. Disposal Options Based on Properties of Pesticides in Soil and Water. In *Pesticide Remediation in Soils and Water*, 1st ed.; Kearney, P., Roberts, T., Eds.; John Wiley & Sons: Hoboken, NJ, USA, 1998.
12. Flores, C.; Morgante, V.; González, M.; Navia, R.; Seeger, M. Adsorption Studies of the Herbicide Simazine in Agricultural Soils of the Aconcagua Valley, Central Chile. *Chemosphere* **2009**, *74*, 1544–1549. [CrossRef]
13. Kalogridi, E.-C.; Christophoridis, C.; Bizani, E.; Drimaropoulou, G.; Fytianos, K. Part I: Temporal and Spatial Distribution of Multiclass Pesticide Residues in Lake Waters of Northern Greece: Application of an Optimized SPE-UPLC-MS/MS Pretreatment and Analytical Method. *Environ. Sci. Pollut. Res.* **2014**, *21*, 7239–7251. [CrossRef]
14. Casado, J.; Santillo, D.; Johnston, P. Multi-Residue Analysis of Pesticides in Surface Water by Liquid Chromatography Quadrupole-Orbitrap High Resolution Tandem Mass Spectrometry. *Anal. Chim. Acta* **2018**, *1024*, 1–17. [CrossRef]
15. Köck-Schulmeyer, M.; Postigo, C.; Farré, M.; Barceló, D.; López de Alda, M. Medium to Highly Polar Pesticides in Seawater: Analysis and Fate in Coastal Areas of Catalonia (NE Spain). *Chemosphere* **2019**, *215*, 515–523. [CrossRef]
16. Gros, M.; Petrović, M.; Barceló, D. Wastewater treatment plants as a pathway for aquatic contamination by pharmaceuticals in the Ebro river basis (Northeast Spain). *Environ. Toxicol. Chem.* **2007**, *26*, 1553. [CrossRef]
17. López-Pacheco, I.Y.; Silva-Núñez, A.; Salinas-Salazar, C.; Arévalo-Gallegos, A.; Lizarazo-Holguin, L.A.; Barceló, D.; Iqbal, H.M.N.; Parra-Saldívar, R. Anthropogenic Contaminants of High Concern: Existence in Water Resources and Their Adverse Effects. *Sci. Total Environ.* **2019**, *690*, 1068–1088. [CrossRef]
18. Rivera-Utrilla, J.; Sánchez-Polo, M.; Ferro-García, M.Á.; Prados-Joya, G.; Ocampo-Pérez, R. Pharmaceuticals as Emerging Contaminants and Their Removal from Water. A Review. *Chemosphere* **2013**, *93*, 1268–1287. [CrossRef]
19. Liu, H.-H.; Wong, C.S.; Zeng, E.Y. Recognizing the Limitations of Performance Reference Compound (PRC)-Calibration Technique in Passive Water Sampling. *Environ. Sci. Technol.* **2013**, *47*, 10104–10105. [CrossRef]
20. Boxall, A.B.A.; Rudd, M.A.; Brooks, B.W.; Caldwell, D.J.; Choi, K.; Hickmann, S.; Innes, E.; Ostapyk, K.; Staveley, J.P.; Verslycke, T.; et al. Pharmaceuticals and Personal Care Products in the Environment: What Are the Big Questions? *Environ. Health Perspect.* **2012**, *120*, 1221–1229. [CrossRef]
21. Hass, U.; Duennbier, U.; Massmann, G. Occurrence and Distribution of Psychoactive Compounds and Their Metabolites in the Urban Water Cycle of Berlin (Germany). *Water Res.* **2012**, *46*, 6013–6022. [CrossRef]
22. Fono, L.J.; Kolodziej, E.P.; Sedlak, D.L. Attenuation of Wastewater-Derived Contaminants in an Effluent-Dominated River. *Environ. Sci. Technol.* **2006**, *40*, 7257–7262. [CrossRef]
23. Li, Y.; Yao, C.; Zha, D.; Yang, W.; Lu, G. Selection of Performance Reference Compound (PRC) for Passive Sampling of Pharmaceutical Residues in an Effluent Dominated River. *Chemosphere* **2018**, *211*, 884–892. [CrossRef]
24. Zhang, R.; Du, J.; Dong, X.; Huang, Y.; Xie, H.; Chen, J.; Li, X.; Kadokami, K. Occurrence and Ecological Risks of 156 Pharmaceuticals and 296 Pesticides in Seawater from Mariculture Areas of Northeast China. *Sci. Total Environ.* **2021**, *792*, 148375. [CrossRef]
25. Lai, W.W.-P.; Lin, Y.-C.; Wang, Y.-H.; Guo, Y.L.; Lin, A.Y.-C. Occurrence of Emerging Contaminants in Aquaculture Waters: Cross-Contamination between Aquaculture Systems and Surrounding Waters. *Water. Air. Soil Pollut.* **2018**, *229*, 249. [CrossRef]
26. Xie, H.; Hao, H.; Xu, N.; Liang, X.; Gao, D.; Xu, Y.; Gao, Y.; Tao, H.; Wong, M. Pharmaceuticals and Personal Care Products in Water, Sediments, Aquatic Organisms, and Fish Feeds in the Pearl River Delta: Occurrence, Distribution, Potential Sources, and Health Risk Assessment. *Sci. Total Environ.* **2019**, *659*, 230–239. [CrossRef]
27. Ismail, N.A.H.; Wee, S.Y.; Kamarulzaman, N.H.; Aris, A.Z. Quantification of Multi-Classes of Endocrine-Disrupting Compounds in Estuarine Water. *Environ. Pollut.* **2019**, *249*, 1019–1028. [CrossRef]
28. Togola, A.; Budzinski, H. Development of Polar Organic Integrative Samplers for Analysis of Pharmaceuticals in Aquatic Systems. *Anal. Chem.* **2007**, *79*, 6734–6741. [CrossRef]
29. Martinezbueno, M.; Hernando, M.; Aguera, A.; Fernandezalba, A. Application of Passive Sampling Devices for Screening of Micro-Pollutants in Marine Aquaculture Using LC–MS/MS. *Talanta* **2009**, *77*, 1518–1527. [CrossRef]
30. Valenzuela, E.F.; Menezes, H.C.; Cardeal, Z.L. Passive and Grab Sampling Methods to Assess Pesticide Residues in Water. A Review. *Environ. Chem. Lett.* **2020**, *18*, 1019–1048. [CrossRef]
31. Miège, C.; Mazzella, N.; Allan, I.; Dulio, V.; Smedes, F.; Tixier, C.; Vermeirssen, E.; Brant, J.; O'Toole, S.; Budzinski, H.; et al. Position paper on passive sampling techniques for the monitoring of contaminants in the aquatic environment—Achievements to date and perspectives. *Trends Environ. Anal. Chem.* **2015**, *8*, 20–26. [CrossRef]
32. Nannou, C.I.; Kosma, C.I.; Albanis, T.A. Albanis. Occurrence of pharmaceuticals in surface waters: Analytical method development and environmental risk assessment. *Int. J. Environ. Anal. Chem.* **2015**, *95*, 1242–1262. [CrossRef]
33. Papadakis, E.N.; Tsaboula, A.; Vryzas, Z.; Kotopoulou, A.; Kintzikoglou, K.; Papadopoulou-Mourkidou, E. Pesticides in the rivers and streams of two river basins in northern Greece. *Sci. Total Environ.* **2018**, *624*, 732–743. [CrossRef] [PubMed]
34. Lambropoulou, D.; Hela, D.; Koltsakidou, A.; Konstantinou, I. Overview of the Pesticide Residues in Greek Rivers: Occurrence and Environmental Risk Assessment. In *The Rivers of Greece. The Handbook of Environmental Chemistry*; Skoulikidis, N., Dimitriou, E., Karaouzas, I., Eds.; Springer: Berlin/Heidelberg, Germany, 2015; Volume 59. [CrossRef]
35. Nannou, C. Modern Analytical Methods for the Determination of Pesticide and Pharmaceutical Residues in Natural Waters and Sediments. Ph.D. Thesis, Department of Chemistry, University of Ioannina, Ioannina, Greece, 2018.

36. European Commission. SANTE/12682/2019, Guidance Document on Analytical Quality Control and Method Validation Procedures for Pesticide Residues and Analysis in Food and Feed. 2019. Available online: https://www.eurl-pesticides.eu/userfiles/file/EurlALL/AqcGuidance_SANTE_2019_12682.pdf (accessed on 30 January 2021).
37. Alvarez, D.A.; Petty, J.D.; Huckins, J.N.; Jones-Lepp, T.L.; Getting, D.T.; Goddard, J.P.; Manahan, S.E. development of a passive, in situ, integrative sampler for hydrophilic organic contaminants in aquatic environments. *Environ. Toxicol. Chem.* **2004**, *23*, 1640. [CrossRef] [PubMed]
38. Thomatou, A.A.; Zacharias, I.; Hela, D.; Konstantinou, I. Passive sampling of selected pesticides in aquatic environment using polar oragnic chemical integrative samplers. *Environ. Sci. Pollut. Res.* **2011**, *18*, 1222–1233. [CrossRef] [PubMed]
39. Vrana, B.; Urík, J.; Fedorova, G.; Švecová, H.; Grabicová, K.; Golovko, O.; Randák, T.; Grabic, R. In Situ Calibration of Polar Organic Chemical Integrative Sampler (POCIS) for Monitoring of Pharmaceuticals in Surface Waters. *Environ. Pollut.* **2021**, *269*, 116121. [CrossRef]
40. Sadutto, D.; Pico, Y. Sample Preparation to Determine Pharmaceutical and Personal Care Products in an All-Water Matrix: Solid Phase Extraction. *Molecules* **2020**, *25*, 5204. [CrossRef]
41. Wu, J.; Qian, X.; Yang, Z.; Zhang, L. Study on the matrix effect in the determination of selected pharmaceutical residues in seawater by solid-phase extraction and ultra-high-performance liquid chromatography–electrospray ionization low-energy collision-induced dissociation tandem mass spectrometry. *J. Chromatogr. A* **2010**, *1217*, 1471–1475.
42. Muñoz, I.; Martínez Bueno, M.J.; Agüera, A.; Fernández-Alba, A.R. Environmental and Human Health Risk Assessment of Organic Micro-Pollutants Occurring in a Spanish Marine Fish Farm. *Environ. Pollut.* **2010**, *158*, 1809–1816. [CrossRef]
43. Ojemaye, C.Y.; Petrik, L. Pharmaceuticals in the marine environment: A review. *Environ. Rev.* **2019**, *27*, 151–165. [CrossRef]
44. Carballeira, C.; De Orte, M.R.; Viana, I.G.; DelValls, T.A.; Carballeira, A. Assessing the toxicity of chemical compounds associated with land-based marine fish farms: The sea urchin embryo bioassay with Paracentrotus lividus and Arbacia lixula. *Arch. Environ. Contam. Toxicol.* **2012**, *63*, 249–261. [CrossRef]
45. EU. Cybutryne EQS Dossier 2011 Prepared by the Sub-Group on Review of the Priority Substances List (under Working Group E of the Common Implementation Strategy for the Water Framework Directive). 2011. Available online: https://circabc.europa.eu/sd/d/1eb5aa3b-bf6c-48ca-8ce0-00488a0c2905/Cybutryne%20EQS%20%20dossier%202011.pdf (accessed on 30 January 2011).

Article

Consumers' Perception on Traceability of Greek Traditional Foods in the Post-COVID-19 Era

Dimitris Skalkos [1,*], Ioanna S. Kosma [1], Eleni Chasioti [1], Thomas Bintsis [2] and Haralabos C. Karantonis [3]

1 Laboratory of Food Chemistry, Department of Chemistry, University of Ioannina, 45110 Ioannina, Greece; i.kosma@uoi.gr (I.S.K.); el.chasioti@uoi.gr (E.C.)
2 School of Science and Technology, Hellenic Open University, 11 Aristotelous, 54624 Thessaloniki, Greece; bintsis.thomas@ac.eap.gr
3 Laboratory of Food Chemistry, Biochemistry and Technology, Department of Food Science and Nutrition, School of the Environment, University of The Aegean, Metropolitan Ioakeim 2, 81400 Mytilene, Greece; chkarantonis@aegean.gr
* Correspondence: dskalkos@uoi.gr; Tel.: +30-2651-008345

Abstract: In the rising new global economic and social period, after the COVID-19 pandemic, traceability is expected to be a critical parameter for the selection of foods by consumers worldwide. Accordingly, traditional foods (TFs) can become the foods of choice in the new era due to their originality, authenticity, unique organoleptic properties, and locality. In this paper, the consumers' perception on traceability regarding Greek TFs and northwest Greek TFs is investigated, in order to find out the specific information they require for the purchase of these foods. Traceability was tested using variables related to package, product, quality, process, and personal information of these foods. A self-response questionnaire survey was carried out in September and October 2021 on a sample of 1707 participants through the Google platform. The results show that the participants consider traceability regarding questions on package information "quite important" and "very important" by an average of 68%, on food information by 64%, on quality information by 69%, on production process information by 78%, and on personal information by 65%. A similar pattern was recorded for the regional northwest Greek TFs for information on production process, personal, and package data, although there was a significant increase in the perception by the participants for data related to food information itself by 87% and more related to quality information by 94%.

Keywords: traditional foods; traceability; package information; product information; quality information; process information; personal information; questionnaire survey; post-COVID-19 era

Citation: Skalkos, D.; Kosma, I.S.; Chasioti, E.; Bintsis, T.; Karantonis, H.C. Consumers' Perception on Traceability of Greek Traditional Foods in the Post-COVID-19 Era. *Sustainability* **2021**, *13*, 12687. https://doi.org/10.3390/su132212687

Academic Editor: Mario D'Amico

Received: 25 October 2021
Accepted: 15 November 2021
Published: 16 November 2021

1. Introduction

Reports show that the economic crisis caused by the COVID-19 pandemic has a major impact on the global economy and significant changes will occur in the long run [1]. It affects all aspects of human life including the consumption of goods. There are signs of a growing anticonsumer movement, distinguished by Philip Kottler, of at least five types of anticonsumers [2]: the degrowth activists who feel that too much time and effort are going into consuming; the life simplifiers who want to eat less and buy less; the climate activists who worry about the damage to the planet through consumption; the food chooser who have turned into vegetarians and vegans; and the conservation activists who plead not to destroy existing goods but to reuse, repair, and redecorate them. These changes in the global dietary patterns introduce changes in the food production and supply processes as well. The highly globalized nature of today's food production and the supply commodities need to move from the world's source of grain supply to where they are consumed [3]. Internet and communication technologies, blockchain in the food supply chain and other industry 4.0 applications, as well as approaches that redefine the way we consume food, are the innovation with the highest potential in the new era [4]. There is also an equally pressing need to exploit social marketing to understand attitudes, perceptions, and barriers that influence the behavior change of consumers and the agrifood industry [4]. Greece is a country which has experienced a sovereign debt crisis, similar to the coming global crisis, the results of which are currently under study at different levels such as the SMEs [5] or employees performance [6]. Subsequently, these changes will contribute to adapting to the new norms forged by the COVID-19 pandemic, where there is a significant gap in knowledge for decision making.

Literature Review

In the post-COVID-19 era, a major issue for the customers will be the traceability in the foods they choose to buy [7]. The history of a food product is the definition of food traceability, and it is important because it ensures valuable data for consumers [8]. Consumers' demand for information about the traceability of food products has increased significantly in the last decade due to market globalization and issues related to food quality, safety, trust, and environmental protection [9–13]. Credible information, good reputation, and, at the same time, the enhancement of consumers' welfare are interrelated aspects of brand performance included in the traceability frame [14]. These concepts are strongly related in consumers' minds, and, therefore, cannot be easily separated in explaining choices [15]. Food traceability can reduce information asymmetry and food safety risks [16–18]. Traceability depends on parameters connected to supply chain and to trade related issues [19–21]. The distrust to the governments worldwide is what makes traceability a valuable tool for increased consumer confidence in food safety [20,22]. A recent study proved that it is possible to affirm that disease/pest and inputs traceability are the elements that increase consumers' trust in food safety [23]. The results of another recent study in six China cities, just before the pandemic, showed that consumers are willing to pay for traceable food with strong evidence of preference heterogeneity and with their valuations differing upon the degree of their trust in government's supervision of food safety and food labels [24]. Another recent report studying the consumers' perspective on food origin traceability in Poland proved that parameters such as food product features, food product packaging information, and shopping place frequency are significant on tracing the food origin [25]. To prioritize drivers to create traceability in the food supply chain after COVID-19 pandemic, 16 drivers were identified and test-grouped into four groups of drivers as informational, environmental, social, and economic [7].

In the new post-COVID-19 era, traditional foods (TFs) can play a vital role as the food of choice for the anticonsumers described above, due to their particular characteristics and properties [26]. They have played an important role in the development of different cultures and regions [27]. Recent study proves that TFs in Europe have a role in food consumption [28]. They reflect cultural inheritance and have left their imprints on the contemporary dietary patterns [29]. The definition of the term "traditional" related to foods is provided by the European Union as "Tradition means proven usage in the community market for a time period showing transmission between generations; at least 25 years" [30]. TFs interfere between the consumers and producers, promoting cultural associations within each area [31]. Sensory attributes, gastronomic heritages, eating habits, and association with certain local areas are more characteristics of TFs [32–35]. The European Union has labeled TFs in three mini categories: PDO, protected designation of origin; PGI, protected geographical indication; and TGI, traditional specialty guaranteed [36]. EU regulation 1151/12 assists producers of TFs to communicate the products' characteristics and farming attributes to buyers and consumers [37]. The definition of the term "traditional" in the above document means proven usage on the domestic market for a period that allows transmission between generations, with the period being at least 30 years.

Greece uses the provisions of the Regulation in the national Legislation with Ministerial Decree (3321/145849) issued by the Hellenic Ministry of Food and Agricultural Development since 2006 [38]. Registered traditional Greek foods by the different types are shown in Table 1. Food and agriculture, producing mainly Greek TFs, make up 3.5% of Greece GDP, the majority of which exported for consumption overseas in Europe, Russia, the US, and elsewhere [39]. This brings an important revenue stream for the Greek economy and keeps many farmers and food producers afloat.

Table 1. Distribution of Greek recognized foods between the different categories.

Type of Food Products	PDOs	PGIs	TSGs
Wine	33	116	
Olive oil	21		
Meat	2		
Cheese	22	1	
Foods of animal origin	2		
Fish	1		
Fruits and vegetables	29	21	1
Others	6		
Total	116	138	1

We studied the consumers' trust in Greek TFs in the post-COVID-19 era and found that they trust them since they "strongly agree" by an average of 20% and "agree" by an average of 50% that TFs are safe, healthy, sustainable, authentic, and tasty [40].

TFs of northwest Greece (namely the region of Epirus) comprise a significant portion of the overall Greek TFs. It is a region with local traditional food products as described elsewhere [41] Selected regional TFs, mainly the PDO cheeses and wines, are exported throughout Europe, thus promoting the regional brand name. Our recent results studying the northwest Greece TFs indicate that the COVID-19 crisis has not interfered in consumers' attitudes and perceptions regarding TFs [41].

We have shown the importance of the traceability of foods in the post-COVID-19 period and the potential of the TFs as the food of choice for the consumers of this new period, namely the anticonsumers. The aim of the present work was to assess the five determinants associated with the consumers' perception on the traceability of Greek TFs in order to identify the key pieces of information required to ensure their future prospect, growth, and development. These five determinants according to the existing literature on food traceability [42–46] are information related to: package, product, quality, process, and personal data. The current study examines these five determinants of consumers' perception on the traceability of Greek TFs and the northwest Greek TFs in the post-COVID-19 period:

(I) Consumers' perception on **package information** of Greek TFs. This involves data regarding characteristics such as nutritional composition and energy value, expiration date, production date, and additional information required for the particular food;
(II) Consumers' perception on **product information** of Greek TFs. This involves data regarding the origin, the producer, the brand name and the price of the food;
(III) Consumers' perception on **quality information** of Greek TFs. This involves data regarding quality label, certification label, safety label, and European origin label;
(IV) Consumers' perception on **process information** of Greek TFs. This involves data regarding the method of production, the level of processing, the raw materials, and the additional ingredients used for the production of food;
(V) Consumers' perception on **personal information** of Greek TFs. This involves data regarding pre-existing knowledge, recommendation by others, pre-existing personal experience, and origin of purchase.

In addition, the five determinants were examined on:

(VI) Consumers' perception on the **traceability of the Northwest Greek TFs** (Epirus' region).

This involves data regarding the package, product, quality, process, and personal data of the food mentioned in I–IV above.

2. Materials and Methods

2.1. Data Collection and Sample Characterization

A questionnaire to investigate the access data related to consumers' perception concerning traceability of Greek TFs, including TFs of Epirus region. The questionnaire included seven parts. The parts were built up using a similar previous study [46]. The first part included questions about the sociodemographic characteristics of the respondents. The second part consisted of four questions designed to assess the perception on the package information of the TFs, namely the nutritional and energy value, the production and expiration date, and the additional information which the participants would like to find in order to purchase them in the post COVID era. The third part included four questions focused on the participants' perception on the product information of the TFs, namely the producer, the geographic origin, the name, and the price of the TFs, which motivates their purchase. In the fourth part, issues concerning the participants' perception on the quality information of TFs, such as quality, certification, safety, and EU labels were assessed through four questions. The fifth part included four questions that approached the preferred process information data, namely the method of production, the level of processing, the raw material, and the ingredients used, of the participants in relation to their preference of TFs. In the sixth part, using four questions, the participants' preference on personal information of the TFs, such as pre-existing knowledge and personal experience, recommendation, and origin of purchase, regarding their perception was assessed. Finally, in the seventh part, using five questions, participants were asked to express their preference on traceability of the northwest Greece TFs, which can direct their purchasing choices. Issues such as package information, product information, quality information, process information, and personal information were taken into consideration. Quality of the data was obtained through the application of the questionnaire to 50 respondents who answered the questions easily. Electronic questionnaire was used. The distribution method chosen was by e-mail following the literature practices [47–49]. A snowball method was used to obtain a large number of participants [50]. The sample of the population is very well distributed among the different demographic characteristics, with participants familiar with the new technologies.

A higher rate for female respondents recorded at 61.7% is similar to the observation by other papers as well [51–54], leading to the conclusion that women respond more willingly to food-related surveys as they are primarily involved in the household organization. The research questionnaire was created through the Google platform and the Google Forms function. The geographical context for the present study was all the Greek regions, divided into five parts: north, west, central, south, and the islands, since the country includes many of them in the Aegean and the Ionian seas. The sample included students, among others. The participants received information explaining the purpose of the research, while obtaining access to the electronic form of the questionnaire through an attached link.

The survey took place during the period September–October 2021 and consisted of 1707 participants (Table 2).

Table 2. Sociodemographic characterization of the sample.

Variable	Groups	(%)
Gender	Male	38.3
	Female	61.7
Age	18–25	38.9
	26–35	10.0
	36–45	12.9
	46–55	22.2
	56+	16.0
Level of education	None/primary school	0.2
	Secondary school	0.4
	High school	7.1
	University	92.3
Civil state	Single	53.7
	Married	42.0
	Divorced	3.7
	Widow/widower	0.5
Job situation	Employed	55.7
	Unemployed	3.5
	Student	37.6
	Retired	3.1
Permanent resident in Greece	NORTH GREECE (regions of Macedonia—Thrace)	27.7
	WEST GREECE (region of Epirus—Etoloakarnania prefecture)	25.2
	CENTRAL GREECE (including Athens)	35.4
	SOUTH GREECE (region of Peloponnese)	3.7
	ISLANDS (Ionian and Aegean)	8.0

From the 1707 participants, 38.3% were male and 61.7% female. Regarding the spatial distribution, 25.2% were permanent residents of west Greece, 35.4% of central Greece (including the capital Athens), 27.7% residents of north Greece, 8.0% residents of the Greek islands, and 3.7% of south Greece, leading to a wide geographic distribution. The majority of the participants were aged between 18–25, 46–55, and 56+ years (38.9%, 22.2%, and 16.0%, respectively), while the other age groups, 26–35 and 36–45, were the least represented (10.0% and 12.9%, respectively). Regarding the level of education, most of the participants had higher education (university, 92.3%), and only 0.6% had completed primary or secondary school, while the employment status category was dominated by employed (55.7%), and students (37.8%) participants. Regarding the civil state of the participants, 42.0% were married, 53.7% were single, 3.7% were divorced, and only 0.5% were widows. It is worth mentioning that there was a significant percentage of young participants (students, at the age of 18–25) in the study which gives a better prospective, value to the results obtained, since the new generation better shows the trends of the future.

2.2. Data Analysis

Basic statistical tools were used for the exploratory analysis of the data. The survey was prepared in Greek and divided into seven parts, as detailed above:

Part I. Sociodemographic data;
Part II. Consumers' perception on the package information of Greek TFs;
Part III. Consumers' perception on the product information of Greek TFs;
Part IV. Consumers' perception on the quality information of Greek TFs;
Part V. Consumers' perception on the process information of Greek TFs;
Part VI. Consumers' perception on the personal information of Greek TFs;
Part VII. Consumers' perception on the traceability of northwest Greek TFs.

In order to measure the respondents' opinion about a set of statements related to TFs, a 5-point Likert scale, ranging from 1 = not at all important, 2 = less important, 3 = moderately important, 4 = quite important, and 5 = very important to me, was used [55].

Details of the statistics performed have recently been described in detail [40]. The Cramer's V coefficient used, ranging from 0 to 1, can be interpreted as follows: $V \approx 0.1$

weak association, V ≈ 0.3 moderate association, and V ≈ 0.5 or over, strong association. In all the tests performed, the level of significance considered was 5% ($p < 0.05$).

3. Results

In the results presented in the tables below, the percentages of not at all important (1) and less important (2) are less than 15% and considered minor, and no specific attention is given to all of them. Table 3 presents the participants' perception on package information of Greek TFs. The results show that the majority of the participants find the information about the expiration date (76.1%) and nutritional value (52.2%) to be very important, while a significant portion finds the information concerning the date of production (38.6%) to be very important, while they are not interested in additional information provided (12.9%).

Table 3. Participants' perception on **package information** of Greek TFs (scale from 1 = not at all important to 5 = very important).

Questions How Important Is the Information on the Food Package to You, Regarding	Answers According to Scale Points (%)				
	1	2	3	4	5
1. The nutritional composition and energy value	1.6	3.4	14.3	28.6	52.2
2. The best before date	0.5	1.7	5.6	16.2	76.1
3. The date of production	4.4	9.3	20.7	27.0	38.6
4. The access to additional information (by the use of a phone number or website)	16.4	19.4	29.3	22.0	12.9

The chi-square test presented in Table 4 shows that there were significant differences between consumers' perceptions on package information of Greek TFs in terms of:

1. Nutritional value of Greek TFs: between age ($x^2 = 78.366$, $p = 0.000$), level of education ($x^2 = 57.565$, $p = 0.000$), civil state ($x^2 = 60.294$, $p = 0.000$), and job situation ($x^2 = 66.550$, $p = 0.000$).
2. Best before date of Greek TFs: between gender ($x^2 = 24.264$, $p = 0.007$), level of education ($x^2 = 62.914$, $p = 0.000$), and civil state ($x^2 = 21.092$, $p = 0.049$).
3. Date of production of Greek TFs: between gender ($x^2 = 10.268$, $p = 0.036$), age ($x^2 = 121.564$, $p = 0.000$), level of education ($x^2 = 25.896$, $p = 0.011$), civil state ($x^2 = 94.153$, $p = 0.000$), and job situation ($x^2 = 93.529$, $p = 0.001$).
4. Access to additional information: between age ($x^2 = 11.918$, $p = 0.018$), age ($x^2 = 118.338$, $p = 0.000$), level of education ($x^2 = 22.240$, $p = 0.035$), civil state ($x^2 = 72.014$, $p = 0.000$), and job situation ($x^2 = 90.281$, $p = 0.002$).

Table 5 presents the participants' perception on product information of Greek TFs. The results show that 55.3% of the participants find the price of the product to be very important and 33.5% find its geographical origin to be very important. The name of the product and the identification of the producer are of less importance (by 21.6% and 20.3% for the very important answer, respectively).

The chi-square test presented in Table 4 showed that there were significant differences between consumers' perceptions on product information of Greek TFs in terms of:

1. Identification of the producer: between age ($x^2 = 83.269$, $p = 0.000$), civil state ($x^2 = 51.557$, $p = 0.000$), and job situation ($x^2 = 72.686$, $p = 0.000$).
2. Geographic origin of the food: between gender ($x^2 = 13.580$, $p = 0.009$), age ($x^2 = 145.520$, $p = 0.000$), civil state ($x^2 = 88.211$, $p = 0.000$), and job situation ($x^2 = 105.982$, $p = 0.000$).
3. Name of the product (branding): between age ($x^2 = 34.549$, $p = 0.005$), civil state ($x^2 = 28.105$, $p = 0.005$), job situation ($x^2 = 25.226$, $p = 0.014$), and residency ($x^2 = 27.619$, $p = 0.035$).
4. Price of the product: between age ($x^2 = 20.552$, $p = 0.000$), age ($x^2 = 64.024$, $p = 0.000$), level of education ($x^2 = 69.607$, $p = 0.000$), job situation ($x^2 = 44.006$, $p = 0.000$), and residency ($x^2 = 29.137$, $p = 0.023$).

Table 4. Association between variables (A) package, (B) product, and (C) quality information of Greek TFs and the sociodemographic variables.

	Gender			Age			Level of Education			Civil State			Job Situation			Residency		
	X^2 *	p **	V ***	X^2	p	V	X^2	p	V	X^2	p	V	X^2	p	V	X^2	p	V
A. Package information of Greek TFs																		
1. The nutritional and energy value				78.366	0.000	0.215	57.565	0.000	0.184	60.294	0.000	0.190	66.550	0.000	0.199			
2. The best before date	24.264	0.000	0.120				62.914	0.000	0.193	21.092	0.049	0.112						
3. The date of production	10.268	0.036	0.078	121.564	0.000	0.268	25.896	0.011	0.124	94.153	0.000	0.237	93.529	0.001	0.236			
4. The access to additional information	11.918	0.018	0.084	118.338	0.000	0.264	22.240	0.035	0.115	72.014	0.000	0.207	90.281	0.000	0.231			
B. Product information of Greek TFs																		
1. The identification of the producer				83.269	0.000	0.221				51.557	0.000	0.175	72.686	0.000	0.207			
2. The geographic origin of the food	13.580	0.009	0.089	145.520	0.000	0.293				88.211	0.000	0.229	105.982	0.000	0.250			
3. The name of the product (branding)				34.549	0.005	0.143				28.105	0.005	0.130	25.226	0.014	0.122	27.619	0.035	0.128
4. The price of the food	20.552	0.000	0.110	64.024	0.000	0.194	69.607	0.000	0.203				44.006	0.000	0.162	29.137	0.023	0.132
C. Quality information of Greek TFs																		
1. The quality label of the product				40.928	0.001	0.155				33.248	0.001	0.141	26.728	0.008	0.126			
2. The certification label/logo	10.759	0.029	0.080	79.525	0.000	0.217				62.227	0.000	0.193	59.218	0.000	0.188			
3. The safety label	14.472	0.006	0.093	102.316	0.000	0.246	25.035	0.015	0.122	77.325	0.000	0.215	95.851	0.000	0.239	29.538	0.021	0.133
4. The European origin label	14.388	0.005	0.092	172.109	0.000	0.319				128.583	0.000	0.277	125.554	0.000	0.273			

* chi-square test, ** level of significance of 5%: $p < 0.05$, *** Cramer's coefficient.

Table 5. Participants' perception on **product information** of Greek TFs (Scale from 1 = not at all important to 5 = very important).

Questions How Important Is the Information of the Food to You, Regarding	Answers According to Scale Points (%)				
	1	2	3	4	5
1. The identification of the producer	5.9	11.2	33.0	29.5	20.3
2. The geographic origin of the food	3.5	8.5	20.8	33.7	33.5
3. The name of the product (branding)	5.3	10.5	30.8	31.9	21.6
4. The price of the food	0.5	2.1	12.2	29.9	55.3

Table 6 presents the participants' perception on quality information of Greek TFs. The results show that consumers find all of the above information to be quite important and very important. Specifically, the safety label as well as the certification logo seem to be a very important information that concerns them by 39.4% and by 37.4%, respectively, followed closely by the quality label and the European origin label (35.2% and 36.7%, respectively, with very important information as an answer).

Table 6. Participants' perception on **quality information** of Greek TFs (Scale from 1 = not at all important to 5 = very important).

Questions How Important Is the Information of the Food Quality to You, Regarding	Answers According to Scale Points (%)				
	1	2	3	4	5
1. The quality label of the product (i.e., retailer quality label, national quality label, quality label of organizations, etc.)	1.9	5.9	20.0	37.0	35.2
2. The certification label/logo (i.e., ECO label, etc.)	4.3	8.1	18.9	31.3	37.4
3. The safety label (i.e., salmonella free, ISO, safety checked, etc.)	5.6	9.0	19.0	27.0	39.4
4. The European origin label (PDO, PGI, and TSG)	5.3	7.9	19.1	31.0	36.7

The chi-square test presented in Table 4 shows that there were significant differences between consumers' perceptions on quality information of Greek TFs in terms of:

1. Quality label of the product: between age ($x^2 = 40.928$, $p = 0.001$), civil state ($x^2 = 33.248$, $p = 0.001$), and job situation ($x^2 = 26.728$, $p = 0.008$).
2. Certification label/logo: between gender ($x^2 = 10.759$, $p = 0.029$), age ($x^2 = 79.525$, $p = 0.000$), civil state ($x^2 = 62.227$, $p = 0.000$), and job situation ($x^2 = 59.218$, $p = 0.000$).
3. Safety label: between gender ($x^2 = 14.472$, $p = 0.006$), age ($x^2 = 102.316$, $p = 0.000$), level of education ($x^2 = 25.035$, $p = 0.015$), civil state ($x^2 = 77.325$, $p = 0.000$), job situation ($x^2 = 95.851$, $p = 0.000$), and residency ($x^2 = 29.538$, $p = 0.021$).
4. European origin label: between gender ($x^2 = 14.388$, $p = 0.005$), age ($x^2 = 172.109$, $p = 0.000$), civil state ($x^2 = 128.583$, $p = 0.000$), and job situation ($x^2 = 125.554$, $p = 0.000$).

Table 7 presents the participants' perception on process information of Greek TFs. The results show that more than 50% of the participants find the information to be very important about the raw materials used (56.8%) and the other ingredients used (58.8%), the additives (58.8%) used for the production process. On the other, participants seem to believe that the method of production and the level of processing is information of less importance (33.4%, and 35.6%, respectively, with very important information as the answer of choice).

The chi-square test presented in Table 8 showed that there were significant differences between consumers' perceptions on process information of Greek TFs in terms of:

1. The used method of production: between age ($x^2 = 147.852$, $p = 0.000$), civil state ($x^2 = 83.269$, $p = 0.000$), and job situation ($x^2 = 103.922$, $p = 0.000$).
2. The level of processing: between age ($x^2 = 61.676$, $p = 0.000$), civil state ($x^2 = 36.345$, $p = 0.000$), and job situation ($x^2 = 49.155$, $p = 0.000$).
3. The raw materials used: between gender ($x^2 = 9.877$, $p = 0.043$), age ($x^2 = 189.659$, $p = 0.000$), civil state ($x^2 = 128.570$, $p = 0.000$), job situation ($x^2 = 158.369$, $p = 0.001$), and residency ($x^2 = 29.258$, $p = 0.022$).

4. The ingredients used: between age (x^2 = 166.051, p = 0.000), civil state (x^2 = 100.133, p = 0.000), and job situation (x^2 = 119.588, p = 0.000).

Table 7. Participants' perception on **process information** of Greek TFs (Scale from 1 = not at all important to 5 = very important).

Questions	**Answers According to Scale Points (%)**				
How Important Is the Information about the Process of the Food to You, Regarding	**1**	**2**	**3**	**4**	**5**
1. The used method of production (e.g., organic production, etc.)	2.8	7.7	19.7	33.4	36.5
2. The level of processing (e.g., whole tomato or tomato soup, etc.)	2.4	7.2	19.5	35.3	35.6
3. The raw materials the food is made from	1.1	3.4	11.0	27.6	56.8
4. The ingredients used	1.4	3.5	10.2	26.1	58.8

Table 9 presents the participants' perception on personal information of Greek TFs. The results show that none of these pieces of information are very important by a major percentage, i.e., more that 50% of the participants. They find by 83.7% the pre-existing experience concerning the TFs as quite and very important (41.7% quite important and 42% very important) and by 77% the pre-existing knowledge (43.2% quite important and 33.8% very important). On the other hand, consumers find the recommendation by others to be moderately important (35.8%) and quite important (37.8%). Finally, the origin of purchase seems to be moderately important for 29.0% and quite important for 30.6% of the participants.

The chi-square test presented in Table 8, showed that there were significant differences between consumers' perceptions on personal information of Greek TFs in terms of:

1. Pre-existing knowledge: between age (x^2 = 79.875, p = 0.000), level of education (x^2 = 26.582, p = 0.009), civil state (x^2 = 53.269, p = 0.000), and job situation (x^2 = 59.155, p = 0.000).
2. Recommendation by friends and family: only between gender (x^2 = 19.569, p = 0.001.
3. Pre-existing personal experience: between gender (x^2 = 11.344, p = 0.023), age (x^2 = 30.045, p = 0.018), level of education (x^2 = 33.175, p = 0.001), and job situation (x^2 = 23.021, p = 0.028).
4. Origin of purchase (e.g., super market, minimarket, grocery store, and market place): only between gender (x^2 = 24.299, p = 0.000).

Table 10 presents the participants' perception on traceability of northwest (the region of Epirus) Greek TFs. The results show that 71.8% of the participants find the information concerning the quality of the food to be very important, while the information of the food itself greatly concerns 55.1%. Package data on the other hand seems to be a moderately important information for 31.6% and quite important for 36.0%. Finally, production process information is very important for 39.5% and personal experience for 37.4% of the participants.

The chi-square test presented in Table 8 shows that there were significant differences between consumers' perceptions on traceability of northwest Greek TFs in terms of:

1. Package data: between civil state (x^2 = 22.322, p = 0.034), job situation (x^2 = 22.391, p = 0.033), and residency (x^2 = 28.887, p = 0.025).
2. The food itself: between age (x^2 = 35.910, p = 0.003), level of education (x^2 = 22.594, p = 0.031), civil state (x^2 = 21.561, p = 0.043), job situation (x^2 = 24.512, p = 0.017), and residency (x^2 = 26.526, p = 0.047).
3. Quality of the food: between gender (x^2 = 14.008, p = 0.007), age (x^2 = 32.754, p = 0.008), level of education (x^2 = 126.505, p = 0.000), civil state (x^2 = 23.125, p = 0.027), and job situation (x^2 = 22.782, p = 0.030).
4. Production process: between age (x^2 = 119.974, p = 0.000), level of education (x^2 = 33.557, p = 0.001), civil state (x^2 = 77.535, p = 0.000), and job situation (x^2 = 91.718, p = 0.000).
5. Personal experience: between gender (x^2 = 23.834, p = 0.000), age (x^2 = 33.107, p = 0.007), level of education (x^2 = 62.662, p = 0.000), civil state (x^2 = 25.440, p = 0.013), and job situation (x^2 = 31.843, p = 0.001).

Table 8. Association between variables (A) process, (B) personal information of Greek TFs, (C) traceability of northwest Greek TFs, and the sociodemographic variables.

	Gender			Age			Level of Education			Civil State			Job Situation			Residency		
	X^2 *	p **	V ***	X^2	p	V	X^2	p	V	X^2	p	V	X^2	p	V	X^2	p	V
A. Process information of Greek TFs																		
1. The used method of production				147.852	0.000	0.296				83.269	0.000	0.223	103.922	0.000	0.248			
2. The level of processing				61.676	0.000	0.191				36.345	0.000	0.147	49.155	0.000	0.171			
3. The raw materials the food is made from	9.877	0.043	0.076	189.659	0.000	0.335				128.570	0.000	0.277	158.369	0.001	0.306	29.258	0.022	0.132
4. The ingredients used				166.051	0.000	0.313				100.133	0.000	0.244	119.588	0.000	0.266			
B. Personal information of Greek TFs																		
1. Pre-existing knowledge				79.875	0.000	0.217	26.582	0.009	0.125	53.269	0.000	0.178	59.155	0.000	0.187			
2. Recommendation by friends and family	19.569	0.001	0.108															
3. Pre-existing personal experience	11.344	0.023	0.082	30.045	0.018	0.133	33.175	0.001	0.140				23.021	0.028	0.117			
4. Product origin of purchase	24.299	0.000	0.120															
C. Traceability information of northwest Greek TFs																		
1. Package data										22.322	0.034	0.115	22.391	0.033	0.115	28.887	0.025	0.131
2. The food itself				35.910	0.003	0.146	22.594	0.031	0.116	21.561	0.043	0.113	24.512	0.017	0.121	26.526	0.047	0.126
3. Quality of the food	14.008	0.007	0.091	32.754	0.008	0.139	126.505	0.000	0.274	23.125	0.027	0.118	22.782	0.030	0.116			
4. Production process				119.974	0.000	0.267	33.557	0.001	0.141	77.535	0.000	0.215	91.718	0.000	0.234			
5. Personal experience with the food	23.834	0.000	0.119	33.107	0.007	0.140	62.662	0.000	0.193	25.440	0.013	0.123	31.843	0.001	0.137			

* chi-square test, ** level of significance of 5%: $p < 0.05$, *** Cramer's coefficient.

Table 9. Participants' perception on **personal information** of Greek TFs (scale from 1 = not at all important to 5 = very important).

Questions How Important Is the Personal Information of the Food to You, Regarding	Answers According to Scale Points (%)				
	1	2	3	4	5
1. Pre-existing knowledge	0.9	3.1	18.9	43.2	33.8
2. Recommendation by friends and family	3.5	12.2	35.8	37.8	10.7
3. Pre-existing personal experience	0.5	1.5	14.2	41.7	42.0
4. Product origin of purchase (e.g., super market, mini market, grocery store, and market place)	8.2	13.4	29.0	30.6	18.8

Table 10. Participants' perception on **traceability** of the northwest Greek TFs (Scale from 1 = not at all important to 5 = very important).

Questions Based on the above 5 Categories, How Important Is the Information of the Northwest Greek TFs to You, Regarding	Answers According to Scale Points (%)				
	1	2	3	4	5
1. Package data	3.4	9.8	31.6	36.0	19.3
2. The food itself	0.5	1.9	10.3	32.2	55.1
3. Quality of the food	0.4	0.6	5.0	22.2	71.8
4. Production process	2.0	4.9	18.6	35.0	39.5
5. Personal experience with the food	0.8	2.6	18.9	40.3	37.4

4. Discussion

In this research, the consumer's perception regarding the five main traceability determinants of TFs, specifically of Greek TFs, after the COVID-19 pandemic is investigated for the first time (package/product/quality/process/personal information). Greek TFs have a long tradition of increased production and use, and it is for this reason that they were chosen for this study. In addition, the northwest Greek TFs, from the region of Epirus, were also chosen for comparison reasons, since this is a typical Greek mountainous, environmentally intact region with increased TFs and significant recognition by Greek consumers, as we have proved recently [41]. The sociodemographic characteristics of the participants of the survey exhibited in accordance to the literature [56]. They were from all different parts of Greece in order to ensure geographical distribution as well.

The package information data chosen in this study had a positive perception by the participants by more than 65%, except for the access to additional information which had only 34.9% (Table 3). The results of the chi-square test indicated that there were significant differences regarding package information between: (a) "gender" regarding the best before date, date of production, and access to additional information with weak association (V = 0.120/0.078/0.084); (b) "age" regarding nutritional value, date of production, and access to additional info with weak to moderate association (V = 0.215/0.268/0.264); (c) "level of education" regarding nutritional value, best before date, date of production, and access to additional info with weak association (V = 0.184/0.193/0.124/0.115); (d) "civil state" regarding nutritional value, best before date, date of production, and access to additional info with weak to moderate association (V = 0.190/0.112/0.237/0.207); and (e) "job situation" regarding nutritional value, date of production, and access to additional info with weak to moderate association (V = 0.199/0.236/0.231). Consumers' recognition and understanding on package information of food, especially regarding production, have been studied for more than a decade and proven to be important for their preference [57]. Nutritional knowledge are broadly helpful improving the accuracy of product choices, regardless of personal factors such as age, education, sex, etc. [58]. The legibility of food package information appears to be an equal challenge for young and elderly consumers [59]. Consumers value the best before date and production date as important information regarding their final decision to throw or not the food away [60]. These results agree with our finding regarding the importance of package information.

Overall, the participants consider the chosen product information to be important, although there is significant difference in their perception as shown in Table 4. The results of the chi-square test indicated that there were significant differences regarding product information between: (a) "gender" regarding the geographic origin of the food and its price with weak association (V = 0.089/0.110); (b) "age" regarding the identification of the producer, the geographic origin of the food, the name of the product, and its price with weak to moderate association (V = 0.221/0.293/0.143/0.194); (c) "level of education" regarding the price of the food with weak to moderate association (V = 0.203); (d) "civil state" regarding the identification of the producer, the geographic origin of the food, and the name of the product with weak to moderate association (V = 0.175/0.229/0.130); and (e) "job situation" regarding the identification of the producer, the geographic origin of the food, the name of the product, and its price with weak to moderate association (V = 0.207/0.250/0.122/0.162); and (f) "residency" regarding the name of the product and its price with weak association (V = 0.128/0.132). The importance of food price by the consumers has been studied extensively in the past [61–63]. A recent study proves that the price of food items is sometimes the only consideration when selecting food products, irrespective of their perceived quality and nutritional value [64]. The literature has examined consumers' preference for food of specific origin mainly country of origin and found a strong positive impact [65,66]. A recent report finds significant consumers' acceptance and preference for the Artic regional food products of Canada [67]. The name of the products (brands) especially for national and private brands of foods as reported this year have the same positive impact [68]. Our data prove the validity of the literature findings on Greek TFs as well.

The positive results regarding participants' perception on quality information of Greek TFs, shown in Table 5, are similar for all the issues addressed, in the range of 66–68%, as shown in Table 5. The results of the chi-square test indicated that there were significant differences regarding quality information between: (a) "gender" regarding the certification label, the safety label, and the European origin label with weak association (V = 0.080/0.093/0.092); (b) "age" regarding the quality label of the product, the certification label, the safety label, and the European origin label with weak to moderate association (V = 0.155/0.217/0.246/0.319); (c) "level of education" regarding the safety label with weak association (V = 0.122); (d) "civil state" regarding the quality label of the product, the certification label, the safety label, and the European origin label with weak to moderate association (V = 0.141/0.193/0.215/0.277); (e) "job situation" regarding the quality label of the product, the certification label, the safety label, and the European origin label with weak association (V = 0.126/0.188/0.239/0.273); and (f) "residency" regarding the safety label with weak association (V = 0.133). Recent reports indicate that different consumer segments have different attitudes and perceptions regarding food quality labels [69] and that PDO and organic labels are considered both labels substitutes by the majority of consumers [70]. These findings are also in agreement with our results regarding Greek TFs as well.

Increased results for traceability were recorded for the process information selected as shown in Table 7 above ranging from 70% to 85%. The results of the chi-square test indicated that there were significant differences regarding process information of the TFs between: (a) "gender" regarding the raw materials used with weak association (V = 0.076); (b) "age" regarding the used method of production, the level of processing, the raw materials, and the ingredients used with weak to moderate association (V = 0.296/0.191/0.335/0.313); (c) "civil state" regarding the used method of production, the level of processing, the raw materials, and the ingredients used with weak to moderate association (V = 0.233/0.147/0.277/0.244); (d) "job situation" regarding the used method of production, the level of processing, the raw materials, and the ingredients used with weak to moderate association (V = 0.248/0.171/0.306/0.266); and (a) "residency" regarding the raw materials used with weak association (V = 0.132).

When it comes to personal information regarding the traceability parameters for Greek TFs of choice, as shown in Table 8, positive results over 50% were recorded for pre-existing knowledge and pre-existing personal experience, while results slightly less than 50% were recorded for the origin of purchase and recommendation by friend and family. The results of the chi-square test indicated that there were significant differences regarding personal information between: (a) "gender" regarding the recommendation by others, the pre-existing personal experience, and the origin of purchase with weak association (V = 0.108/0.082/0.120); (b) "age" regarding the pre-existing knowledge, and the pre-existing personal experience with weak association (V = 0.217/0.133); (c) "level of education" regarding the pre-existing knowledge and the pre-existing personal experience with weak association (V = 0.125/0.140); (d) "civil state" regarding the pre-existing knowledge with weak association (V = 0.178); and (e) "job situation" regarding the pre-existing knowledge and the pre-existing personal experience with weak association (V = 0.187/0.177). The reports in the literature on personal knowledge of food preference specify them on different items such as safety, hygiene, etc. [71–74], therefore cannot be compared with our TFs findings.

The participants' perception on the traceability parameters of the Epirus' Greek TFs, as shown in Table 9, compared with their perception for Greek TFs followed the similar pattern for most of the five parameters tested, except the quality information and the food information, as shown in Table 9. The results of the chi-square test indicated that there were significant differences regarding traceability information on northwest Greek TFs between: (a) "gender" regarding the quality of the food and personal experience with weak association (V = 0.091/0.119); (b) "age" regarding the food itself, quality of the food, the production process, and personal experience with weak association (V = 0.146/0.139/0.267/0.140); (c) "level of education" regarding the food itself, quality of the food, the production process, and personal experience with weak to moderate association (V = 0.116/0.274/0.141/0.193); (d) "civil state" regarding package data, the food itself, quality of the food, the production process, and personal experience with weak association (V = 0.115/0.113/0.118/0.215/0.123); (e) "job situation" regarding package data, the food itself, quality of the food, the production process, and personal experience with weak association (V = 0.115/0.121/0.116/0.234/0.137); and (f) "residency" regarding package data and the food with weak association (V = 0.131/0.126). Recent results suggest that COVID-19 psychological pressure was associated with an impulsive approach to buying food [75]. Consequently, it is of major importance to predict whether or not the food-purchasing behavior reverts to pre-COVID-19 habits when the emergency is over or it takes another path in the new rising economy.

5. Conclusions

This research work explores the consumers' perception on the five main determinants of traceability of Greek TFs at the beginning of the new post-COVID-19 era. The study applied these parameters on food traceability of the TFs in the Greek consumers' mind in order to find the parameters that are significant to their preference for information regarding the purchase of TFs. A questionnaire was completed by 1707 Greek participants conducted in September and October 2021. The present pandemic is causing major changes on consumers' mind and preferences, which is leading to changes of their selection of foods in an unprecedented way under investigation currently. With a relevant degree of uncertainty, it is believed that people will be more selected on food, especially the new generation of anticonsumers, purchasing it in a personalized way, with a focus on the environmental, health, and safety effects. Our results show that the participants of this study appreciate, in the order of importance, the information regarding production > process > quality > package data > personal > food itself available as traceability characteristics in order to consider them the food of choice in the future. Participants express their satisfaction with the package of these five characteristics associated with the TFs.

In order to evaluate the possible regional originalities and characteristics of the consumers' evaluation on the traceability of Greek TFs, a regional TF group of products, namely the northwest Greece (region of Epirus), TFs were used at the end of the same survey with the same participants. The results showed that customers perceived in a similar manner the Epirus' TFs, information on production process, personal data, and package data, while the quality and the food itself data were considered more important as compared to the corresponding issues of the overall Greek TFs.

In the study, more women, educated, and employed participants, as well as young students, took place in the survey, and this can be considered a limitation of the study, even though the number of responses obtained is considered adequate. In addition, a limitation of the study is the use of Greek TFs only without the use of TFs by other countries which can have a different impact to the consumers. Finally, a limitation of the study is also the use of Greek participants only and not from other countries as well. Different cultures, especially outside the Mediterranean area, are expected to have minor differences on the traceability perception of TFs. This is the first study on understanding the traceability parameters of TFs for purchase and consumption in the new period after the pandemic crisis from the consumers' point of view.

Despite the importance of our findings, additional studies are needed in order to investigate further the parameters of traceability in the TFs, the long-lasting effects, and adaptations behavior to the "new normality". The findings contribute further to the main objective, which is the integration of TFs into the daily food consumption in the countries where there is the potential for increased production such as Greece. They also contribute to economic policies interventions required aimed at supporting increased production of TFs in Greece and elsewhere as they are important key factors for regional and territorial development, especially in inner and marginal areas. Further studies should expand in two different directions: studying TFs of other countries EU primarily, either themselves or in comparison, and studying the concept of traceability through the in-depth investigation of other pieces of information for Greek TFs perceived positively by the consumers.

Author Contributions: Conceptualization, supervision, and methodology, D.S.; writing—original draft preparation, D.S. and I.S.K.; investigation, E.C.; review and editing, D.S., T.B. and H.C.K. All authors have read and agreed to the published version of the manuscript.

Funding: This research received no external funding.

Institutional Review Board Statement: Not applicable.

Informed Consent Statement: Not applicable.

Conflicts of Interest: The authors declare no conflict of interest.

References

1. Garad, A.; Budiyanto, G.; Ansi, A.M.A.L. Impact of covid-19 pandemic on the global economy and future prospects: A systematic review of global reports. *J. Theor. Appl. Inf. Technol.* **2021**, *99*, 1–15.
2. Kotler, P. The Consumer in the Age of Coronavirus. *J. Creating Value* **2020**, *6*, 12–15. [CrossRef]
3. FAO Q&A COVID-19 Pandemic—Impact on Food and Agriculture. Available online: https://www.fao.org/2019-ncov/q-and-a/impact-on-food-and-agriculture/en/ (accessed on 15 October 2021).
4. Galanakis, C.M.; Rizou, M.; Aldawoud, T.M.S.; Ucak, I.; Rowan, N.J. Innovations and technology disruptions in the food sector within the COVID-19 pandemic and post-lockdown era. *Trends Food Sci. Technol.* **2021**, *110*, 193–200. [CrossRef]
5. Dimitropoulos, P.; Koronios, K.; Thrassou, A.; Vrontis, D. Cash holdings, corporate performance and viability of Greek SMEs: Implications for stakeholder relationship management. *EuroMed J. Bus.* **2019**, *15*, 333–348. [CrossRef]
6. Koronios, K.; Kriemadis, A.; Dimitropoulos, P.; Papdopoulos, A. A values framework for measuring the influence of ethics and motivation regarding the performance of employees. *Bus. Entrep. J.* **2019**, *8*, 1–19.
7. Patidar, A.; Sharma, M.; Agrawal, R. Prioritizing drivers to creating traceability in the food supply chain. *Procedia CIRP* **2021**, *98*, 690–695. [CrossRef]
8. Olsen, P.; Borit, M. How to define traceability. *Trends Food Sci. Technol.* **2013**, *29*, 142–150. [CrossRef]
9. Britwum, K.; Yiannaka, A. Consumer willingness to pay for food safety interventions: The role of message framing and issue involvement. *Food Policy* **2019**, *86*, 101726. [CrossRef]
10. Wongprawmas, R.; Canavari, M. Consumers' willingness-to-pay for food safety labels in an emerging market: The case of fresh produce in Thailand. *Food Policy* **2017**, *69*, 25–34. [CrossRef]
11. Van Loo, E.J.; Caputo, V.; Nayga, R.M.; Seo, H.-S.; Zhang, B.; Verbeke, W. Sustainability labels on coffee: Consumer preferences, willingness-to-pay and visual attention to attributes. *Ecol. Econ.* **2015**, *118*, 215–225. [CrossRef]

12. Lusk, J.L.; Tonsor, G.T.; Schroeder, T.C.; Hayes, D.J. Effect of government quality grade labels on consumer demand for pork chops in the short and long run. *Food Policy* **2018**, *77*, 91–102. [CrossRef]
13. Liao, P.-A.; Chang, H.-H.; Chang, C.-Y. Why is the food traceability system unsuccessful in Taiwan? Empirical evidence from a national survey of fruit and vegetable farmers. *Food Policy* **2011**, *36*, 686–693. [CrossRef]
14. Lassoued, R.; Hobbs, J. Consumer confidence in credence attributes: The role of brand trust. *Food Policy* **2015**, *52*, 99–107. [CrossRef]
15. Van Rijswijk, W.; Frewer, L.J. Consumer perceptions of food quality and safety and their relation to traceability. *Br. Food J.* **2008**, *110*, 1034–1046. [CrossRef]
16. Dandage, K.; Badia-Melis, R.; Ruiz-Garcia, L. Indian perspective in food traceability: A review. *Food Control.* **2017**, *71*, 217–227. [CrossRef]
17. Xu, L.; Yang, X.; Wu, L.; Chen, X.; Chen, L.; Tsai, F.S. Consumers' Willingness to Pay for Food with Information on Animal Welfare, Lean Meat Essence Detection, and Traceability. *Int. J. Environ. Res. Public Health* **2019**, *16*, 3616. [CrossRef] [PubMed]
18. Wu, L.; Wang, S.; Zhu, D.; Hu, W.; Wang, H. Chinese consumers' preferences and willingness to pay for traceable food quality and safety attributes: The case of pork. *China Econ. Rev.* **2015**, *35*, 121–136. [CrossRef]
19. Pekkirbizli, T.; Almadani, M.I.; Theuvsen, L. Food safety and quality assurance systems in Turkish agribusiness: An empirical analysis of determinants of adoption. *Econ. Agro-Aliment.* **2015**, *17*, 31–55. [CrossRef]
20. Rimpeekool, W.; Seubsman, S.-A.; Banwell, C.; Kirk, M.; Yiengprugsawan, V.; Sleigh, A. Food and nutrition labelling in Thailand: A long march from subsistence producers to international traders. *Food Policy* **2015**, *56*, 59–66. [CrossRef]
21. Wongprawmas, R.; Bravo, C.A.P.; Lazo, A.; Canavari, M.; Spiller, A. Practitioners' perceptions of the credibility of food quality assurance schemes: Exploring the effect of country of origin. *Qual. Assur. Saf. Crop. Foods* **2015**, *7*, 789–799. [CrossRef]
22. Chang, A.; Tseng, C.-H.; Chu, M.-Y. Value creation from a food traceability system based on a hierarchical model of consumer personality traits. *Br. Food J.* **2013**, *115*, 1361–1380. [CrossRef]
23. Matzembacher, D.E.; Stangherlin, I.D.C.; Slongo, L.A.; Cataldi, R. An integration of traceability elements and their impact in consumer's trust. *Food Control.* **2018**, *92*, 420–429. [CrossRef]
24. Liu, R.; Gao, Z.; Nayga, R.M.; Snell, H.A.; Ma, H. Consumers' valuation for food traceability in China: Does trust matter? *Food Policy* **2019**, *88*, 101768. [CrossRef]
25. Walaszczyk, A.; Galińska, B. Food Origin Traceability from a Consumer's Perspective. *Sustainability* **2020**, *12*, 1872. [CrossRef]
26. Skalkos, D. Traditional Foods in Europe: Perceptions & Prospects in the New Business Era. *Mod. Concepts Dev. Agron.* **2021**, *8*, 787–789. [CrossRef]
27. Almli, V.L. Consumer Acceptance of Innovations in Traditional Food. Attitudes, Expectations and Perception. Ph.D. Thesis, Norwegian University of Life Sciences, Ås, Norway, 2012.
28. Guiné, R.P.; Florença, S.G.; Barroca, M.J.; Anjos, O. The duality of innovation and food development versus purely traditional foods. *Trends Food Sci. Technol.* **2021**, *109*, 16–24. [CrossRef]
29. Trichopoulou, A.; Soukara, S.; Vasilopoulou, E. Traditional foods: A science and society perspective. *Trends Food Sci. Technol.* **2007**, *18*, 420–427. [CrossRef]
30. EU. *Council Regulation (EC) No 509/2006 on Agricultural Products and Foodstuffs as Traditional Specialties Guaranteed*; Official Journal of the European Union L 93/1; European Union: Brussels, Belgium, 2006.
31. Bojnec, S.; Petrescu, D.C.; Petrescu-Mag, R.M.; Radulescu, C.V. Locally Produced Organic Food: Consumer Preferences. *Amfiteatru Econ.* **2019**, *21*, 209–227. [CrossRef]
32. Serrano-Cruz, M.R.; Espinoza-Ortega, A.; Sepúlveda, W.S.; Vizcarra-Bordi, I.; Thomé-Ortiz, H. Factors associated with the consumption of traditional foods in central Mexico. *Br. Food J.* **2018**, *120*, 2695–2709. [CrossRef]
33. Pieniak, Z.; Verbeke, W.; Vanhonacker, F.; Guerrero, L.; Hersleth, M. Association between traditional food consumption and motives for food choice in six European countries. *Appetite* **2009**, *53*, 101–108. [CrossRef]
34. Guerrero, L.; Guàrdia, M.D.; Xicola, J.; Verbeke, W.; Vanhonacker, F.; Zakowska-Biemans, S.; Sajdakowska, M.; Sulmont-Rossé, C.; Issanchou, S.; Contel, M.; et al. Consumer-driven definition of traditional food products and innovation in traditional foods. A qualitative cross-cultural study. *Appetite* **2009**, *52*, 345–354. [CrossRef]
35. Verbeke, W.; Guerrero, L.; Almli, V.L.; Vanhonacker, F.; Hersleth, M. European Consumers' Definition and Perception of Traditional Foods. In *Traditional Foods*; Springer: Boston, MA, USA, 2016; pp. 3–16. [CrossRef]
36. EU. *Certificates of Specific Character for Agricultural Products and Food Status*; Official Journal of the European Union L 2082/1992; European Union: Brussels, Belgium, 1992; pp. 9–14.
37. EU. *Quality Schemes for Agricultural Products and Food Stuffs*; Official Journal of the European Union L 1151/2012; European Union: Brussels, Belgium, 2012; pp. 31–59.
38. Hellenic Ministry of Agricultural Development and Food. Greek Traditional Products (PDO-PGI-TSG). 2021. Available online: http://www.minagric.gr/index.php/el/for-farmer-2/2012-02-02-07-52-07 (accessed on 15 October 2021).
39. ELSTAT. Greek Statistics. 2021. Available online: www.statistics.gr/en/the-greek-economy (accessed on 15 October 2021).
40. Skalkos, D.; Kosma, I.S.; Vasiliou, A.; Guine, R.P.F. Consumers' Trust in Greek Traditional Foods in the Post COVID-19 Era. *Sustainability* **2021**, *13*, 9975. [CrossRef]
41. Skalkos, D.; Kosma, I.; Chasioti, E.; Skendi, A.; Papageorgiou, M.; Guiné, R. Consumers' Attitude and Perception toward Traditional Foods of Northwest Greece during the COVID-19 Pandemic. *Appl. Sci.* **2021**, *11*, 4080. [CrossRef]

42. Montet, D.; Dey, G. History of Food Traceability: Analytical Techniques. In *Food Traceability and Authenticity*; Chapman and Hall/CRC: Boca Raton, FL, USA, 2017; pp. 1–30.
43. Opara, L.U.; Vol, E.; Opara, L.U.; Vol, E. Traceability in agriculture and food supply chain: A review of basic concepts, technological implications, and future prospects. *Food Agric. Environ.* **2003**, *1*, 101–106.
44. Mora, C.; Menozzi, D. Benefits of traceability in food markets: Consumers' perception and action. *Food Econ.—Acta Agric. Scand. Sect. C* **2008**, *5*, 92–105. [CrossRef]
45. Jin, S.; Zhou, L. Consumer interest in information provided by food traceability systems in Japan. *Food Qual. Prefer.* **2014**, *36*, 144–152. [CrossRef]
46. Van Rijswijk, W.; Frewer, L.J.; Menozzi, D.; Faioli, G. Consumer perceptions of traceability: A cross-national comparison of the associated benefits. *Food Qual. Prefer.* **2008**, *19*, 452–464. [CrossRef]
47. Palmieri, N.; Perito, M.A.; Macrì, M.C.; Lupi, C. Exploring consumers' willingness to eat insects in Italy. *Br. Food J.* **2019**, *121*, 2937–2950. [CrossRef]
48. Palmieri, N.; Suardi, A.; Pari, L. Italian Consumers' Willingness to Pay for Eucalyptus Firewood. *Sustainability* **2020**, *12*, 2629. [CrossRef]
49. Palmieri, N.; Perito, M.A.; Lupi, C. Consumer acceptance of cultured meat: Some hints from Italy. *Br. Food J.* **2020**, *123*, 109–123. [CrossRef]
50. Palmieri, N.; Perito, M. Consumers' Willingness to Consume Sustainable and Local Wine in Italy. *Ital. J. Food Sci.* **2020**, *32*, 222–233.
51. De Leeuw, A.; Valois, P.; Ajzen, I.; Schmidt, P. Using the theory of planned behavior to identify key beliefs underlying pro-environmental behavior in high-school students: Implications for educational interventions. *J. Environ. Psychol.* **2015**, *42*, 128–138. [CrossRef]
52. Pappalardo, G.; Lusk, J.L. The role of beliefs in purchasing process of functional foods. *Food Qual. Prefer.* **2016**, *53*, 151–158. [CrossRef]
53. Chinnici, G.; d'Amico, M.; Pecorino, B. A multivariate statistical analysis on the consumers of organic products. *Br. Food J.* **2002**, *104*, 187–199. [CrossRef]
54. Giampietri, E.; Verneau, F.; del Giudice, T.; Carfora, V.; Finco, A. A Theory of Planned behaviour perspective for investigating the role of trust in consumer purchasing decision related to short food supply chains. *Food Qual. Prefer.* **2018**, *64*, 160–166. [CrossRef]
55. Likert, R. A technique for the measurement of attitudes. *Arch. Psychol.* **1932**, *140*, 44–53.
56. Petrescu-Mag, R.; Vermeir, I.; Petrescu, D.; Crista, F.; Banatean-Dunea, I. Traditional Foods at the Click of a Button: The Preference for the Online Purchase of Romanian Traditional Foods during the COVID-19 Pandemic. *Sustainability* **2020**, *12*, 9956. [CrossRef]
57. Hoogland, C.T.; de Boer, J.; Boersema, J.J. Food and sustainability: Do consumers recognize, understand and value on-package information on production standards? *Appetite* **2007**, *49*, 47–57. [CrossRef]
58. Miller, L.M.S.; Cassady, D.L.; Beckett, L.A.; Applegate, E.A.; Wilson, M.D.; Gibson, T.N.; Ellwood, K. Misunderstanding of Front-Of-Package Nutrition Information on US Food Products. *PLoS ONE* **2015**, *10*, e0125306. [CrossRef]
59. Droulers, O.; Amar, J. The legibility of food package information in France: An equal challenge for young and elderly consumers? *Public Health Nutr.* **2016**, *19*, 1059–1066. [CrossRef]
60. Kavanaugh, M.; Quinlan, J.J. Consumer knowledge and behaviors regarding food date labels and food waste. *Food Control.* **2020**, *115*, 107285. [CrossRef]
61. Epstein, L.H.; Jankowiak, N.; Nederkoorn, C.; Raynor, H.A.; French, S.A.; Finkelstein, E. Experimental research on the relation between food price changes and food-purchasing patterns: A targeted review. *Am. J. Clin. Nutr.* **2012**, *95*, 789–809. [CrossRef] [PubMed]
62. Horgen, K.B.; Brownell, K.D. Comparison of price change and health message interventions in promoting healthy food choices. *Health Psychol.* **2002**, *21*, 505–512. [CrossRef] [PubMed]
63. Thow, A.M.; Downs, S.; Jan, S. A systematic review of the effectiveness of food taxes and subsidies to improve diets: Understanding the recent evidence. *Nutr. Rev.* **2014**, *72*, 551–565. [CrossRef]
64. Koen, N.; Wentzel-Viljoen, E.; Blaauw, R. Price rather than nutrition information the main influencer of consumer food purchasing behaviour in South Africa: A qualitative study. *Int. J. Consum. Stud.* **2018**, *42*, 409–418. [CrossRef]
65. Lim, K.H.; Hu, W.; Maynard, L.J.; Goddard, E.U.S. Consumers' Preference and Willingness to Pay for Country-of-Origin-Labeled Beef Steak and Food Safety Enhancements. *Can. J. Agric. Econ. Can. Agroecon.* **2013**, *61*, 93–118. [CrossRef]
66. Balcombe, K.; Bradley, D.; Fraser, I.; Hussein, M. Consumer preferences regarding country of origin for multiple meat products. *Food Policy* **2016**, *64*, 49–62. [CrossRef]
67. Yang, Y.; Hobbs, J.E.; Natcher, D.C. Assessing consumer willingness to pay for Arctic food products. *Food Policy* **2020**, *92*, 101846. [CrossRef]
68. Van Loo, E.J.; Minnens, F.; Verbeke, W. Consumer Preferences for Private Label Brand vs. National Brand Organic Juice and Eggs: A Latent Class Approach. *Sustainability* **2021**, *13*, 7028. [CrossRef]
69. Sadílek, T. Consumer preferences regarding food quality labels: The case of Czechia. *Br. Food J.* **2019**, *121*, 2508–2523. [CrossRef]
70. Pérez, L.P.Y.; Gracia, A.; Barreiro-Hurlé, J. Not Seeing the Forest for the Trees: The Impact of Multiple Labelling on Consumer Choices for Olive Oil. *Foods* **2020**, *9*, 186. [CrossRef]
71. Ismail, F.H.; Chik, C.T.; Muhammad, R.; Yusoff, N.M. Food Safety Knowledge and Personal Hygiene Practices amongst Mobile Food Handlers in Shah Alam, Selangor. *Procedia—Soc. Behav. Sci.* **2016**, *222*, 290–298. [CrossRef]

72. Rahman, A.; Tosepu, R.; Karimuna, S.R.; Yusran, S.; Zainuddin, A.; Junaid, J. Personal Hygiene, Sanitation and Food Safety Knowledge of Food Workers at the University Canteen in Indonesia. *Public Health Indones.* **2018**, *4*, 154–161. [CrossRef]
73. Farragher, T.; Wang, W.C.; Worsley, A. The associations of vegetable consumption with food mavenism, personal values, food knowledge and demographic factors. *Appetite* **2016**, *97*, 29–36. [CrossRef] [PubMed]
74. Hardiah, M.; Nabawiyah, H.; Pibriyanti, K. Correlation between Knowledge and Attitudes to the Behavior of Personal Hygiene Food Handlers in Nutrient Department. *Sport Nutr. J.* **2020**, *2*, 17–24. [CrossRef]
75. Russo, C.; Simeone, M.; Demartini, E.; Marescotti, M.E.; Gaviglio, A. Psychological pressure and changes in food consumption: The effect of COVID-19 crisis. *Heliyon* **2021**, *7*, e06607. [CrossRef]

Article

Adapting Open Innovation Practices for the Creation of a Traceability System in a Meat-Producing Industry in Northwest Greece

Agapi Dima [1], Eleni Arvaniti [2], Chrysostomos Stylios [3,4,*], Dimitrios Kafetzopoulos [5] and Dimitris Skalkos [1,6]

[1] Computer Technology Institute and Press "Diophantus" (CTI), 26504 Patras, Greece; agapidima@cti.gr (A.D.); dskalkos@uoi.gr (D.S.)
[2] Department of Environmental Engineering, University of Patras, 26504 Patras, Greece; eleniarvanit@upatras.gr
[3] Industrial Systems Institute, Athena RC, 26504 Patras, Greece
[4] Department of Informatics and Telecommunications, University of Ioannina, 47100 Arta, Greece
[5] Department of Business Administration, University of Macedonia, 54636 Thessaloniki, Greece; dimkafe@uom.edu.gr
[6] Laboratory of Food Chemistry, Department of Chemistry, University of Ioannina, 45110 Ioannina, Greece
* Correspondence: stylios@isi.gr

Abstract: Traceability is becoming an essential tool for both the industry and consumers to confirm the characteristics of food products, leading industries to implement traceability to their merchandise. In order for the Computer Technology Institute and Press "Diophantus" (CTI) to help small and medium-sized enterprises (SMEs) implement traceability systems based on open innovation, principles were introduced. This paper presents market research that was carried out in order to determine the significant concerns of the Greek consumers about pork meat and pork products, their opinion on traceability information, and their preferences regarding how they would like to receive this information. The survey was conducted online and took place from mid-February to mid-March 2021 on a sample of 224 participants. The market research showed a very high interest concerning traceability, especially on the expiry date of the meat (87.9%), while the way and conditions of transport of the meat products follow (79%). Furthermore, consumers showed that they believe that the quality and safety of pork products would be improved with traceability (70.1%) and (79%) would prefer to buy traceable compared with untraceable pork, signifying the importance of traceability for consumers. Additionally, it was found that consumers and SMEs have common concerns regarding traceability. The information gathered from this market research will be used to adapt the traceability system to consumers' needs.

Keywords: traceability; pork meat; market research; open innovation

Citation: Dima, A.; Arvaniti, E.; Stylios, C.; Kafetzopoulos, D.; Skalkos, D. Adapting Open Innovation Practices for the Creation of a Traceability System in a Meat-Producing Industry in Northwest Greece. *Sustainability* **2022**, *14*, 5111. https://doi.org/10.3390/su14095111

Academic Editor: Filippo Giarratana

Received: 22 March 2022
Accepted: 21 April 2022
Published: 24 April 2022

1. Introduction

1.1. Traceability

In the last years, the food scandals in Europe and China unveiled the importance of an all-encompassing food traceability system. The UK mad cow disease, the 2013 horsemeat scandal in the European Union (EU), and the 2008 melamine scandal in China [1] were only a few of them and showed the need for new regulations and procedures about food fraud and food safety. Situations such as these not only erode the reputation of companies and have economic impacts [2] but can also be dangerous for consumers' health, whether they are intentional or not [3].

Food traceability is the ability to access specific information about a food product that has been captured and integrated with the product's recorded identification throughout the supply chain.

In any case, traceability refers to a system that can continuously track a food product and its history and location. The main principles of traceability were defined by Codex Alimentarius Commission (CAC) as the ability to follow the movement of a food through specified stage(s) of production, processing, and distribution [4]. This is one of the main definitions but several others have been proposed depending on standardization, organizations, legislations, and the academic literature [5].

Practically, a system such as this encompasses the creation of identification for each product in all the stages of the supply chain (farm to fork). This ID is coded on the product and corresponds to a file containing information about the history of the product and its components, both in the previous and next stages of the chain (sequential traceability) and in the current stage (internal traceability).

Traceability shows the complete history of the product, which is very advantageous, especially during crisis management, as a defective product can be located and recalled at any step. Furthermore, traceability can also provide information that allows better control of all the processes (e.g., optimal use of raw materials, inventory control, production planning, troubleshooting should an issue arise, quality control, etc.). Traceability can also be used at any time in order to substantiate the company's claims about the characteristics of its products (e.g., quality, origin, GMOs, etc.) [5–8].

Obviously, an effective traceability system must be very complex to include all the information needed for each product and all the procedures for making said product. However, it is difficult for small and medium-sized enterprises (SMEs) to implement a traceability system on their own, mainly due to lack of funds or know-how [9].

Traceability systems are also critical in terms of commerce. An implemented traceability system can allow seamless global trade of products that have verified origins [10]. The EU, China, Canada, India, and other countries heavily promote traceability. However, since there will inadvertently be variations of what these traceability systems contain due to differences across geographies, cultures, and products, there is a need to make traceability components compliant with standards shared by all partners, which will, in turn, make it easier to share and compare information [10]. Recently, consumers' perception of Greek traditional foods using variables related to package, product, quality, process, and personal information was investigated [11]. The results show that consumers considered questions on package information "quite important" and "very important" by an average of 68%, on food information by 64%, on quality information by 69%, on the production process by 78%, and on personal information by 65%.

Many different techniques have been used to prove traceability and food authenticity. Analytical methods such as GC, HLPC, and several spectroscopic and DNA analysis techniques are being used to confirm traceability and determine if a product has been adulterated [12,13].

The leaps made in technology have helped to overcome several issues concerning traceability. For example, portable spectroscopy devices can provide rapid, on-site, easy-to-use, and cost-effective food analysis throughout the whole food supply chain. Furthermore, blockchain technology could improve traceability throughout the entire supply chain combined with other approaches. However, such platforms are impossible to be used in several parts of the world. If digitization is not advanced enough, and is not in many instances, any potential system will be unavailable and unusable to companies and end-users alike [14]. Cell phones have abilities that can be used to make traceability more efficient and make traceability information more available to consumers. By scanning a code (QR code, bar code, etc.) or using NFC technology, consumers can trace information and can be easily connected to databases that have all the available information (food origin, feeds, and date of slaughtering for meat, pesticides that were used on fruits and vegetables, etc.) from the point it was produced to the point of sale [15].

There is also a high interest in specifically implementing blockchain in traceability systems. It has been shown that blockchains and other distributed ledger technologies can help implement traceability and sharing of information for small and medium enterprises [16]. Blockchain's capacity of irreversibly storing data, as well as other features it has, has shown promising results as it can create a secure string of information that can be shared with everyone (industry, consumers, and authorities) [17]. However, as of now, only a few programs have been launched concerning traceability, making it too soon to tell whether this is something that will be genuinely beneficial to the industry as well as consumers [18,19]. Furthermore, many challenges need to be addressed as to how an implementation such as this one will happen [20], considering it requires costly infrastructure changes [21].

Nevertheless, some severe issues concern consumers and the meat-producing industries considering meat traceability. As of now, there is no unifying framework concerning meat traceability, which leads to confusion as different principles and guidelines exist simultaneously [22]. However, to this end, new research efforts are being made so that all participants follow the same procedures, while new flexible and user-friendly traceability systems are also being proposed [6,22,23].

1.2. Current Status of Pork Production in Greece

Pork meat consumption fluctuates annually in Greece. Consumption fell slightly by 7.3%, reaching 884 thousand tons in 2017, remaining almost at the same level as in 2016. The degree of self-sufficiency in the domestic meat market has remained stable at about 52% over the last five years [24].

Pork outperforms consumer preferences over other types of meat:

- The per capita consumption of pork has remained stable at 27.2 kg/person in recent years;
- Poultry meat follows, with an annual per capita consumption of 26.9 kg/person in 2017;
- Beef/veal (15.4 kg/person);
- Sheep and goat meat (6.8 kg/person).

Pork consumption amounted to 292.5 thousand tons in 2017. A percentage of 53% of the pork in the domestic market was imported from 2013–2017. On the other hand, pork-meat exports were very limited, covering only 9–11% of production. The degree of self-sufficiency in pork has shrunk by 9% over the last seven years [24].

1.3. Open Innovation

SMEs have limited resources and capabilities, which restricts them significantly [25]. This can be helped by implementing open innovation to provide a viable solution. Open innovation is a relatively new term; it first appeared in the 2003 book of Chesbrough, and it proposed that companies combining internal and external ideas when innovating would benefit more than by adhering to the traditional research and development model [26]. The prevailing definition for open innovation is purposive inflows and outflows of knowledge to accelerate internal innovation and expand the markets for external use of innovation, respectively [27]. This means practically that, instead of a closed-off research and development process with an in-house team, collaborations are promoted between different partners combining forces. As a result, diverse collaborators such as companies, research centers, universities, and even people working on the project have a positive effect as the different perspectives and backgrounds of everyone involved lead to the creation of better products, services, or research [28].

The research institutes participating in open innovation reaped many benefits as it was found that it strengthened the position of the public research institute, increased internal networking, and broadened and improved the capabilities and knowledge of the involved researchers [29]. Furthermore, research consistently shows that open innovation has been beneficial to firms that have used it [30]. More specifically, there have been many success stories when open innovation was used in the food industry, creating fascinating results [31].

As new technology appears to need multidisciplinary development, open innovation can help, as a single organization struggles to provide what is needed, especially if it is a smaller one [32].

For a successful open collaboration to provide the necessary results, finding the right partners for the project is vital. Computer Technology Institute and Press "Diophantus" (CTI) was approached for this project. CTI Diophantus has the know-how to create and implement a complete traceability system. This system will allow consumers, using various types of smartphones, etc., to access all information available for the products through an integrated data center.

This collaboration is beneficial as an SME would not have been able to spend funds and time to create or license a traceability system just for themself. On the other hand, CTI Diophantus can further proceed with the research work created from all the steps towards implementing the traceability system in real life. This is quite significant, as SMEs are usually more directed to using open innovation during the commercialization and seldom during the research phase, as in this case [33].

The first step in this direction was conducting market research concerning pork, pork products, and the information available to consumers through traceability. This market research was critical for the next steps, especially since there is a substantial lack of research concerning the consumer behavior toward pork meat and pork meat products [34].

1.4. Aim of the Investigation

This work aims to study and analyze the data collected, obtain some insights into consumers' preferences and identify opportunities or problems that could be present through the implementation and use of the innovative traceability system. More specifically, this market research examined:

How important it is for the consumer to know specific information about pork meat and pork products that can be provided through traceability.

How consumers think that traceability will affect the pork-producing industries in terms of quality and safety.

How consumers think that traceability will affect their buying habits concerning pork and pork products.

How consumers prefer to receive traceability information.

Besides identifying consumer trends concerning pork meat and products, this market research will also be used to improve the corresponding traceability system. Using this information about what consumers want to know about the products' characteristics/quality, the traceability system will be adapted accordingly. In conclusion, besides identifying consumer trends, the market research helps define the requirements and standards that the Integrated Traceability Information System will have.

2. Materials and Methods

2.1. Research Methodology

The survey took place from mid-February to mid-March 2021. The survey was conducted online and shared on social media for maximum exposure. In total, 224 questionnaires were completed by all age groups and by both sexes. The participants originated from the regions of Western and Central Macedonia. In addition, data about how a traceability system would affect an industry were collected from pork-meat and pork-product producing SMEs. These SMEs incorporate several stages of the supply chain that are of interest in this study.

An interview protocol was developed from which the collection of information was based. Data analysis was performed using the IBM Statistical Package for Social Sciences (SPSS) for Windows, version 24 (SPSS Inc., Chicago, IL, USA)

The present research results could be a starting point for further future research on the same topics.

2.2. General Information

The data of the replied questionnaires were gathered and then processed using SPSS, version 24. The first six questions were introductory and contained demographic questions (gender, age, educational level, personal income, and marital status) and pork frequency consumption status.

The next three sections included only multiple-choice questions with a single answer option. The first one was to describe how important consumers deemed information about pork meat that can be provided through traceability (questions 1–16). All questions of the first section were drawn up following a scale of seven points: 1 = Definitely unimportant; 2 = Probably not important; 3 = Slightly important; 4 = Maybe important; 5 = Probably important; 6 = Very important; 7 = Definitely important. The second section focused on identifying how the public reacts to the traceability information provided for the sold pork.

All questions of the second section were drawn up following a scale of seven points: 1 = Definitely not; 2 = Probably not; 3 = Maybe not; 4 = Maybe; 5 = Maybe yes; 6 = Probably yes; 7 = Definitely yes. The last section was about the preferred ways consumers could access traceability, and three different ways were presented to consumers.

3. Results

3.1. Profile of the Sample

As shown in Table 1, most of the participants were women (75%). This is mainly because women are more sensitive and interested in health and nutrition issues as they are the primary household decision makers, and consequently were more responsive to the research.

Table 1. Sociodemographic profile of the sample, N = 224%.

Variable	Groups	(%)
Gender	Male	25
	Female	75
Age	18–30	17
	31–45	21.4
	46–55	42
	>55	19.6
Educational level	High school	34.8
	University	58.9
	Postgraduate	6.3
Personal monthly income EUR	<500	9.8
	500–1000	29.9
	1000–1500	25.9
	1500–2000	18.8
	>2000	15.6
Marital status	Single	17.4
	Married	75.9
	Divorced	4.9
	Widowed	1.8
Pork consumption frequency	twice a week	18.8
	once a week	68.3
	once or twice a month	12.9

As for the ages of the participants, it was found that people of all ages completed the questionnaire. A total of 17% of the participants were 18–30 years old, 21.4% were 31–45 years old, 42% were 46–55 years old, and 19.6% were over 55 years old. Therefore, the most significant sample of consumers who replied to the questionnaire was the 46–55- year-old one.

Figure 1 shows the answers about which information was the most important for consumers. Consumers expressed high interest in learning information about the way and the conditions of transport of the meat products, the expiry date of the pork products, and the results of inspections made by the appropriate health services.

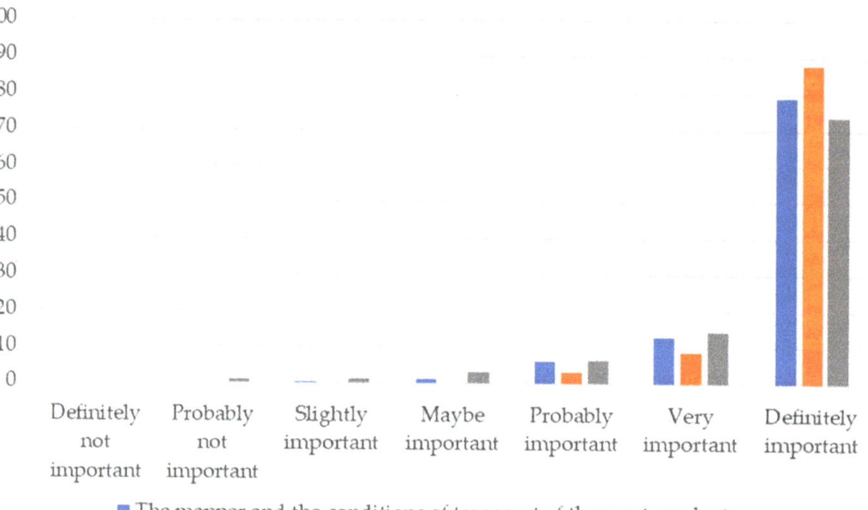

Figure 1. Consumers' trends concerning information about the hygiene of the animal, the transport, and the expiry date of the pork products.

In Table 3, the answers about how consumers think that traceability information would affect the pork meat industry and their consumer habits are shown.

A total of 70.1% of the participants believe that the quality and safety of pork products would be improved with traceability, and 79% would prefer to buy traceable pork compared with untraceable pork.

It also appears that consumers consider that traceability has significant benefits in improving the public confidence concerning pork meat products (40.2%), reducing the consumption of unsuitable products (39.3%), making the quality–price comparison better (34.2%), and protecting public health (percentage 33.0%). It is also important to mention that only a meager percentage of the participants answered that traceability would not substantially impact the pork they are buying (less than 10% on most questions).

In Figure 2, it can be seen more clearly what consumers consider will change due to traceability. As said before, they believe that the safety of the products will be improved, the risk of consuming unsuitable products will be reduced, and they will prefer traceable meat over untraceable. It is also observed that only a fraction of consumers thought that traceability would have little effect on the products or their shopping habits. This shows that the public has a high interest in pork and pork-product traceability.

Table 3. Public perceptions about how meat traceability will affect consumers. N = 224%.

If You Have the above Information about the Pork You Are Buying, Do You Consider That:	Definitely Not	Probably Not	Maybe Not	Maybe	Maybe Yes	Probably Yes	Definitely Yes
	1	2	3	4	5	6	7
17. Would the quality and safety of pork products be improved with the use of traceability?	0.4	1.3	0.9	4	9.8	13.4	70.1
18. Would your confidence in the hygiene and safety of pork products increase?	0	1.3	2.7	8	15.6	32.1	40.2
19. Would public health be more effectively protected?	0	0.4	2.2	5.8	21	37.5	33
20. Could you make a better quality–price comparison?	0	3.1	3.6	7.1	18.3	33.5	34.4
21. Would the risk of consuming unsuitable products be reduced?	0.4	1.3	2.2	11.2	11.6	33.9	39.3
22. Would you prefer to buy pork characterized by traceability over some other not-traceable meat?	0	0	0	3.1	4.9	12.9	79

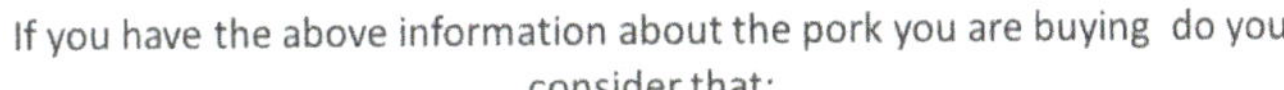

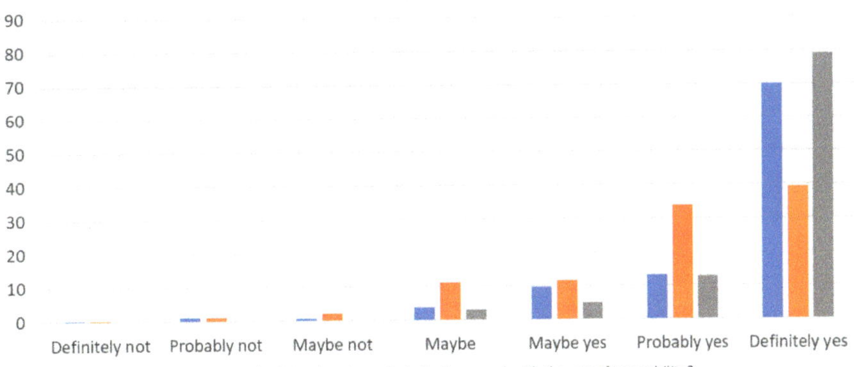

Figure 2. Consumers' perceptions about the effects of traceability information.

3.4. Methods of Receiving Information

The responses considering the means of providing traceability information to consumers were quite interesting. As seen in Figure 3, most consumers prefer a screen in the marketplace providing this information, with the mobile app as the second choice. This can be explained by the need of consumers to be provided with traceability information as they are trying to select which pork products to buy. Visiting the company's website is the least preferred method, even though it is a choice that is accessible 24 h a day, seven days a week.

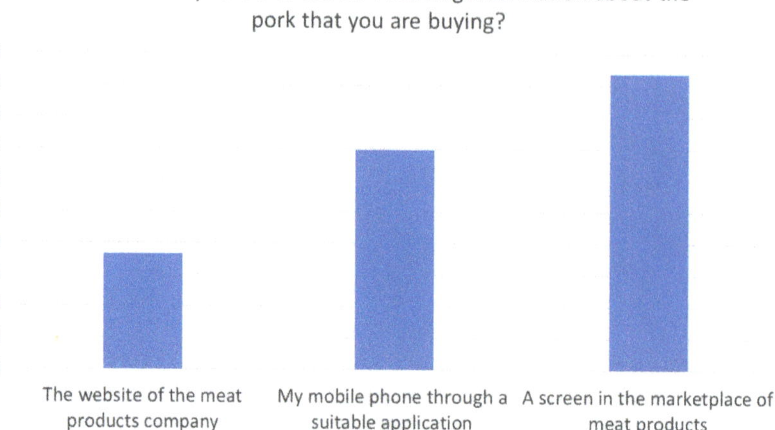

Figure 3. Consumers' preference for receiving traceability information about pork.

Consumers seem to prefer the informational screen in the market as it is a direct method to provide information that they are already familiar with.

3.5. SMEs and Traceability

A pork and pork-product-producing SME was selected to gain useful information about the factory plant and the conditions the pigs were raised in, the logistics, and other details that would affect traceability. The piglets are born by sows and are raised on the farm. After weaning, they are transferred to stables. They are fed with the appropriate food for about 5.5 months before being led to the slaughterhouse during their stay there. Every month, a veterinary inspection is conducted certifying the health of the animals and recording the condition of the animals, their vaccinations, possible diseases, treatments, etc. These inspections are of the utmost importance. Deworming is also conducted regularly at the stables.

The type and quality of feed are determined on an annual basis. Feed samples are tested for antibiotics, aflatoxins, etc., by the incharge health services. Possible modification of the feed composition is determined according to the cost and the content of nutrients. The incharge health department also carries out a veterinary check during the slaughter of animals.

Storage conditions (temperature and humidity) are considered necessary for the safety of the products and the delivery date to the retailer. The slaughter date and expiration of meat products are also considered essential for ensuring their safety. The meat is transported to the retailers by privately owned refrigerator trucks, in which the necessary transport conditions are observed. The trucks are maintained regularly, and their temperature and hygiene are checked before loading them with products.

The information that was considered most important by the SME and is concerned with the traceability system is:

- The use of a code to identify each pork product or with the use of a batch number for all the animals;
- The health of the animal and the hygiene of the carcass, which is ensured by the necessary veterinary checks, under the responsibility of both the appropriate health authorities and as well as the company;
- The breeding of the animal with nutritious and safe feeds;
- The yield of useful meat by measuring the initial weight of the animal, the breeding time, the weight before slaughter, and the weight of the useful meat;

- The storage conditions of the meat products before they are sent to the retailers and the date they received them;
- The expiration date indicated on the packaged products is data that certifies that the distributed products are safe for consumption.

As the market research showed, there is a high overlap between what the enterprise and consumers deem essential information about meat and meat products.

As for the advantages of the application of the advanced traceability system, it has been determined that there will be many benefits that help the company in many ways, such as:

- Standardizing procedures and having more effective overseeing of the overall process.
- More effective monitoring will help in troubleshooting and lead to immediate problem management, minimizing the number of errors associated with all stages of production. This also minimizes the creation of problematic and /or unsafe batches and makes them more easily identified and withdrawn.
- Livestock management will also be more effective, leading to a higher meat yield, resulting in increased productivity.

Furthermore, the production of safe products will be ensured, resulting in increased consumer confidence and reduced complaints. This leads to increased sales and, therefore, more revenue. Moreover, monitoring and evaluating suppliers with this system makes it easier to guarantee the future purchase of healthy animals.

Consequently, it is determined that integrating the innovative traceability system provided by CTI Diophantus is in the company's best interest.

4. Discussion

Analysis of the Results of the Market Research

By presenting market research results, useful conclusions can be extracted from consumers' preferences concerning pork meat and its products. This can be of great insight to any company concerning their products. In total, 68% of consumers have pork once a week, which means that this meat is an integral part of their diet, and there is a relatively large sample in this category for the preferences of consumers to be reliably exhibited. Moreover, since the risk of consuming inappropriate products is high, the company should have the appropriate tools to manage possible incidents concerning the meat produced and its meat products. An advanced innovative meat traceability system based on RFID technology that can be used with HACCP implementation can help with these incidents and create a framework for traceability for pork or pork-meat products [22]. Considering all the above, the proposed advanced innovative traceability system will help supply-chain members upload the information they want or need to share with the relevant authorities or consumers and have the traceability information stored in a form that can serve this purpose.

Another significant finding from the market research is that most consumers prefer pork products with traceability information (90%). This should push pork-producing companies to invest directly in the innovative advanced traceability system. It would fulfill consumers' need and requirement to access the information on the pork products they buy, benefiting both consumers and businesses. Increased sales offset the investment costs due to increased consumer confidence, as the end-user is provided with safer products of higher quality. This research results also show that consumers would like to learn as much information as possible about meat products. Additionally, consumers show interest not only in specific traceability information, but in the entire life cycle of meat products, from animal husbandry to the final products that come into their hands, giving more importance to meat safety (expiry date, storage conditions, and country of origin) and meat hygiene (results of tests by health services, results of chemical and/or microbiological tests, and information on the health and breeding of the animal). Furthermore, even though the legislation in the European Union is considered complete and covers all stages of the supply chain, there is not a standardized way of recording all this information, leaving companies on their own to decide the best way to record the necessary information in a manner so that it can be available if needed. By implementing a traceability system, the company gains all the previously established benefits and a competitive advantage against other companies that do not possess one.

As mentioned before, an SME would have been unable to perform the task of implementing a new traceability system on its own, and this is why open innovation was used. In this case, the utilization of open innovation does not stop after installing the system. The means (mobile apps, site, and screen in the market) used for communicating the information to the consumer can create ways of interaction between consumers and the company. This way, consumers are able to send feedback to the company, state their preferences, review the products, and even send ideas for new products.

5. Conclusions

In this market research, Greek consumers and their behavior towards pork meat and pork products was examined, as well as the information pertaining to these products. This information was used to see consumers' attitudes about pork meat and pork-meat products and their interest in traceability and the information they would like to be available regarding said products. The research showed that consumers are very interested in the information provided to them and that it can affect their buying habits to a high degree. This high interest showed that a traceability system would benefit the company by giving it an edge in the market. Furthermore, it is shown that between an SME and consumers, there is a very high overlap regarding the information about pork and pork products that they both consider of high importance, and that by implementing a traceability system, this information can be provided to them both.

Author Contributions: Conceptualization, A.D. and E.A.; methodology, A.D. and E.A.; validation, A.D., E.A. and C.S.; formal analysis, A.D. and E.A.; investigation, E.A. and D.K.; resources, E.A. and D.K.; data curation, D.K.; writing—original draft preparation, A.D. and E.A.; writing—review and editing C.S. and D.K.; visualization, A.D.; supervision, C.S.; project administration, C.S. and D.K.; funding acquisition, C.S., D.K. and D.S. All authors have read and agreed to the published version of the manuscript.

Funding: "This research is funded by Operational Program of Region of Western Macedonia 2014–2020 under the project "Ko-MEAT-IT: Developing a modern system of advanced traceability of the Kozani meat to improve productive performance, quality and hygiene using intelligent information systems" cofinanced by the European Union—European Regional Development Fund (ERDF) and National Resources.

Institutional Review Board Statement: Not applicable.

Informed Consent Statement: Not applicable.

Data Availability Statement: Not applicable.

Conflicts of Interest: The authors declare no conflict of interest.

References

1. Pei, X.; Tandon, A.; Alldrick, A.; Giorgi, L.; Huang, W.; Yang, R. The China melamine milk scandal and its implications for food safety regulation. *Food Policy* **2011**, *36*, 412–420. [CrossRef]
2. Li, S.; Wang, Y.; Tacken, G.M.L.; Liu, Y.; Sijtsema, S.J. Consumer trust in the dairy value chain in China: The role of trustworthiness, the melamine scandal, and the media. *J. Dairy Sci.* **2021**, *104*, 8554–8567. [CrossRef] [PubMed]
3. Visciano, P.; Schirone, M. Food frauds: Global incidents and misleading situations. *Trends Food Sci. Technol.* **2021**, *114*, 424–442. [CrossRef]
4. FAO Codex Alimentarius. Principles for traceability/product tracing as a tool within a food inspection and certification system. *Cac/Gl* **2018**, *1*, 1–423.
5. Islam, S.; Cullen, J.M. Food traceability: A generic theoretical framework. *Food Control.* **2021**, *123*, 107848. [CrossRef]
6. Charalampous, V.; Margariti, S.V.; Salmas, D.; Stylios, C.; Kafetzopoulos, D.; Skalkos, D. Design and develop cloud-based system for meat traceability. *CEUR Workshop Proc.* **2020**, *2761*, 475–484.
7. Galimberti, A.; Casiraghi, M.; Bruni, I.; Guzzetti, L.; Cortis, P.; Berterame, N.M.; Labra, M. From DNA barcoding to personalized nutrition: The evolution of food traceability. *Curr. Opin. Food Sci.* **2019**, *28*, 41–48. [CrossRef]

8. Espiñeira, M.; Santaclara, F.J. *What Is Food Traceability?* Elsevier Ltd.: Amsterdam, The Netherlands, 2016; ISBN 978-0-08-100321-3.
9. Curto, J.P.; Gaspar, P.D. Traceability in food supply chains: Review and SME focused analysis—Part 1. *AIMS Agric. Food* **2021**, *6*, 679–707. [CrossRef]
10. Qian, J.; Ruiz-Garcia, L.; Fan, B.; Robla Villalba, J.I.; McCarthy, U.; Zhang, B.; Yu, Q.; Wu, W. Food traceability system from governmental, corporate, and consumer perspectives in the European Union and China: A comparative review. *Trends Food Sci. Technol.* **2020**, *99*, 402–412. [CrossRef]
11. Skalkos, D.; Kosma, I.S.; Vasiliou, A.; Guine, R.P.F. Consumers' trust in greek traditional foods in the post covid-19 era. *Sustainubility* **2021**, *13*, 9975. [CrossRef]
12. Bianchi, F.; Giannetto, M.; Careri, M. Analytical systems and metrological traceability of measurement data in food control assessment. *TrAC-Trends Anal. Chem.* **2018**, *107*, 142–150. [CrossRef]
13. Wadood, S.A.; Boli, G.; Xiaowen, Z.; Hussain, I.; Yimin, W. Recent development in the application of analytical techniques for the traceability and authenticity of food of plant origin. *Microchem. J.* **2020**, *152*, 104295. [CrossRef]
14. McVey, C.; Elliott, C.T.; Cannavan, A.; Kelly, S.D.; Petchkongkaew, A.; Haughey, S.A. Portable spectroscopy for high throughput food authenticity screening: Advancements in technology and integration into digital traceability systems. *Trends Food Sci. Technol.* **2021**, *118*, 777–790. [CrossRef]
15. Ma, T.; Wang, H.; Wei, M.; Lan, T.; Wang, J.; Bao, S.; Ge, Q.; Fang, Y.; Sun, X. Application of smart-phone use in rapid food detection, food traceability systems, and personalized diet guidance, making our diet more health. *Food Res. Int.* **2022**, *152*, 110918. [CrossRef] [PubMed]
16. Hashimy, L.; Treiblmaier, H.; Jain, G. Distributed ledger technology as a catalyst for open innovation adoption among small and medium-sized enterprises. *J. High Technol. Manag. Res.* **2021**, *32*, 100405. [CrossRef]
17. Iftekhar, A.; Cui, X. Blockchain-based traceability system that ensures food safety measures to protect consumer safety and COVID-19 free supply chains. *Foods* **2021**, *10*, 1289. [CrossRef] [PubMed]
18. Feng, H.; Wang, X.; Duan, Y.; Zhang, J.; Zhang, X. Applying blockchain technology to improve agri-food traceability: A review of development methods, benefits and challenges. *J. Clean. Prod.* **2020**, *260*, 121031. [CrossRef]
19. Galvez, J.F.; Mejuto, J.C.; Simal-Gandara, J. Future challenges on the use of blockchain for food traceability analysis. *TrAC-Trends Anal. Chem.* **2018**, *107*, 222–232. [CrossRef]
20. Pearson, S.; May, D.; Leontidis, G.; Swainson, M.; Brewer, S.; Bidaut, L.; Frey, J.G.; Parr, G.; Maull, R.; Zisman, A. Are Distributed Ledger Technologies the panacea for food traceability? *Glob. Food Sec.* **2019**, *20*, 145–149. [CrossRef]
21. Creydt, M.; Fischer, M. Blockchain and more—Algorithm driven food traceability. *Food Control.* **2019**, *105*, 45–51. [CrossRef]
22. Kafetzopoulos, D.; Stylios, C.; Skalkos, D. Managing traceability in the meat processing industry: Principles, guidelines and technologies. *CEUR Workshop Proc.* **2020**, *2761*, 302–308.
23. Haleem, A.; Khan, S.; Khan, M.I. Traceability implementation in food supply chain: A grey-DEMATEL approach. *Inf. Process. Agric.* **2019**, *6*, 335–348. [CrossRef]
24. *ICAP CRIF Business Information Services | ICAP CRIF in Greece MEAT*; ICAP CRIF: Athens, Greece, 2018.
25. Gamage, S.K.N.; Ekanayake, E.M.S.; Abeyrathne, G.A.K.N.J.; Prasanna, R.P.I.R.; Jayasundara, J.M.S.B.; Rajapakshe, P.S.K. A review of global challenges and survival strategies of small and medium enterprises (SMEs). *Economies* **2020**, *8*, 79. [CrossRef]
26. Chesbrough, H.W. Open innovation. In *The Routledge Companion to Innovation Management*; Routledge: London, UK, 2019; p. 24. [CrossRef]
27. Chesbrough, H. *Open Innovation: The New Imperative for Creating and Profiting from Technology*; Harvard Business Press: Boston, MA, USA, 2006; pp. 1–9.
28. Tang, T.Y.; Fisher, G.J.; Qualls, W.J. The effects of inbound open innovation, outbound open innovation, and team role diversity on open source software project performance. *Ind. Mark. Manag.* **2021**, *94*, 216–228. [CrossRef]
29. Van Lancker, J.; Wauters, E.; Van Huylenbroeck, G. Open innovation in public research institutes -success and influencing factors. *Int. J. Innov. Manag.* **2019**, *23*, 1950064. [CrossRef]
30. Lu, Q.; Chesbrough, H. Measuring open innovation practices through topic modelling: Revisiting their impact on firm financial performance. *Technovation* **2021**, 102434. [CrossRef]
31. Miglietta, N.; Battisti, E.; Campanella, F. Value maximization and open innovation in food and beverage industry: Evidence from US market. *Br. Food J.* **2017**, *119*, 2477–2492. [CrossRef]
32. Uribe-Echeberria, R.; Igartua, J.I.; Lizarralde, R. Implementing open innovation in research and technology organisations: Approaches and impact. *J. Open Innov. Technol. Mark. Complex.* **2019**, *5*, 91. [CrossRef]
33. Bertello, A.; Ferraris, A.; De Bernardi, P.; Bertoldi, B. Challenges to open innovation in traditional SMEs: An analysis of pre-competitive projects in university-industry-government collaboration. *Int. Entrep. Manag. J.* **2022**, *18*, 89–104. [CrossRef]
34. Mondéjar-Jiménez, J.A.; Sánchez-Cubo, F.; Mondéjar-Jiménez, J. Consumer Behaviour towards Pork Meat Products: A Literature Review and Data Analysis. *Foods* **2022**, *11*, 307. [CrossRef] [PubMed]

Article

Development and Validation of a Measurement Instrument for Sustainability in Food Supply Chains

Theofilos Mastos, Katerina Gotzamani and Dimitrios Kafetzopoulos *

Department of Business Administration, University of Macedonia, 54636 Thessaloniki, Greece; tmastos@uom.edu.gr (T.M.); kgotza@uom.edu.gr (K.G.)
* Correspondence: dimkafe@uom.edu.gr

Abstract: The purpose of this paper is to develop a measurement instrument for sustainable supply chain management (SSCM) critical factors, practices and performance and validate it in the food industry. A literature review was conducted in order to identify pertinent variables and propose relevant measuring items. An email survey was carried out in 423 Greek companies in the food and beverage sector. The questionnaire was sent by e-mail in the Google Forms format and it was requested to be answered by a representative of the company. The collected data was processed using exploratory factor analysis in order to extract the latent constructs of the SSCM critical factors, practices and performance measures. The validity of the proposed instrument was confirmed through confirmatory factor analysis. The extracted SSCM critical factors are "firm-level sustainability critical factors" and "supply chain sustainability critical factors". The extracted SSCM practices are "supply chain collaboration" and "supply chain strategic orientation". The extracted SSCM performance factors are "economic performance", "social performance" and "environmental performance". The three developed constructs constitute a measurement instrument that can be used both by practitioners who desire to implement SSCM and by researchers who can apply the proposed scales in other research projects or use them as assessment tools.

Keywords: sustainable supply chain management; measurement instrument; critical factors; practices; performance; Greece

Citation: Mastos, T.; Gotzamani, K.; Kafetzopoulos, D. Development and Validation of a Measurement Instrument for Sustainability in Food Supply Chains. *Sustainability* **2022**, *14*, 5203. https://doi.org/10.3390/su14095203

Academic Editor: Attila Gere

Received: 28 March 2022
Accepted: 23 April 2022
Published: 26 April 2022

Publisher's Note: MDPI stays neutral with regard to jurisdictional claims in published maps and institutional affiliations.

1. Introduction

Sustainable supply chain management (SSCM) is one of the key sustainability concepts receiving significant attention during the last two decades [1,2]. SSCM involves the management of material, information and capital flows as well as the cooperation among all companies in the supply chain, considering all three dimensions of sustainable development, i.e., economic, environmental and social [2]. SSCM involves practices (SSCM-PRA) related to environmental, social and economic activities which often have a positive influence on SSCM performance (SSCM-PER) [3]. These practices might be enabled or inhibited by various contingent factors that are critical for the successful implementation of SSCM. Different industries address these SSCM critical factors (SSCM-CF) from several perspectives based on their size, organizational culture, geographical location and their stakeholders. SSCM has been investigated in several sectors, such as oil and gas [4], the automotive industry [5], energy [6] and the food industry [7]. The food industry, in particular, is one of the sectors facing significant sustainability challenges due to the special biological processes employed, the perishability and bulkiness of food products and environmental and social concerns such as climate change and food safety, respectively [7–10]. At the same time, factors such as globalization, advanced technology and transportation affect food supply chain sustainability [11,12], since changes or re-configurations in one stage of the supply chain are expected to affect other stages of the supply chain as well. In addition, during the last two years, food supply chains have been heavily influenced by the COVID-19 pandemic and as a result, SSCM has become even more important in the face of increasing demand and disruptive events that boost uncertainty [13].

Numerous studies have highlighted the importance of SSCM critical success factors, SSCM practices and SSCM performance. While previous research, especially in the food industry, has offered valuable results [14–16], the literature is still limited regarding the common conceptualization of SSCM critical success factors, SSCM practices and SSCM performance across food supply chains [1,7,17]. Regarding the SSCM critical factors, ref. [18] has identified a set of key enablers and inhibitors for implementing SSCM in small Greek enterprises. In [13], which explored SSCM critical factors during the COVID-19 pandemic, it was found that information sharing, food safety and innovation are only some of the driving forces that companies need to take into account in order to develop sustainable food supply chains during uncertainty. In [7], a conceptual set of SSCM practices were proposed, highlighting the need to evaluate the practices in more depth. Furthermore, SSCM performance has been investigated in the literature in relationship to SSCM practices [3,19,20]. What is common in the abovementioned studies is that different factors are used to describe each construct, indicating a lack of agreement on how these factors should be used in the field of SSCM. In addition to the above, the validation of SSCM critical factors, SSCM practices and SSCM performance needs to be investigated in more depth [1,7,17].

Based on the above arguments, the aim of this study is to empirically validate the theoretical scales of three key SSCM concepts i.e., SSCM critical factors, SSCM practices and SSCM performance in the Greek food industry. The discussion concerning the measurement instrument of the three key SSCM concepts is important because it provides an enhanced understanding of the complexity of the SSCM implementation in the food sector. Identifying these scales, and their related critical factors and measurement items, is crucial both for practitioners and researchers. Their identification will help practitioners in the food industry; first, to secure, provide and promote the necessary resources for effective SSCM, both within their companies and along their supply chains; second, to recognize and apply the necessary practices for the implementation of SSCM; and third, to use the appropriate measures for SSCM performance appraisal and improvement. In addition, the identification of these scales will help researchers to advance theory in SSCM and to test various research hypotheses regarding their relationships within the food industry in particular. The proposed measurement instrument contributes to the development of knowledge on the operationalization of the three key SSCM concepts. Exploratory and confirmatory factor analyses are deployed for the purpose of this study. With this approach, this study addresses the identified challenge that highlights the application of quantitative research methods (surveys) in order to test the reliability and validity of the developed SSCM theory [1]. Another gap that is addressed in this work is the limited work that has been conducted on the investigation of industry- and location-specific factors [21,22]. It is proposed that future research should identify industry-specific and geographically significant factors of SSCM [21]. Hence, this study will do so by exploring these factors in the Greek food industry. It is expected that the developed measurement instrument will offer useful guidance for SSCM critical factors, practices and performance measurement and provide a stepping-stone for future research in the field.

The rest of the paper is organized as follows. The Section 2 presents an overview of the related literature for the development of the three constructs. The research methodology is described in Section 3, while the results are presented in Section 4. Finally, in Section 5, the results are discussed in conjunction with previous research and conclusions are drawn, including the study limitations as well as future research paths.

2. Literature Review

2.1. SSCM Critical Factors

In the extensive literature on SSCM, it is supported that several factors are responsible for the success or failure of the implementation of SSCM [1]. Indeed, many researchers have described a number of factors (enablers, inhibitors, drivers, firm level strengths, barriers, etc.) that may impact the implementation of SSCM practices [18,23–27]. In this research work, these factors are named critical factors (CFs). In order to detect the SSCM critical factors, one should identify the enablers, drivers, success factors, motives, as well as barriers and inhibiting factors, that may influence the adoption and implementation of SSCM practices. The investigation of CFs is mainly analysed in two dimensions: the firm level and the external level, which includes the supply chain dimension as well. Among the most common firm level CFs are top management commitment [23,24,26], customer demands [15,24,25], and knowledge and expertise about sustainability [15,25]. Government policy [21,24,25,28,29], international/national regulatory frameworks [15,23,25], pressure and interaction with stakeholders, competitors and investors [15,24], and food incidents [15] are identified as some of the most common external CFs. On the external level, factors with a supply chain focus are also identified as critical. Information sharing [13,21,23,26,30] and building trustful relationships are two of the most critical supply chain factors for implementing SSCM in the food industry [16,30].

As already noted, several researchers have studied the critical factors for implementing SSCM. However, there is a scarcity of research related to the operationalisation and validation of the SSCM-CF construct. Hence, in line with these arguments, the following research hypothesis is generated:

Hypothesis 1 (H1). *SSCM critical factors (SSCM-CF) in the food industry can be reflected by firm level critical sustainability factors and external critical sustainability factors.*

2.2. SSCM Practices

Based on the SSCM definition given in the introduction and on [31]'s definition on supply chain management practices, SSCM practices are characterized "as a set of sustainability (i.e., economic, environmental and social) activities undertaken in an organization in cooperation with each stakeholders, to promote effective sustainability management of its supply chain". SSCM practices span from green supply chain management practices, such as environmental management and eco-design [5,20,32], to logistics social responsibility practices, such as socially responsible purchasing, sustainable transportation, reverse logistics, sustainable packaging and sustainable warehousing [33]. SSCM practices may also include land management and recycling activities [19] as well as codes of conduct and social audits [34]. Among the most common SSCM practices, especially in the food industry, are strategic orientation, supply chain continuity, collaboration, risk management, and proactivity [7]. Despite the fact that these practices are tested in the context of Chinese manufacturing firms from several sectors [35], the validation of these practices exclusively in the food industry is still limited. Hence, in order to address the need to further evaluate these practices [7], this study adopts them and posits the following hypothesis:

Hypothesis 2 (H2). *SSCM practices (SSCM-PRA) in the food industry can be reflected by strategic orientation, supply chain continuity, collaboration, risk management, and proactivity.*

2.3. SSCM Performance

SSCM performance refers to how well a supply chain achieves its environmental, economic and social goals. The literature mainly focuses on the economic and environmental facet of performance. The social dimension and the integration of the three sustainability dimensions are still lagging behind [2]. In [36], a rising interest in the aforementioned gap was revealed but more research is still needed in the field. SSCM performance is usually analysed as a three-dimensional concept including environmental, economic and social aspects. The ultimate goal of the implementation of SSCM practices is to improve overall SSCM performance. Among the most frequently used environmental performance measures are the reduction or avoidance of hazardous/harmful/toxic materials, water and energy consumption, recycled materials, Life Cycle Analysis (LCA), and environmental penalties [17,36]. Energy efficiency, air emissions, and greenhouse gas emissions are also highly cited indicators in the extant body of literature [36]. Regarding economic performance, the most frequent measures are quality, including quality of products provided by suppliers [17] or quality of the production process [37]. Sales, market share, and profits, as well as delivery time and customer satisfaction, are frequently used as financial performance indicators. The social dimension of SSCM performance is the hardest to measure due to the qualitative nature of the social issues. For example, the supply chain impact on customer experience or social welfare is difficult to be quantified; hence, the development of quantitative metrics is of crucial importance. Several researchers have identified a few measures that are frequently used in the SSCM literature. Among the most common measures are recordable accidents, training and education, and labour practices. So far, researchers have investigated SSCM performance measurement and more than 2500 unique metrics have been identified, indicating a lack of agreement and demonstrating that it is not yet clear how SSCM performance should be measured [17]. In addition to that, several studies have investigated sustainability performance as a multi-dimensional concept. For example, [3] investigated sustainability performance as a four-dimensional concept including operational, economic, environmental, and social outcomes, while [17] found 13 key characteristics for measuring performance of SSCM, including economic, environmental, social, volunteer, resilience, long-term, stakeholder, flow, coordination, relationship, value, efficiency, and performance indicators. In the food industry, researchers have investigated sustainability performance in terms of efficiency, flexibility, responsiveness, and product quality [8,14]. In [19], the performance of social and environmental sustainability practices in the food industry was found to be reflected by quality, cost, and environmental outcomes. Despite the fact that the definition of SSCM clearly states that the dimensions of sustainability are the economic, environmental, and social, empirical studies show that sustainability performance is a multidimensional concept. So far, as noticed in the literature, most papers concentrate on the performance measurement of one or two sustainability dimensions, mainly the environmental and economic [36]. Based on the above, it is obvious that there is a need to develop a valid and reliable construct to measure the performance of SSCM, especially in the food industry. In line with these arguments and SSCM theory, the following research hypothesis is generated:

Hypothesis 3 (H3). *The three dimensions of SSCM performance (SSCM-PER) (environmental, economic, and social) in the food industry reflect the measured indicators identified in the literature.*

3. Research Methodology

3.1. Research Instrument

The data of this study were collected through a structured survey questionnaire based on the literature review on SSCM [7,15,17,25,26,29,30,35,38]. Similar studies in the field have also conducted surveys demonstrating the relevance of this method in SSCM research [32]. The survey questionnaire was structured in four sections. The first section included questions regarding the critical factors for effective SSCM implementation. The second section included questions regarding the implementation of various SSCM practices. The third section included questions regarding the SSCM performance and the fourth section included questions regarding the profile of the companies and the respondents. The content validity of the questionnaire was ensured through extensive literature review that resulted in an initial list of 80 items. In order to further validate the questionnaire's content, a pilot study was conducted with 10 experts from the food industry. The experts' comments and suggestions were incorporated during the questionnaire pretesting phase [39,40], in order to improve the questions regarding the clarity of expression, the explanation of terms and items, the research scope, and the expected results. The draft version of the questionnaire was also reviewed and revised by four academics/researchers [41], resulting in a final list of 68 items. A seven-point Likert scale (with 1 "strongly disagree" and 7 "strongly agree") was used in order to allow respondents to report the extent to which they agree or disagree with each of the 68 items of the questionnaire. The statistical packages SPSS 24 and AMOS 21 were used for data processing.

3.2. Research Sample

The population of the survey consisted of firms in the food and beverage sector included in Greek sustainability databases, such as "CSR HELLAS NET", "Sustainable Greece 2020", "CSR Index GR", etc. Apart from the sustainability databases, the population included firms from other business databases, such as the "Federation of Hellenic Food Industries" and ICAP (the largest business information and consulting firm in Greece), among others, totaling an initial sample of 904 companies. Due to the COVID-19 pandemic restrictions, all companies were contacted via e-mail. The questionnaire was addressed to the personnel responsible for the supply chain and administered via e-mail by the authors and university students, who participated in a training session related to the research scope and content. The e-mail included a cover letter that assured the confidentiality of the submitted answers. The respondents were also advised to provide their contact details in case they were interested in the research results. Responses were collected from May 2020 to August 2021. By the end of the survey, 423 completed questionnaires were collected, yielding a response rate of 46.8%. This response rate was considered acceptable, as compared to other similar studies [35].

3.3. Non-Response Bias and Common Method Bias

In order to examine the dataset for non-response bias, the sample was divided into early and late respondents, where late respondents represent the theoretical non-respondents [42]. Comparisons between the two groups were made with use of the Mann–Whitney U test and no statistically significant differences were found, indicating that non-response bias is not an issue in this study.

Furthermore, the common method bias, which is another critical validity risk in behavioral research [43], is tested. To avoid this phenomenon, the Harman's single factor test was applied to test whether a single factor explained more than 50 percent of the variance in the data. All items were loaded in one single factor and the total variance explained was 35.198 percent, way below 50 percent, assuring the absence of common method bias in this study.

4. Results

4.1. Company Profiles

The sample included companies that belong to several food industry sub-sectors, covering the entire food supply chain network and ensuring that the findings do not relate only to specific supply chain members. Of the companies involved, 39.5% of the firms participate in more than one supply chain activities and 21.7% operate in the food services sub-sector; 16.1% operate in the retail sector and 10.9% in food manufacturing. The rest of the companies operate in other sub-sectors such as wholesale (4.5%), crop and animal production (2.6%), beverage industry (2.6%), and transportation and storage (2.1%) (Figure 1). Regarding the size, based on the number of employees, the responding companies were grouped as follows: 51.5% were very small enterprises (1–10 employees), 22.5% small enterprises (11–50 employees), 13.2% medium enterprises (51–250 employees), and 12.8%were large enterprises (>250 employees) (Figure 2).

4.2. Exploratory and Confirmatory Factor Analysis

Before performing the EFA and CFA, the items were examined individually in order to identify unique or extreme observations. The SPSS's boxplot is applied in order to define the extreme observations that are greater than 1.5 quartiles away from the end of the box [44]. Defined as outliers, 37 observations were deleted from the analysis as they were very likely to influence the outcome of any multivariate analysis [44]. The SSCM critical factors, SSCM practices, and SSCM performance items identified in the literature were used as measured variables in the following analysis. EFA was applied in order to extract the latent constructs of SSCM Critical Factors. Two latent factors (constructs) were extracted with the following values: KMO: 0.855, Bartlett's test of Sphericity: 985.769, df: 28, p: 0.000, eigen-value > 1, 0.907 = MSA $\geq$ 0.814, 0.800 = factor loadings $\geq$ 0.603, explaining 60.604% of the total variance. The factors were named after the items that were loaded on them, as follows: "firm-level critical sustainability factors" and "supply chain critical sustainability factors". EFA was also applied to extract the latent constructs of the SSCM Practices. Two latent factors were extracted (KMO: 0.872, Bartlett's test of Sphericity: 1181.660, df: 28, p: 0.000, eigen-value > 1, 0.909 = MSA $\geq$ 0.815, 0.863 = factor loadings $\geq$ 0.672), explaining 63.547% of the total variance, and were named as follows: "supply chain collaboration" and "supply chain strategic orientation". Eleven items related to supply chain continuity, risk management, and proactivity demonstrated cross-loadings greater than 0.4 on more than one latent construct; hence, they were dropped, since they do not provide clear measures of a specific factor [44]. One item from these practices was also dropped, since it demonstrated a factor loading below 0.5, which is not considered practically significant [44]. Finally, EFA was applied on SSCM performance, extracting three latent factors (KMO: 0.865, Bartlett's test of Sphericity: 3028.048, p: 0.000, eigen-value > 1, 0.937 = MSA $\geq$ 0.792, 0.937 = factor loadings $\geq$ 0.652), namely "economic performance", "social performance", and "environmental performance", explaining 70.780% of the total variance. The reliability of the extracted factors was confirmed by using Cronbach's alpha coefficient, recognized as a good direct measure of internal consistency. In each latent construct, the alpha value exceeds 0.7 [42,43], indicating that all factors are measured by reasonably reliable items.

In order to estimate the level of SSCM critical factors' adoption, the SSCM implementation practices, as well as the SSCM performance, perceived by the respondents, the mean scores of the three constructs were computed and analyzed. From the following three tables (Tables 1–3), it is evident that the companies have a high level of SSCM-CF adoption and SSCM-PRA implementation. Furthermore, the mean value of the social performance reached 6.24 with a standard deviation of 0.99, indicating the positive level of social performance of the participating companies. In general, it can be argued that the same level of importance has been given to all aspects of SSCM.

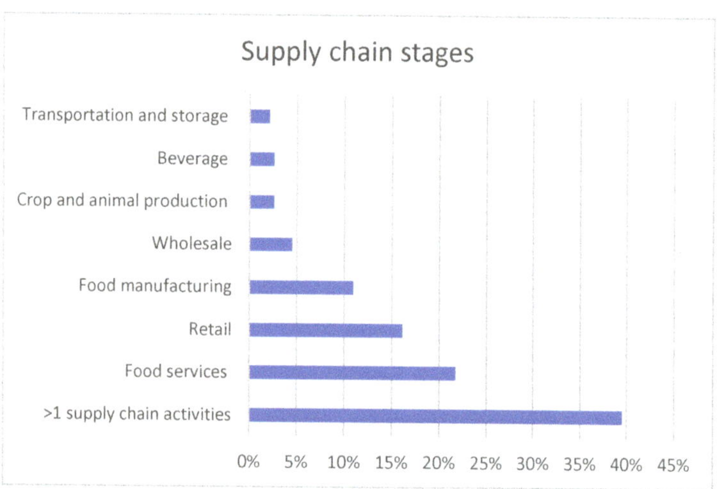

Figure 1. Supply chain stages.

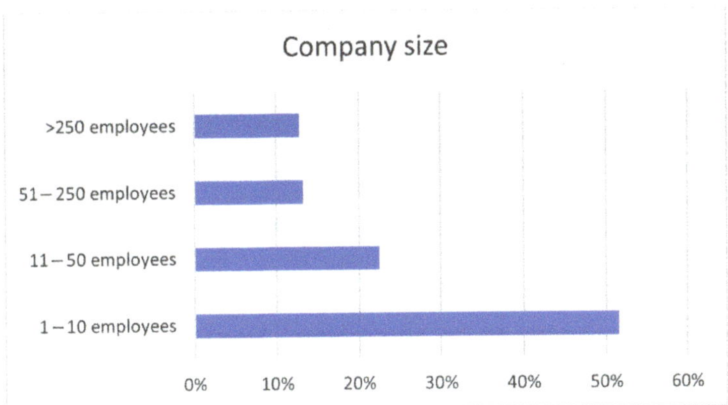

Figure 2. Company size.

Table 1. Descriptive analysis of SSCM-CF construct.

Factors	Items	Mean	SD
Firm-Level Critical Sustainability Factors	Sustainability knowledge and expertise	6.11	1.09
	Sustainability training	5.93	1.10
	Top management commitment to sustainability	5.28	1.47
	Customer needs and requirements for sustainability	6.14	0.96
	Mean value	5.87	1.15
Supply Chain Critical Sustainability Factors	Trust and commitment between supply chain partners	6.25	0.91
	Information sharing between supply chain partners	5.63	1.31
	Preventive measures regarding food scares, incidents and scandals of supply chain partners	6.51	0.85
	Mean value	6.13	1.02

In order to determine whether the empirical data fit the extracted latent factors of the EFA, CFA (maximum likelihood estimation technique) was performed for each of the three constructs (SSCM critical factors, SSCM practices and SSCM performance). The extracted latent factors of the three constructs show acceptable fit to the empirical data. The goodness of fit of the three constructs to the measured data is presented in Table 4. It is evident that the findings of this study consistently support the structure of the latent factors of the three developed constructs.

Table 2. Descriptive analysis of SSCM-PRA construct.

Factors	Items	Mean	SD
Supply chain collaboration	Technical integration of supply chain partners	5.15	1.35
	Monitoring supply chain partners	5.32	1.34
	Knowledge, information and resources sharing (upstream and downstream)	5.32	1.36
	Training and discussing sustainability issues with suppliers	4.67	1.60
	Mean value	5.12	1.41
Supply chain strategic orientation	Sustainability strategic goal setting	5.97	1.03
	Equal importance on environmental, social and economic issues	5.88	1.14
	Understanding sustainable development issues	5.45	1.23
	Mean value	5.77	1.13

Table 3. Descriptive analysis of SSCM-PER construct.

Factors	Items	Mean	SD
Economic performance	Profit growth rate	5.04	1.49
	Profit margin	5.01	1.52
	Cash flow	5.03	1.46
	Return on investment (ROI)	5.13	1.44
	Mean value	5.05	1.48
Environmental performance	Water consumption	5.12	1.52
	Waste reduction	5.60	1.38
	Energy efficiency	5.63	1.24
	Mean value	5.45	1.38
Social performance	Accidents per employee	6.26	1.02
	Accidents related to environment	6.36	0.96
	Environmental penalties	6.27	1.06
	Health and safety	6.27	0.91
	Product safety	6.20	0.97
	Hazardous/harmful/toxic materials	6.06	1.05
	Mean value	6.24	0.99

Table 4. The goodness of fit of the three constructs of the measurement instrument.

Fit Indices	SSCM Critical Factors	SSCM Practices	SSCM Performance	Acceptable Fit Indices
Absolute fit indices				
Chi-square (CMIN or χ^2)	37.516	29.980	144.476	$0 \leq \chi^2 \leq 2$df
Degrees of freedom (df)	13.000	13.000	60.000	
Probability level	0.000 *	0.000 *	0.000 *	$p > 0.05$
Root mean square residual (RMR)	0.042	0.054	0.060	<0.08
Root mean square of approximation (RMSEA)	0.070	0.058	0.060	<0.08
Incremental fit indices				
Incremental fit index (IFI)	0.972	0.983	0.972	>0.90
Tucker–Lewis coefficient (TLI)	0.954	0.973	0.963	>0.90
Comparative fit index (CFI)	0.972	0.983	0.972	>0.90
Parsimonious fit indices				
Chi-square/degrees of freedom (χ^2/df)	2.886	2.306	2.408	Between 1 and 3
Normed fit index (NFI)	0.958	0.971	0.953	>0.50
Goodness of fit index (GFI)	0.975	0.979	0.948	>0.50
Adjusted goodness of fit index (AGFI)	0.945	0.955	0.921	>0.50

Note: * acceptable when n > 250, the number of the measured variables range between 12 and 30, RMR < 0.08, RMSEA < 0.07, and CFI > 0.92 ([44]).

The construct validity of the latent factors is confirmed by calculating the convergent validity (AVE > 0.5), the discriminant validity (AVE > Corr^2) [45,46], the face-content validity (questionnaire feedback from food industry experts), and the nomological validity (significant correlations among the extracted latent factors) [47]. The convergent validity of the latent factors is confirmed by assessing the factor loadings (>0.606), the average variance extracted (AVE) (>0.431), and the construct reliability (CR) (>0.694) in all constructs [44]. It has to be mentioned that AVE value for Supply Chain Critical Sustainability Factors is found less than 0.50. If AVE is between 0.4 and 0.5, but composite reliability (CR) is higher than 0.6, the convergent validity of the construct is still adequate [48,49]. In order to assess discriminant validity, the AVE is compared with the highest squared correlation between the factor of interest and the remaining latent factors [47]. As shown in Table 5, the AVE is greater than the Corr^2, confirming the discriminant validity [44]. The items reflecting the three SSCM constructs, along with their standardised regression weights, are represented in Tables 6–8. The results of CFA confirmed the three constructs revealed by EFA and demonstrate that the extracted latent factors show acceptable fit to the empirical data.

Table 5. Constructs validity and reliability.

Latent Factors	CR	AVE	Cronbach's Alpha	Corr^2
SSCM Critical Factors				
Firm-level Critical Sustainability Factors (FLCSF)	0.829	0.549	0.817	0.244
Supply Chain Critical Sustainability Factors (SCCSF)	0.694	0.431	0.706	0.251
SSCM Practices				
Supply Chain Collaboration (SCC)	0.835	0.561	0.824	0.087
Supply Chain Strategic Orientation (SC)	0.753	0.504	0.776	0.251
SSCM Performance				
Economic Performance (FIN)	0.929	0.765	0.932	0.030
Social Performance (SOC)	0.873	0.536	0.877	0.234
Environmental Performance (ENV)	0.774	0.535	0.770	0.110

Table 6. CFA and standardised regression weights for SSCM Critical Factors.

Factors	Items	Components	
		1	2
Firm-level Critical Sustainability Factors	Sustainability knowledge and expertise	0.747	
	Sustainability training	0.714	
	Top management commitment to sustainability	0.743	
	Customer needs and requirements for sustainability	0.759	
Supply Chain Critical Sustainability Factors	Trust and commitment between supply chain partners		0.679
	Information sharing between supply chain partners		0.629
	Preventive measures regarding food scares, incidents and scandals of supply chain partners		0.659

Table 7. CFA and standardised regression weights for SSCM Practices.

Factors	Items	Components	
		1	2
Supply chain collaboration	Technical integration of supply chain partners	0.830	
	Monitoring supply chain partners	0.788	
	Knowledge, information and resources sharing (upstream and downstream)	0.754	
	Training and discussing sustainability issues with suppliers	0.606	
Supply chain strategic orientation	Sustainability strategic goal setting		0.689
	Equal importance on environmental, social and economic issues		0.718
	Understanding sustainable development issues		0.723

Table 8. CFA and standardised regression weights for SSCM Performance.

Factors	Items	Components		
		1	2	3
Economic performance	Profit growth rate	0.956		
	Profit margin	0.898		
	Cash flow	0.840		
	Return on investment (ROI)	0.824		
Environmental performance	Water consumption		0.729	
	Waste reduction		0.805	
	Energy efficiency		0.653	
Social performance	Accidents per employee			0.777
	Accidents related to environment			0.768
	Environmental penalties			0.669
	Health and safety			0.758
	Product safety			0.749
	Hazardous/harmful/toxic materials			0.661

5. Discussion, Implications and Concluding Remarks

This study developed and validated a measurement instrument comprised of three key SSCM constructs: one for SSCM critical factors, one for SSCM practices, and one for SSCM performance. The confirmation and validation of the three constructs supports the theory that firm-level and supply chain critical sustainability factors may be responsible for the success or failure of the implementation of SSCM practices that influence sustainability performance. The exploratory and confirmatory factor analysis revealed two latent constructs that reflect SSCM critical factors, two factors that represent SSCM practices, and three factors that measure SSCM performance. The extracted SSCM critical factors reflect the internal and external environment of an organization and indicate the focus that should be given to factors that are under the control of the company, such as providing sustainability knowledge and expertise or ensuring top management commitment, and on factors that are not 100% under the control of a company, such as the preventive measures regarding food scares, incidents, and scandals that are related to the supply chain as a whole. These factors have been labeled: "firm-level critical sustainability factors" and "supply chain critical sustainability factors". This distinction is in line with prior research that has acknowledged that there are critical factors for the implementation of SSCM [1] that can be categorized into firm-level and external, including also the supply chain level [15,50].

Regarding the SSCM practices, two factors have been extracted reflecting the practices that companies adopt to manage and control their supply chains. The first latent factor is named "supply chain collaboration", reflecting the importance of technical integration of supply chain partners and the sharing of information, knowledge, and resources upstream and downstream of the supply chain. This is not a surprise, since supply chain collaboration is one of the most commonly applied practices, especially in the food sector [7,51]. The second latent factor is named "supply chain strategic orientation". This is a more intangible factor since it involves measures such as sustainability strategic goal setting and placing equal importance on environmental, social, and economic issues. The results are in line with previous studies, which show that a strategic SSCM focus on all business decisions, even the ones that are directly related to the company's operations, is critical for the successful management of supply chains [2,52]. It is worth mentioning that the respondents did not explicitly consider supply chain continuity (establishing long-term relationships), risk management (adoption of standards and certifications) or proactivity practices (innovation capability or stakeholder management) [7] to be significant. This is an interesting finding, since it would have been expected that responses given during the pandemic, which caused severe disruptions to the food supply chains, would have considered risk management, supply chain continuity, and proactive practices as highly significant. This can be justified by the fact that Greek companies are usually reluctant to adopt and endorse SSCM practices and appear to be less proactive [53]. However, the recent study of Kafetzopoulos et al. [54] found that an agri-food company's knowledge orientation, collaboration, and quality orientation are factors that drive innovation. Hence, the two identified practices of supply chain collaboration and strategic orientation are expected to act as significant enablers for innovation in the future.

Regarding SSCM performance, three factors have been extracted and validated, addressing the three dimensions of sustainability i.e., environmental, social, and economic performance. The first factor is named "economic performance", since it involves key economic measures. Profit seems to prevail in this construct and this is not a surprise since this is one of the key measures that companies try to improve both in the short and long term. In line with this, ref. [14] also identified gross profit margin as one of the key sustainability performance indicators in the food industry (dairy sector). The second factor of the SSCM-PER construct is named "environmental performance". This construct involves key environmental measures such as water consumption, waste reduction, and energy efficiency, which are also supported by prior literature [55]. The third factor identified through the EFA and CFA is "social performance". This factor involves items that measure accidents per employee, health and safety, and product safety, which are commonly used in the description of the social dimension of sustainability [56]. Interestingly, some indicators in the social dimension cover environmental issues as well. For example, the management of hazardous materials, environmental accidents, and penalties are indicators that can be found in the environmental performance dimension. The hazardous materials are included in the social dimension because companies are responsible for treating these materials safely in order to avoid health and safety incidents that would hurt their social image. The social image of a company can also be affected by environmental accidents and penalties. In addition to the above, the majority of Greek firms have limited awareness of sustainability performance, especially in their supply chains, and they operate based on a low-cost/cost-cutting strategy [53]; hence, accidents and penalties related to environmental issues are expected to increase costs.

The results of the present work fully support the first and third research hypotheses. Regarding the second research hypothesis, there is evidence that the level of adoption of SSCM practices such as supply chain continuity, risk management, and proactivity should be further increased. Based on the above and on the evaluation process of the food firms' sample against the SSCM practices construct, the second research hypothesis is partially accepted.

5.1. Research and Practical Implications

The present work contributes to the existing research by providing a measurement instrument comprised of three key SSCM concepts, i.e., SSCM critical factors, SSCM practices, and SSCM performance. The first construct identifies the factors that are critical for companies that desire to successfully implement SSCM. The second identifies the factors related to SSCM practices and the third identifies the factors that need to be taken into account in order to improve SSCM performance. The three constructs of the measurement instrument offer insights into the nature of SSCM critical factors, practices, and performance in the food industry in the Greek business context. This paper answers the question of what factors to measure in order to implement SSCM practices and improve SSCM performance. The developed measurement instrument can be used both by practitioners and researchers. Supply chain practitioners can apply the three scales individually or together both in the firm level and the supply chain level. Furthermore, these constructs may be exploited by managers who desire to implement SSCM and allocate resources in order to improve supply chain sustainability performance. The proposed measurement instrument offers the opportunity to supply chain professionals to appropriately align their supply chain strategy towards positive environmental and social outcomes. The environmental and social performance items are the key aspects that should be taken into consideration for improving SSCM. Both practitioners and researchers may take advantage of the proposed scales and use them as assessment frameworks, benchmarking tools, or guidelines for the design of future strategies or research projects. Last but not least, the proposed measurement instrument was developed during the pandemic; hence, it may be beneficial for managers that wish to develop SSCM during uncertain times.

5.2. Limitations and Future Steps

It is recognized that there are several limitations to this work that can be used as future research propositions. First, in this study, the three constructs were tested separately. Based on the research findings, it is proposed that future studies should emphasise investigation of the relationships among the three constructs in order to provide a deeper understanding of how SSCM critical factors, SSCM practices, and SSCM performance relate to each other. The second limitation is related to the characteristics of the food firms' sample. The suggested measurement instrument is valid in the food industry and especially in the Greek business context. Future studies may examine the way the instrument's validity replicates in other sectors and countries. Another future opportunity would be to develop a SEM-PLS model, especially if the sample is smaller than this study. The analysis with SEM-PLS could be compared with the SPSS analysis to test if the developed measurement instrument is confirmed or not. Finally, this study was designed before the outbreak of the pandemic and was conducted during the pandemic. An interesting future research opportunity would be to repeat this survey in the post- COVID 19 era and compare the findings.

Author Contributions: Conceptualization, T.M. and K.G.; methodology, T.M., K.G. and D.K.; validation, T.M., K.G. and D.K.; formal analysis, T.M.; investigation, T.M.; resources, T.M.; data curation, T.M. and D.K.; writing—original draft preparation, T.M.; writing—review and editing, T.M., K.G. and D.K.; visualization, T.M.; supervision, K.G.; project administration, T.M. All authors have read and agreed to the published version of the manuscript.

Funding: This research received no external funding.

Institutional Review Board Statement: Not applicable.

Informed Consent Statement: Not applicable.

Data Availability Statement: Not applicable.

Conflicts of Interest: The authors declare no conflict of interest.

References

1. Ansari, Z.N.; Kant, R. A state-of-art literature review reflecting 15 years of focus on sustainable supply chain management. *J. Clean. Prod.* **2017**, *142*, 2524–2543. [CrossRef]
2. Seuring, S.; Müller, M. From a literature review to a conceptual framework for sustainable supply chain management. *J. Clean. Prod.* **2008**, *16*, 1699–1710. [CrossRef]
3. Zailani, S.; Jeyaraman, K.; Vengadasan, G.; Premkumar, R. Sustainable supply chain management (SSCM) in Malaysia: A survey. *Int. J. Prod. Econ.* **2012**, *140*, 330–340. [CrossRef]
4. Matos, S.; Hall, J. Integrating sustainable development in the supply chain: The case of life cycle assessment in oil and gas and agricultural biotechnology. *J. Oper. Manag.* **2007**, *25*, 1083–1102. [CrossRef]
5. Zhu, Q.; Sarkis, J. Relationships between operational practices and performance among early adopters of green supply chain management practices in Chinese manufacturing enterprises. *J. Oper. Manag.* **2004**, *22*, 265–289. [CrossRef]
6. Vance, L.; Cabezas, H.; Heckl, I.; Bertok, B.; Friedler, F. Synthesis of sustainable energy supply chain by the P-graph framework. *Ind. Eng. Chem. Res.* **2012**, *52*, 266–274. [CrossRef]
7. Beske, P.; Land, A.; Seuring, S. Sustainable supply chain management practices and dynamic capabilities in the food industry: A critical analysis of the literature. *Int. J. Prod. Econ.* **2014**, *152*, 131–143. [CrossRef]
8. Aramyan, L.; Ondersteijn, C.; Kooten, O.; Oude Lansink, A. Performance indicators in agri-food production chains. In *Quantifying the Agri-Food Supply Chain*; Springer: Dordrecht, The Netherlands, 2006; Chapter 5; pp. 49–66.
9. Carter, C.R.; Liane Easton, P. Sustainable supply chain management: Evolution and future directions. *Int. J. Phys. Distrib. Logist. Manag.* **2011**, *41*, 46–62. [CrossRef]
10. Reisch, L.; Eberle, U.; Lorek, S. Sustainable food consumption: An overview of contemporary issues and policies. *Sustain. Sci. Pract. Policy* **2013**, *9*, 7–25. [CrossRef]
11. McNeely, J.A.; Scherr, S.J. *Ecoagriculture: Strategies to Feed the World and Save Wild Biodiversity*; Island Press: Washington, DC, USA, 2003.
12. Yakovleva, N. Measuring the sustainability of the food supply chain: A case study of the UK. *J. Environ. Policy Plan.* **2007**, *9*, 75–100. [CrossRef]
13. Kazancoglu, Y.; Ozbiltekin-Pala, M.; Sezer, M.D.; Ekren, B.Y.; Kumar, V. Assessing the Impact of COVID-19 on Sustainable Food Supply Chains. *Sustainability* **2022**, *14*, 143. [CrossRef]
14. Bourlakis, M.; Maglaras, G.; Gallear, D.; Fotopoulos, C. Examining sustainability performance in the supply chain: The case of the Greek dairy sector. *Ind. Mark. Manag.* **2014**, *43*, 56–66. [CrossRef]
15. Chkanikova, O.; Mont, O. Corporate supply chain responsibility: Drivers and barriers for sustainable food retailing. *Corp. Soc. Responsib. Environ. Manag.* **2015**, *22*, 65–82. [CrossRef]
16. Golini, R.; Moretto, A.; Caniato, F.; Caridi, M.; Kalchschmidt, M. Developing sustainability in the Italian meat supply chain: An empirical investigation. *Int. J. Prod. Res.* **2017**, *55*, 1183–1209. [CrossRef]
17. Ahi, P.; Searcy, C. An analysis of metrics used to measure performance in green and sustainable supply chains. *J. Clean. Prod.* **2014**, *86*, 360–377. [CrossRef]
18. Mastos, T.; Gotzamani, K. Sustainable supply chain management in the food industry: A case study of enablers and inhibitors from Greek small enterprises. *World Rev. Entrep. Manag. Sustain. Dev.* 2020; in press.
19. Pullman, M.E.; Maloni, M.J.; Carter, C.R. Food for thought: Social versus environmental sustainability practices and performance outcomes. *J. Supply Chain. Manag.* **2009**, *45*, 38–54. [CrossRef]
20. Herrmann, F.F.; Barbosa-Povoa, A.P.; Butturi, M.A.; Marinelli, S.; Sellitto, M.A. Green Supply Chain Management: Conceptual Framework and Models for Analysis. *Sustainability* **2021**, *13*, 8127. [CrossRef]
21. Saeed, M.A.; Kersten, W. Drivers of sustainable supply chain management: Identification and classification. *Sustainability* **2019**, *11*, 1137. [CrossRef]
22. Sajjad, A.; Eweje, G.; Tappin, D. Sustainable supply chain management: Motivators and barriers. *Bus. Strategy Environ.* **2015**, *24*, 643–655. [CrossRef]
23. Faisal, N.M. Sustainable supply chains: A study of interaction among the enablers. *Bus. Process Manag. J.* **2010**, *16*, 508–529. [CrossRef]
24. Walker, H.; Jones, N. Sustainable supply chain management across the UK private sector. *Supply Chain Manag. Int. J.* **2012**, *17*, 15–28. [CrossRef]
25. Giunipero, L.C.; Hooker, R.E.; Denslow, D. Purchasing and supply management sustainability: Drivers and barriers. *J. Purch. Supply Manag.* **2012**, *18*, 258–269. [CrossRef]
26. Wittstruck, D.; Teuteberg, F. Understanding the success factors of sustainable supply chain management: Empirical evidence from the electrics and electronics industry. *Corp. Soc. Responsib. Environ. Manag.* **2012**, *19*, 141–158. [CrossRef]
27. Mastos, T.; Gotzamani, K. Enablers and inhibitors for implementing Sustainable Supply Chain Management practices: Lessons from SMEs in the food industry. In Proceedings of the EurOMA2018 Proceedings of the 25th International EurOMA Conference, Budapest, Hungary, 24–26 June 2018.
28. Gopalakrishnan, K.; Yusuf, Y.Y.; Musa, A.; Abubakar, T.; Ambursa, H.M. Sustainable supply chain management: A case study of British Aerospace (BAe) Systems. *Int. J. Prod. Econ.* **2012**, *140*, 193–203. [CrossRef]

29. Diabat, A.; Kannan, D.; Mathiyazhagan, K. Analysis of enablers for implementation of sustainable supply chain management–A textile case. *J. Clean. Prod.* **2014**, *83*, 391–403. [CrossRef]
30. Grimm, J.H.; Hofstetter, J.S.; Sarkis, J. Critical factors for sub-supplier management: A sustainable food supply chains perspective. *Int. J. Prod. Econ.* **2014**, *152*, 159–173. [CrossRef]
31. Li, S.; Rao, S.S.; Ragu-Nathan, T.S.; Ragu-Nathan, B. Development and validation of a measurement instrument for studying supply chain management practices. *J. Oper. Manag.* **2005**, *23*, 618–641. [CrossRef]
32. Sellitto, M.A.; Hermann, F.F. Influence of green practices on organizational competitiveness: A study of the electrical and electronics industry. *Eng. Manag. J.* **2019**, *31*, 98–112. [CrossRef]
33. Ciliberti, F.; Pontrandolfo, P.; Scozzi, B. Logistics social responsibility: Standard adoption and practices in Italian companies. *Int. J. Prod. Econ.* **2008**, *113*, 88–106. [CrossRef]
34. Awaysheh, A.; Klassen, R.D. The impact of supply chain structure on the use of supplier socially responsible practices. *Int. J. Oper. Prod. Manag.* **2010**, *30*, 1246–1268. [CrossRef]
35. Hong, J.; Zhang, Y.; Ding, M. Sustainable supply chain management practices, supply chain dynamic capabilities, and enterprise performance. *J. Clean. Prod.* **2018**, *172*, 3508–3519. [CrossRef]
36. Beske-Janssen, P.; Johnson, M.P.; Schaltegger, S. 20 years of performance measurement in sustainable supply chain management–what has been achieved? *Supply Chain Manag. Int. J.* **2015**, *20*, 664–680. [CrossRef]
37. Rao, P.; Holt, D. Do green supply chains lead to competitiveness and economic performance? *Int. J. Oper. Prod. Manag.* **2005**, *25*, 898–916. [CrossRef]
38. Tajbakhsh, A.; Hassini, E. Performance measurement of sustainable supply chains: A review and research questions. *Int. J. Prod. Perform. Manag.* **2015**, *64*, 744–783. [CrossRef]
39. Malhotra, M.K.; Grover, V. An assessment of survey research in POM: From constructs to theory. *J. Oper. Manag.* **1998**, *16*, 407–425. [CrossRef]
40. Oksenberg, L.; Cannell, C.; Kalton, G. New strategies for pretesting survey questions. *J. Off. Stat.* **1991**, *7*, 349–365.
41. Yan, T.; Kreuter, F.; Tourangeau, R. Evaluating survey questions: A comparison of methods. *J. Off. Stat.* **2012**, *28*, 503–529.
42. Armstrong, J.S.; Overton, T.S. Estimating nonresponse bias in mail surveys. *J. Mark. Res.* **1977**, *14*, 396–402. [CrossRef]
43. Podsakoff, P.M.; MacKenzie, S.B.; Lee, J.-Y.; Podsakoff, N.P. Common method biases in behavioral research: A critical review of the literature and recommended remedies. *J. Appl. Psychol.* **2003**, *88*, 879–903. [CrossRef] [PubMed]
44. Hair JFJr Black, W.C.; Babin, B.J.; Anderson, R.E.; Tatham, R.L. *Multivariate Data Analysis*, 5th ed.; Pearson Prentice-Hall: New York, NY, USA, 2006.
45. Singh, P. Empirical assessment of ISO 9000 related management practices and performance relationships. *Int. J. Prod. Econ.* **2008**, *113*, 40–59. [CrossRef]
46. Kim, D.Y.; Kumarb, V.; Kumarb, U. Relationship between quality management practices and innovation. *J. Oper. Manag.* **2012**, *30*, 295–315. [CrossRef]
47. Singh, P.; Powera, D.; Chuong, S. A resource dependence theory perspective of ISO9000 in managing organizational environment. *J. Oper. Manag.* **2011**, *29*, 49–64. [CrossRef]
48. Fornell, C.; Larcker, D.F. Evaluating structural equation models with unobservable variables and measurement error. *J. Mark. Res.* **1981**, *18*, 39–50. [CrossRef]
49. Lam, L.W. Impact of competitiveness on salespeople's commitment and performance. *J. Bus. Res.* **2012**, *65*, 1328–1334. [CrossRef]
50. Walker, H.; Di Sisto, L.; McBain, D. Drivers and barriers to environmental supply chain management practices: Lessons from the public and private sectors. *J. Purch. Supply Manag.* **2008**, *14*, 69–85. [CrossRef]
51. Matopoulos, A.; Vlachopoulou, M.; Manthou, V.; Manos, B. A conceptual framework for supply chain collaboration: Empirical evidence from the agri-food industry. *Supply Chain. Manag. Int. J.* **2007**, *12*, 177–186. [CrossRef]
52. Pagell, M.; Wu, Z. Building a more complete theory of sustainable supply chain management using case studies of 10 exemplars. *J. Supply Chain. Manag.* **2009**, *45*, 37–56. [CrossRef]
53. Skouloudis, A.; Evangelinos, K.; Nikolaou, I.; Filho, W.L. An overview of corporate social responsibility in Greece: Perceptions, developments and barriers to overcome. *Bus. Ethics Eur. Rev.* **2011**, *20*, 205–226. [CrossRef]
54. Kafetzopoulos, D.; Vouzas, F.; Skalkos, D. Developing and validating an innovation drivers' measurement instrument in the agri-food sector. *Br. Food J.* **2020**, *122*, 1199–1214. [CrossRef]
55. Varsei, M.; Soosay, C.; Fahimnia, B.; Sarkis, J. Framing sustainability performance of supply chains with multidimensional indicators. Supply Chain Management. *Int. J.* **2014**, *19*, 242–257.
56. Ahi, P.; Searcy, C. Measuring social issues in sustainable supply chains. *Meas. Bus. Excell.* **2015**, *19*, 33–45. [CrossRef]

 sustainability

Perspective

Constructed Wetlands as Nature-Based Solutions in the Post-COVID Agri-Food Supply Chain: Challenges and Opportunities

Vasileios Takavakoglou [1,2,*], Eleanna Pana [2] and Dimitris Skalkos [3]

1 Soil and Water Resources Institute, Hellenic Agricultural Organization "DEMETER", Thermi, 57001 Thessaloniki, Greece
2 Laboratory of General and Agricultural Hydraulics and Land Reclamation, Department of Agriculture, Aristotle University of Thessaloniki, 54124 Thessaloniki, Greece; epana@agro.auth.gr
3 Laboratory of Food Chemistry, Department of Chemistry, University of Ioannina, 45110 Ioannina, Greece; dskalkos@uoi.gr
* Correspondence: v.takavakoglou@swri.gr; Tel.: +30-231-047-3429

Abstract: The COVID-19 crisis has highlighted the interchangeable link between human and nature. The health and socioeconomic impacts of COVID-19 are directly or indirectly linked to the natural environment and to the way that agri-food systems interact with nature. Although the pandemic continues to evolve and there are still many uncertainties, important issues about the future of the agri-food sector and the need for a sustainable and environmentally friendly reformation are beginning to arise in society. Nature-based Solutions (NbSs) encompass a broad range of practices that can be introduced in the agri-food supply chain and address multiple environmental challenges of the COVID-19 and post-COVID-19 era while providing economic and societal benefits. In this perspective, the design and establishment of multifunctional constructed wetlands as NbSs opens a portfolio of eco-innovative options throughout the agri-food supply chain, offering a realistic and promising way towards the green regeneration of the post-COVID-19 economy and the welfare of society. The aim of this work is to explore the potential role of constructed wetlands as Nature-based Solutions in the agri-food supply chain of the forthcoming post-COVID-19 era. More specifically, this work aims to reveal application opportunities of constructed wetlands in the different segments of the agri-food supply chain, identify linkages with societal challenges and EU policies, and discuss their potential limitations, future challenges, and perspectives.

Keywords: agri-food supply chain; environment; NbSs; eco-innovation; constructed wetlands; post-COVID-19

Citation: Takavakoglou, V.; Pana, E.; Skalkos, D. Constructed Wetlands as Nature-Based Solutions in the Post-COVID Agri-Food Supply Chain: Challenges and Opportunities. *Sustainability* **2022**, *14*, 3145. https://doi.org/10.3390/su14063145

Academic Editor: Antonio Boggia

Received: 11 January 2022
Accepted: 7 March 2022
Published: 8 March 2022

1. Introduction

The agri-food sector entails a wide and complex network of feedbacks and tradeoffs between environment, economic activities, transport, trade, livelihoods, and human health. Since its first wave in 2019, the outbreak of COVID-19 is still having an unparalleled effect on the agri-food sector. The health and socioeconomic impacts of the pandemic have been linked to the natural environment and to the way that agri-food systems are organized and operate [1]. The crisis that the agri-food sector is facing today requires adapting transformative changes in technological, economical, and socio-ecological activities to address human needs while preserving Earth's systems in the post-COVID-19 era [2,3]. Nature-based Solutions (NbSs) are gaining importance as solutions that integrate societal challenges and nature conservation across scales and landscapes. In this perspective, they have the potential to offer long-term transformative pathways to agri-food supply chains towards sustainability [4]. Constructed and natural wetlands are at the epicenter of NbSs [5,6]. Unfortunately, constructed wetlands attract attention mainly as natural wastewater treatment systems, while other important ecosystem services that they provide are usually overlooked or are simply considered ancillary [7]. Thus, the multifunctional role of constructed wetlands in the different segments of the food supply chain is often underestimated and their contribution as NbSs to the post-COVID-19 resilience and sustainability of the agri-food sector is not fully assessed.

The aim of this work is to explore the potential role of constructed wetlands as Nature-based Solutions in the agri-food supply chain of the forthcoming post-COVID-19 era. More specifically, this work aims to reveal application opportunities of constructed wetlands in the different segments of the agri-food supply chain, identify the linkages with societal challenges and EU policies, and discuss their potential limitations and future challenges.

For this purpose, a literature review was conducted based on emergent qualitative analysis (deductive and inductive) [8]. This dual approach combines critical elements without relying completely either on existing literature or on the data themselves [9]. The analysis follows a stepwise approach, in which first the environmental aspects of agri-food sector in the post-COVID-19 era are discussed to set the framework for the analysis. Focus is given on the environmental challenges of the agri-food supply chain, as these have been affected by the pandemic and evolved according to the changes in public perceptions and attitudes after the outbreak. In answer to the identified challenges, the authors present the concept of NbSs in the agri-food sector followed by a chapter that presents the relevance and classification of constructed wetlands as NbSs. The next chapter analyzes in detail the present and perspectives of constructed wetlands in the agri-food supply chain by demonstrating existing applications and potential opportunities, analyzing linkages with policies, and discussing limitations and future challenges.

2. Environmental Aspects of the Agri-Food Supply Chain in the Post-COVID-19 Era: A Consumers' Driven Approach

The COVID-19 pandemic has led the agri-food systems into a novel reality with multiple challenges that need to be addressed. In this perspective, issues related to sustainability and environment are of primary importance for the agri-food supply chain [10,11]. Although the pandemic continues to evolve and there are still many uncertainties, important issues about the future of the agri-food sector and the need for a sustainable and environmentally friendly reformation are beginning to arise in society. In this perspective, Kotler (2020) [12] pointed out the emergence and growing importance of five consumer types in the post-COVID-19 era, which are interestingly all related directly or indirectly to environmental issues. These types include:

- Degrowth activists, who worry about the carrying capacity of the earth in relation to the consumption of goods and natural resources and call for nature conservation and the reduction of human material needs;
- Climate activists, who are concerned about climate change and the future of our planet while aiming to reduce the human carbon footprint and the degradation of natural resources;
- Sane food choosers, who are persons who have turned into vegans or vegetarians, are abstaining from the use of meat or animal products, and are opposed to industrial farming of animals for ethical and environmental reasons, including high methane emissions and the increased water footprint of raising livestock;
- Conservation activists, mainly environmentalists with social concerns, who promote the philosophy of repair–reuse–recycle;
- Life simplifiers, who are less interested in owning goods, and in order to cover temporal needs they prefer renting instead of owning.

Considering that consumers may regulate market growth, competitiveness, and economic integration, the assessment of consumers' preferences and behavior is of primary importance in planning post-COVID-19 strategies and measures towards green and sustainable agri-food supply chains [13]. The shifting of consumer preferences can unlock a multitude of both health and environmental benefits such as combating biodiversity and climate threats and crises, relieving environmental stresses, and contributing to sustainable socioeconomic schemes and healthier lifestyles with tangible long-term impacts on the livelihood of human society.

Following the evolution of public perceptions, needs, attitudes, and intentions [14], the environmental aspects of the agri-food supply chain are becoming of primary importance in the agenda of the food industry, decision-makers, and scientists, as these are related to human and environmental health and safety issues. As identified in several recent studies [15–19], challenges related to the environment in the COVID-19 and post-COVID-19 era of the agri-food sector are mainly related to (1) emerging greener consumer behavior, (2) climate change, (3) environmental pollution, (4) resource efficiency, (5) health and hygiene concerns, (6) green energy transition, (7) conservation of biodiversity, and (8) systems resilience and sustainability (Figure 1).

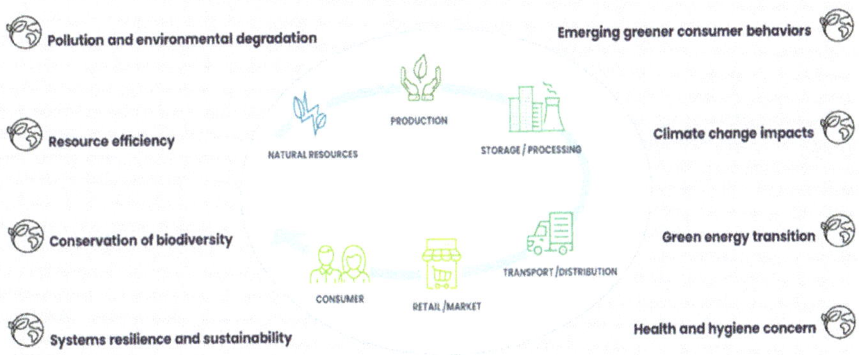

Figure 1. Environment-related challenges of the agri-food supply chain in the COVID-19 and post-COVID-19 era.

Although these challenges are not new in principle, the pandemic resulted in a reorientation of priorities and the urgent need for integrated solutions in respect to multiple societal needs and the changes of citizens' behavioral patterns. Key issues of environmental interest as well as associated challenges affected by pandemic are the following:

- Climate-related issues have improved during COVID-19. Nitrogen and carbon emissions decreased significantly because of the restrictions in transportation and mobility, the decreased usage of electricity, and the ceased industrial production. However, based on projected changes in climate and upcoming socioeconomic developments, most climate change impacts are expected to rebound and maybe increase even more in the coming decades across Europe [20]. Based on these projections and given the sharp rising of fossil fuel prices, as was recently recorded, the gradual transition to green energy is necessary in order to safeguard both the viability of the agri-food sector and climate health in the post-COVID-19 era [21]. Investment in renewable energy sources (e.g., wind, solar, and bioenergy) along with interventions for energy efficiency (e.g., insulation retrofits, green buildings, and infrastructure) are indispensable parts of the armory against the global energy crisis in the years to come. This is of primary importance for the agri-food sector in which the cost of energy and the reliability of supply is critical (e.g., greenhouses, storage, and processing facilities). Green energy solutions may contribute to autonomous and safe operations even in case of emergencies and unexpected events (e.g., COVID-19 outbreak). In a win–win scenario, countries and business in the agri-food sector may benefit from a robust green energy economy and the cutting down of spending over more expensive and less reliable fossil-based sources of energy [22].

- In terms of pollution, water-quality issues related to emerging pollutants and microplastics are of growing importance. The extensive use of personal protective equipment (e.g., masks and gloves) that become waste and the inappropriate use of chemical substances to control pests and/or prevent the transmission of diseases may raise important environmental problems [23]. Soil degradation issues became more intense during COVID-19 lockdown because of the increased quantities of municipal food wastes, the suspension of recycling programs, and restrictions on sustainable waste management practices [17].
- The lockdown measures were found to drive an important shift towards the "circular economy" approach, which aims to maintain the value of products and resources through time while minimizing the generation of waste [23]. According to FAO (2021) [18], the main opportunities and challenges are related to the treatment and reuse of wastewater as well as the recycling of irrigation water, the precision agriculture, and the optimization of agricultural inputs, biofertilizers, and bioenergy. In this direction, the G20 encourages eco-design that permits products and resources to be repaired–recycled–reused and the uptake of relevant business models for economic recovery [24].
- According to the Intergovernmental Platform on Biodiversity and Ecosystem Services (IPBES), the emergence of zoonotic diseases, as well as changes in land use, the expansion of agriculture, and urbanization, could be associated with more than 30% of emerging diseases. Furthermore, it was emphasized that birds, mammals (primates, bats, and rodents), and livestock (e.g., poultry, pigs) could act as reservoirs of pathogens that may have pandemic potential [25]. Thus, multiple biodiversity-related issues arise in the COVID-19 and post-COVID-19 era, including the interconnections between agriculture, biodiversity, and infectious diseases; the trade and consumption of wildlife; the importance of climate change on biodiversity and eventually on the emergence of diseases through the spatiotemporal distribution of potential reservoirs and vectors; the degradation of ecosystem functions and the loss of habitats; and the impact of land-use change on biodiversity from deforestation for agricultural purposes to landscape fragmentation due to transport networks and other human infrastructure development [26,27].

3. The Growing Importance of Nature-Based Solutions

Environmental sustainability may contribute to the prevention as well as coping of potential future pandemics and their impacts [28]. The concept of Nature-based Solutions (NbSs) introduces an alternative pathway towards sustainability through balanced socioecological adaptation and resilience [29]. The European Commission defined NbSs as solutions that are inspired and supported by nature, are designed to address societal challenges, are cost-effective, provide environmental and socioeconomic benefits, and help build resilience [30,31]. In this perspective, the concept of "innovating with nature" may effectively contribute to more sustainable and resilient societies through green growth and job creation [32]. NbSs as an umbrella concept may range from the wise management of natural ecosystems to the establishment of new ecosystem functions and processes [33]. From the perspective of degrees of intervention, NbSs can be divided into three broad categories, as presented in Table 1 [34,35].

NbSs are acknowledged among policy/decision-makers and major European Policy frameworks and strategies, such as the European Green Deal, the EU Health Strategy, and the EU Biodiversity Strategy, all considering their potential to increase health and well-being. They are considered credible means able to address key societal issues (e.g., impact of climate change, natural disasters, and loss of biodiversity) [35], and this was acknowledged at the high-level ministerial panel on NbSs in green recovery held by the IUNC in March 2021. A prominent outcome of the panel was the commitment of involved parties to increase efforts and investment to allocate NbSs a larger role in COVID-19 stimulus plans, acknowledging the great cost–benefit ratio of NbSs, their potential for speedy and streamlined implementation, and their contribution to sustainability and citizens' well-being [29].

Table 1. Types of Nature-based Solutions.

	NbSs Typology
Type I	Minimal (or no) intervention in ecosystems. Aim to sustain protected/natural ecosystems, improve the conservation status and increase environmental awareness, and enhance or restore their functional role and ecosystem health (e.g., ecosystem conservation and restoration strategies).
Type II	Partial interventions in ecosystems. Aim to improve selected ecosystem functions and services by contributing to sustainable, multi-functional ecosystems (e.g., sustainable forestry and agriculture, multifunctional rural landscapes, application of agroecological practices, or strengthening of forest resilience to extreme events through biodiversity enhancement).
Type III	Interventive management of ecosystems (extensive/intrusive) or establishment of new ecosystems. Aim to draw benefits from newly established assemblages of organisms and natural processes while also linked to the concepts of green and blue infrastructure (e.g., green roofs or walls to mitigate city warming or air pollution; natural systems such as constructed wetlands for water pollution control and non-conventional water supply, bio/phyto remediation of heavily polluted or degraded areas).

In agri-food supply chains, NbSs encompass a wide range of promising practices and potential solutions that can be introduced, addressing multiple environmental challenges of the COVID-19 and post-COVID-19 era.

In agricultural production, NbSs can be deployed directly in the context of food and fiber production on agricultural lands, in rural landscapes, or regarding water resources that are used for production [36]. At the farm level, examples of NbSs include agroforestry; intercropping; cultivar mixtures; biological pest control; rehabilitation of soil functions and improvement of soil quality; erosion control measures; biological nitrogen fixation; multifunctional field margins; precision agriculture and smart irrigation techniques for the reduction of inputs; natural systems of water management and recycling of nutrients, waste, and energy; and avoiding contaminants to ensure food safety [37].

In terms of urban sustainability transformation, the concept of edible cities (urban food production and local distribution) can be seen as a multifunctional NbSs [38]. Simultaneously, circular economy-related initiatives can stretch to connect urban with rural areas, as in the case of Kitakyushu city in Japan, where the adoption of a food-recycling loop allows compost from urban areas to be used as fertilizer or an energy source in rural areas [24].

From the consumers' point of view, the COVID-19 outbreak introduced multiple changes in the daily life of people by affecting the foundations of our societies and economies. This is characteristically reflected in the behavior and attitude of consumers [39]. According to Durante and Laran (2016) [40], in stressful situations such as the pandemic, consumers tend to save money and spend strategically on necessary products in order to restore their sense of control. In this perspective, NbSs that reduce the stress of citizens may contribute to more rational consumer behavior and the gradual rebound of the economy in the post-COVID-19 era and potentially lead to greener consumption patterns with considerable environmental benefits. A gradual shift to plant-based diets, for example, may contribute to sequestration from 332 Gt to 574 Gt CO_2 [41].

Furthermore, the use of NbSs for improving townscapes and favoring social cohesion [42] may contribute to the mitigation of climate change impacts (e.g., urban heat shocks) and associated heart and respiratory diseases [43], which is of critical importance under the threat of COVID-19.

In addition to the above, NbSs may also provide employment opportunities, which is critical in the post-COVID-19 era, especially in disadvantaged and climate-vulnerable areas [44]. It is highlighted that approximately 1.2 billion jobs globally are dependent on ecosystem health (e.g., agriculture, forestry, fisheries, tourism), with considerable societal and economic importance [45,46]. NbSs can be used to sustain or enhance the jobs and productivity of those working in the agri-food sector and thus contribute to social justice goals of reduced inequalities, decent employment, equal opportunities, social safety, and cohesion.

A common element in the NbSs approach is that based on nature's paradigm, the establishment of healthy and resilient ecosystems may deliver valuable services that contribute to human well-being while simultaneously addressing environmental and socioeconomic goals [35].

4. Constructed Wetlands as Nature-Based Solutions

Constructed or artificial wetlands are engineered ecosystems that combine the core structural components of natural wetland ecosystems (e.g., water, vegetation, and soil/substrate) in such a way as to mimic and perform selected functions of natural wetlands and thus deliver a range of monetary and non-monetary services. In this perspective, wetland systems are an important tool in the armory of NbSs. The services of constructed wetlands, as in case of natural wetland ecosystems, may include [47]:

- Supporting services (e.g., nutrient cycling, food-web support);
- Regulating services (e.g., water-quality improvement, water-flow regulation, groundwater recharge, and climate regulation);
- Provisioning services (e.g., food, fiber, and water supply, including non-conventional water);
- Cultural services (e.g., education, recreation, aesthetic, spiritual).

The design of constructed wetlands is adjusted according to the targeted services and the purpose of their establishment [48]. Depending on their main functions and the targeted wetland ecosystem services, constructed wetlands may be established as:

a. Natural wastewater (black or grey) treatment systems focused on water-quality improvement;
b. Blue–green areas focused on cultural services;
c. Food and fiber production systems focusing on provisioning services;
d. Building interventions such as wet roofs and green walls with a focus on climate regulation services;
e. Landscape interventions for water-flow regulation and flood control in urban, rural, and mountainous areas;
f. Biodiversity enhancement areas focused on food-web support.

Based on their structural and functional characteristics, constructed wetlands are referred to as green infrastructure, classified under Type III of NbSs. Green infrastructure, as part of NbSs' armory, is defined by the European Commission [49] as a network of areas that are strategically planned, designed, and managed in order to deliver a range of ecosystem services, such as, for example, the improvement of water and/or air quality, mitigation of/adaptation to climate change, recreational areas, and natural risk or disaster attenuation.

Although historically, constructed wetland technology was mainly focused on pollution control and wastewater treatment, a broader approach has recently evolved. According to this, constructed wetlands are part of a wider picture that involves multiple integrative technologies to address sustainability issues in water, energy, and food [50]. Nowadays, and within the concept of circular economy and the water–food–energy nexus, the designers of constructed wetlands are aiming to build multifunctional systems that are able to deliver multiple services with associated benefits for society. A characteristic example in urban areas is the establishment of wet roofs for grey-water treatment, climate regulation, the improvement of energy efficiency, and the reuse of wastewater for primary production and/or landscape amelioration. In parallel, the introduction of constructed wetlands in rural areas as multifunctional wet field margins may contribute to non-point source pollution control and floodwater management, while at the same time creating wildlife habitats and enhancing the biodiversity of the rural landscape [6]. In this perspective, it is underlined that the typologies of NbSs, as presented in Table 1, are dynamic benchmarks and not static classifications of possible NbSs interventions, since many hybrid NbSs may exist along the gradients. For example, constructed wetlands established initially as green infrastructure of Type III will be subsequently managed as Type I systems [34,36].

5. Constructed Wetlands in Agri-Food Supply Chains: Challenges and Opportunities

5.1. Applications and Opportunities

Constructed wetlands, given their nature and multidimensional role, are at the center of NbSs since they constitute cost-effective solutions based on and supported by nature, able to provide multiple environmental and socioeconomic benefits [5]. Considering the increased priority given by the international community to NbSs, constructed wetlands are gaining attention as potential promising solutions to important challenges of the agri-food supply chain that are related to the environment.

The considerable progress of constructed wetland eco-technology in deploying selected ecosystem functions and services opens a portfolio of options throughout the agri-food supply chain. These offer a realistic and promising way forward for addressing conservation, climate, and economic as well as social challenges, while maintaining healthy and resilient agri-food systems in the post-COVID-19 era. Constructed wetlands and NbSs in general can often operate as standalone solutions. However, there is also a recent tendency to integrate them with gray infrastructure, creating hybrid solutions that are able to address complex challenges and meet increasing demands from different sectors of the broader water–food–energy nexus [50]. The applications of constructed wetlands cover the entire range from farm to fork by addressing multiple challenges, from pollution control, green energy transition, biodiversity conservation, and resource efficiency, to social welfare and post-COVID-19 economic regeneration. The potential role and applications of constructed wetlands in the agri-food supply chain in relevance with key societal challenges are presented in Table 2. These societal challenges were selected based on a review of the SDGs of NbSs frameworks [51], the EEA report on NbSs [20], and the outline of the societal challenges of the Horizon 2020 research programme [52].

Table 2. The potential role and applications of constructed wetlands in the agri-food supply chain in relevance with key societal challenges.

Agri-Food Supply Chain	Potential Role and Applications of Constructed Wetlands	Linked Societal Challenges
Production	• Non-point source pollution control in agricultural areas [53] • Non-conventional water supply for irrigation through the reuse of reclaimed wastewater [54] • Green energy production (e.g., from wetland vegetation biomass or through microbial fuel cells) [55,56] • Raw materials for the production of agri-food products (e.g., biomass as substrate for mushroom production) [57] • Food production, including fish farming [58] • Promoting circular economy within the water–soil–waste nexus [59] • Creation of habitats and increase in rural biodiversity [60,61] • Mitigating climate change impacts, including erosion, desertification, depletion of groundwater aquifers, wildfires, floods, etc. [62] • Multifunctional landscapes, including wet field margins for pollution control, water flow regulation, and biodiversity enhancement [63,64] • Remediation of polluted soils and sensitivity to degradation areas [65]	SC 1. SC 2. SC 3. SC 5. SC 6. SC 7. SC 8. SC 10.
Storage and Processing	• Wastewater treatment in food-processing industries [66] • Wetland roofs and green walls to improve buildings' energy efficiency [67] • Non-conventional water supply for industrial use (cooling, landscape amelioration, firefighting, etc.) [68] • Green energy production (e.g., from wetland vegetation biomass or through microbial fuel cells) [69,70]	SC 1. SC 2. SC 3. SC 5. SC 7. SC 10.

Table 2. *Cont.*

Agri-Food Supply Chain	Potential Role and Applications of Constructed Wetlands	Linked Societal Challenges
Transport and Distribution	• Treatment of runoff waters from road and transport networks [71] • Carbon sequestration and CO_2 sinks [72–74] • Providing ecological niches and mitigating ecological impacts of habitat loss or fragmentation [75,76] • Seawater pollution control, including petroleum hydrocarbons in the marine environment, using floating wetlands [77]	SC 4. SC 5. SC 8. SC 10.
Retail and Markets	• Wetland roofs and green walls to improve the energy efficiency of commercial and market buildings [64] • Decentralized water treatment in public and central markets [78] • Food-waste pollution control through landfill leachate treatment [79] • Social cohesion and environmental responsibility strengthening, as in the case of Dumaguete city's public market, where the funds collected from the public restrooms cover the operational and maintenance expenses of a wetland system for wastewater treatment [80]	SC 1. SC 3. SC 5. SC 6. SC 9. SC 10.
Customers and Consumption	• Greywater treatment and reuse for urban landscape amelioration [81] • Enhancement of urban biodiversity and mitigating climate change impacts in urban areas, including heat stress, stormwater management, etc. [82,83] • Urban agriculture and support of short local food supply chains (e.g., floating wetlands for vegetable cultivation in urban areas) [84] • Establishment of educational and environmental awareness areas [85] • Blue–green spaces for people to feel connected with nature and enhance psychological well-being in line with emerging consumer behaviors in the post-COVID-19 era [86]	SC 1. SC 2. SC 3. SC 5. SC 6. SC 7. SC 8. SC 9. SC 10.

(SC 1: Public health and well-being; SC 2: Food security, sustainable agriculture, and forestry; SC 3: Secure, clean, and efficient energy; SC 4: Smart, green, and integrated transport; SC 5: Climate action and resilience to extreme weather- and climate-related events; SC 6: Inclusive, innovative, reflective, and resilient societies; SC 7: Sustainable economic development and decent employment (including green jobs); SC 8: Preserving habitat, reducing biodiversity loss, and increasing green and blue spaces; SC 9: Making cities and human settlements inclusive, safe, resilient, and sustainable; SC 10: Environmental quality (including air quality, water, and waste management), resource efficiency, and raw materials).

The complexity of tackling the environmental challenges in the agri-food sector in the post-COVID-19 era requires an effective transformative change across a wide range of political, technological, and socioeconomic factors [87]. Working with nature is considered today a promising path to this transformative change [88]. In this context, innovative tools for the design, implementation, and assessment of multifunctional constructed wetlands and NbSs in general are required, along with effective processes that are able to effectively support stakeholders' participation [69].

Critical steps in this process are both the deployment of relevant research initiatives as well as the realization of large-scale demonstrative actions and the active mobilization of stakeholders and local champions as lighthouses for the transfer of knowledge and innovation at an operational level across borders. In this direction, the EU already supports several flagship projects, such as WaterLANDS and HYDROUSA through H2020, AQUACYCLE through ENI CBC Med, and MARA-MEDITERRA through the PRIMA Foundation, which not only promote the research and innovation in natural and constructed wetlands as NbSs, but also demonstrate their operational applications and capitalization potential to address environmental, economic, and societal challenges towards a sustainable future.

The new EU Framework Programme pays special attention to transferring the developed knowledge and innovations at an operational level, where they can generate tangible results and serve citizens. Thus, according to EU political priorities and the COVID-19 recovery plan, four key strategic orientations for EU research have been established [89]:

A. To promote open strategic autonomy through the development of key digital, enabling, and emerging technologies, sectors, and value chains;
B. To restore biodiversity and ecosystems as well as to sustainably manage natural resources in order to ensure food security and environmental health;
C. To set Europe as a protagonist in a digitally enabled sustainable, climate-neutral, and circular economy;
D. To establish a resilient, inclusive, and democratic society with high-quality health care, EU citizens empowered to act in green and digital transitions, and an increased level of readiness against disasters and threats.

In this perspective, the introduction of constructed wetland ecotechnology in multiple segments of the agri-food supply chain may have a significant contribution to the achievement of the expected impacts outlined in the Horizon Europe Strategic Plan 2021–2024 [89]. Based on a literature review of constructed wetlands' applications and services, a preliminary assessment of their potential contribution to the abovementioned impacts was performed (Table 3). These ecosystems may provide solutions in the agricultural production phase in respect to climate change adaptation/mitigation, pollution control, biodiversity enhancement, sustainable management of natural resources, and alternative primary and green energy production. Constructed wetlands may also play an important role at the consumer level, especially in urban areas, by addressing environmental challenges for a healthier environment and by promoting social transition towards responsible resource management, consumption of safer products, and eventually the flourishment of green economy and the well-being of society.

Table 3. Potential contribution of constructed wetlands to the expected impacts of Horizon Europe 2021–2024 key strategic orientations, per stage of the agri-food supply chain.

Agri-Food Supply Chain	Expected Impacts of the Key Strategic Orientations														
	A1	A2	A3	A4	B1	B2	B3	C1	C2	C3	C4	D1	D2	D3	D4
Production															
Storage and Processing															
Transport and Distribution															
Retail and Markets															
Customers and Consumption															

(A1: Competitive and secure data economy; A2: Industrial leadership in key and emerging technologies that work for people; A3: Secure and cybersecure digital technology; A4: High-quality digital services for all; B1: Enhancing ecosystems and biodiversity on land and in water; B2: Clean and healthy air, water, and soil; B3: Sustainable food systems from farm to fork on land and sea; C1: Climate change mitigation and adaptation; C2: Affordable and clean energy; C3: Smart and sustainable transport; C4: Circular and clean economy; D1: A resilient EU prepared for emerging threats; D2: A secure, open, and democratic EU society; D3: Good health and high-quality accessible healthcare; D4: Inclusive growth and new job opportunities. Colors: Green-High, Yellow-Moderate; Pink-Low; Grey-N/A).

Overall, constructed wetlands appear to be a promising ecotechnology with a potential significant contribution to EU policy objectives and strategies and especially the EU Green Deal (2019), the EU's Biodiversity Strategy for 2030 (2020), the new EU Strategy on Adaptation to Climate Change (2021), and the Bioeconomy Strategy and its recent update [29]. Especially concerning the agri-food sector, constructed wetlands are expected to play a key role in the Farm to Fork Strategy (2020) and the new Common Agricultural Policy 2023–2027 as potential measures supported by Eco-Schemes (e.g., measures to reduce and prevent water and soil pollution from excess nutrients, creation of nutrient traps and buffer strips, semi-natural habitat creation, rewetting wetlands/peatlands), aiming to provide stronger incentives for climate- and environmentally friendly farming practices [90].

5.2. Shortcomings and Challenges

Although constructed wetlands and NbSs in general appear to have a promising future, there are still several limitations and challenges to be addressed in order to fulfil their potential. In terms of technological maturity, some of the potential applications presented in Table 2 are already applied at an operational level (e.g., wastewater treatment), whereas for others (e.g., microbial fuel cells), further research and testing is needed to reach the necessary Technology Readiness Level (TRL) for commercial exploitation. Furthermore, performance models need to be developed in the case of conventional solutions. These should consider the potential long-term climate implications and scenarios, as well as the eco-evolutionary mechanisms that underpin the capacity of the ecosystems under study to perform and recover or adapt to major perturbations [88]. Another important issue is the documentation of their socio-economic and environmental contribution/benefits. This is directly related not only to decision-making in terms of planning, but also to their effectiveness and impact as part of policy measures and strategies. The social and cultural implications should be studied and analyzed before designing such systems. However, the problem with current evidence for the cost-effectiveness of constructed wetlands as NbSs is that appraisals underestimate the economic benefits of working with nature, especially over the long term. In this perspective, non-monetary benefits (e.g., carbon sequestration, education) are difficult to monetize, or there is high uncertainty about their non-market value. Furthermore, appraisals rarely factor in trade-offs among different interventions and ecosystem services or between stakeholder groups, which may experience the costs and benefits of NbSs differently. For example, the importance of wet field margins is different for farmers, local civil society, visitors, etc., reflecting differences in the extent of dependency on natural resources [77]. Additional challenges appear also in terms of governance. NbSs often involve multiple actions taking place over broad landscapes and seascapes, crossing jurisdictional boundaries. For example, constructed wetlands as buffer zones for non-point source pollution control in rural landscapes require collective action across different levels of decision-making (e.g., local and regional) and among multiple ministries (e.g., agriculture, environment, finance). Therefore, such efforts require cooperation and coordination between stakeholders whose priorities, interests, or values may not align [91].

6. Conclusions

The COVID-19 pandemic reminded people that the pressures of humankind on the planet has disrupted the balance and resilience of natural systems [92]. The interconnections of the pandemic to human and environmental health, including food systems, indicate the need to increase the levels of resilience and the preparedness against disturbances [93]. The pandemic revealed several structural shortcomings regarding the production and access to healthy products, the resilience of agri-food systems, and their relation to environmental health and sustainability [11]. On the other hand, the COVID-19 pandemic opened an opportunity window for the reformation of economies and the transition towards a greener model of development. As massive programs and mitigation measures are launched for the recovery of economies after the pandemic [94], it is important to ensure that sustainability plays a central role in the post-COVID-19 era [95,96]. This represents a critical intervention point in which NbSs could be effectively embedded within strategies and policies regarding sustainable land-use planning and development, resource efficiency, and environmental management, as well as social interventions in support of a green economy. In light of the transformation process towards a green and sustainable post-COVID-19 economy, there are several potential applications for constructed wetlands in the agri-food supply chains as Nature-based Solutions, with multiple environmental and socioeconomic benefits. In this perspective, relevant research activities should be further strengthened to address post-COVID-19 environmental challenges within a broader water–food–energy nexus framework, and social and economic aspects of their operational application as multifunctional systems should be further explored.

Author Contributions: Conceptualization, V.T.; investigation, E.P.; resources, E.P. and D.S.; writing—review and editing, V.T., E.P. and D.S; supervision, V.T. All authors have read and agreed to the published version of the manuscript.

Funding: This research was funded by the Hellenic Foundation for Research and Innovation (HFRI) and the General Secretariat for Research and Innovation (GSRI) under grant agreement No. 1928.

Institutional Review Board Statement: Not applicable.

Informed Consent Statement: Not applicable.

Data Availability Statement: Not applicable.

Conflicts of Interest: The authors declare no conflict of interest.

References

1. Markandya, A.; Salcone, J.; Hussain, S.; Mueller, A.; Thambi, S. COVID, the Environment and Food Systems: Contain, Cope and Rebuild Better. *Front. Environ. Sci.* **2021**, *9*, 674432. [CrossRef]
2. Gillard, R.; Gouldson, A.; Paavola, J.; Van Alstine, J. Transformational responses to climate change: Beyond a systems perspective of social change in mitigation and adaptation. *Wiley Interdisc. Rev. Clim. Chang.* **2016**, *7*, 251–265. [CrossRef]
3. Feola, G. Societal transformation in response to global environmental Change: A review of emerging concepts. *Ambio* **2015**, *44*, 376–390. [CrossRef] [PubMed]
4. Murti, R.; Sheikholeslami, D. *Nature-Based Solutions for Recovery—Opportunities, Policies and Measures*; Technical Paper No. 2; IUCN Nature-Based Recovery Initiative: Gland, Switzerland, 2020.
5. United Nations (UN). Wetlands at the Center of Nature-Based Solutions. 2018. Available online: https://www.unwater.org/wetlands-at-the-center-of-nature-based-solutions/ (accessed on 22 November 2021).
6. Nagabhatla, N.; Metcalfe, C.D. (Eds.) *Multifunctional Wetlands, Environmental Contamination Remediation and Management*; Springer International Publishing: Cham, Switzerland, 2018.
7. Ghermandi, A.; Fichtman, E. Cultural ecosystem services of multifunctional constructed treatment wetlands and waste stabilization ponds: Time to enter the mainstream? *Ecol. Eng.* **2015**, *84*, 615–623. [CrossRef]
8. Bazeley, P. *Qualitative Data Analysis with NVivo*; SAGE: Los Angeles, CA, USA; London, UK, 2007.
9. Bernard, H.R.; Ryan, G.W. *Analyzing Qualitative Data: Systematic Approaches*; Corwin: Thousand Oaks, CA, USA, 2010.
10. Meemken, E.M.; Barrett, C.B.; Michelson, H.C.; Qaim, M.; Reardon, T.; Sellare, J. Sustainability standards in global agrifood supply chains. *Nat. Food* **2021**, *2*, 758–765. [CrossRef]
11. Barrett, C.B.; Fanzo, J.; Herrero, M.; Mason-D'Croz, D.; Mathys, A.; Thornton, P.; Wood, S.; Benton, T.G.; Fan, S.; Lawson-Lartego, L.; et al. COVID-19 pandemic lessons for agri-food systems innovation. *Environ. Res. Lett.* **2021**, *16*, 101001. [CrossRef]
12. Kotler, P. The Consumer in the Age of Coronavirus. *J. Creat. Value* **2020**, *6*, 12–15. [CrossRef]
13. Mehta, S.; Saxena, T.; Purohit, N. The New Consumer Behaviour Paradigm amid COVID-19: Permanent or Transient? *J. Health Manag.* **2020**, *22*, 291–301. [CrossRef]
14. Vázquez-Martínez, U.J.; Morales-Mediano, J.; Leal-Rodríguez, A.L. The impact of the COVID-19 crisis on consumer purchasing motivation and behavior. *Eur. Res. Manag. Bus. Econ.* **2021**, *27*, 100166. [CrossRef]
15. FAO. *Climate Change and Food Security: Risks and Responses*; Food and Agriculture Organization of the United Nations (FAO): Rome, Italy, 2015; pp. 1–122; ISBN 978-92-5-108998-9. Available online: https://www.fao.org/3/i5188e/i5188e.pdf (accessed on 10 December 2021).
16. Yoro, K.O.; Daramola, M.O. CO_2 emission sources, greenhouse gases, and the global warming effect. In *Advances in Carbon Capture*; Elsevier: Amsterdam, The Netherlands, 2020; pp. 3–28.
17. Zambrano-Monserrate, M.A.; Ruanob, M.A.; Sanchez-Alcalde, L. Indirect effects of COVID-19 on the environment. *Sci. Total Environ.* **2020**, *728*, 138813. [CrossRef] [PubMed]
18. FAO. *Circular Economy: Waste-to-Resource & COVID-19*; Info Sheet; Food and Agriculture Organization of the United Nations (FAO): Rome, Italy, 2021. Available online: https://www.fao.org/land-water/overview/covid19/circular/en/ (accessed on 12 November 2021).
19. Mu, W.; van Asselt, E.D.; van der Fels-Klerx, H.J. Towards a resilient food supply chain in the context of food safety. *Food Control* **2021**, *125*, 107953. [CrossRef]
20. European Environment Agency (EEA). *Nature-Based Solutions in Europe: Policy, Knowledge and Practice for Climate Change Adaptation and Disaster Risk Reduction*; European Environment Agency Report 01/2021; Publications Office of the European Union: Luxembourg, 2021. [CrossRef]
21. Usman, M.; Husnain, M.; Riaz, A.; Riaz, A.; Yameen, A. Climate change during the COVID-19 outbreak: Scoping future perspectives. *Environ. Sci. Pollut. Res.* **2021**, *28*, 49302–49313. [CrossRef] [PubMed]
22. Hoang, A.T.; Nižetić, S.; Olcer, A.I.; Ong, H.C.; Chen, W.H.; Chong, C.T.; Sabu, T.; Suhaib, A.B.; Nguyen, X.P. Impacts of COVID-19 pandemic on the global energy system and the shift progress to renewable energy: Opportunities, challenges, and policy implications. *Energy Policy* **2021**, *154*, 112322. [CrossRef] [PubMed]

23. Yang, M.; Chen, L.; Msigwa, G.; Ho, K.; Tang, D.; Yap, P.S. Implications of COVID-19 on global environmental pollution and carbon emissions with strategies for sustainability in the COVID-19 era. *Sci. Total Environ.* **2021**, *809*, 151657. [CrossRef]
24. G20. Towards a More Resource-Efficient and Circular Economy. Background Report Prepared for the 2021 G20 Presidency of Italy. 2021. Available online: https://www.oecd.org/environment/waste/OECD-G20-Towards-a-more-Resource-Efficient-and-Circular-Economy.pdf (accessed on 11 December 2021).
25. Morand, S.; Lajaunie, C. Biodiversity and COVID-19: A report and a long road ahead to avoid another pandemic. *One Earth* **2021**, *4*, 920–923. [CrossRef] [PubMed]
26. Damarad, T.; Bekker, G.J. *COST 341—Habitat Fragmentation Due to Transportation Infrastructure: Findings of the COST Action 341*; Office for Official Publications of the European Communities: Luxembourg, 2003.
27. Ramos, J.S.; Feria-Toribio, J.M. Assessing the effectiveness of protected areas against habitat fragmentation and loss: A long-term multi-scalar analysis in a Mediterranean region. *J. Nat. Conserv.* **2021**, *64*, 126072. [CrossRef]
28. Lehmann, P.; Beck, S.; de Brito, M.M.; Gawel, E.; Groß, M.; Haase, A.; Lepenies, R.; Otto, D.; Schiller, J.; Strunz, S.; et al. Environmental Sustainability Post-COVID-19: Scrutinizing Popular Hypotheses from a Social Science Perspective. *Sustainability* **2021**, *13*, 8679. [CrossRef]
29. Dumitru, A.; Wendling, L. (Eds.) *Evaluating the Impact of Nature-Based Solutions: A Handbook for Practitioners*; European Commission Directorate-General for Research and Innovation Healthy Planet—Climate and Planetary Boundaries, Publications Office of the European Union: Luxembourg, 2021.
30. European Commission (EC). *Horizon 2020 Work Programme 2016-2017–12: Climate Action, Environment, Resource Efficiency and Raw Materials*; European Commission: Brussels, Belgium, 2016.
31. Raymond, C.M.; Frantzeskaki, N.; Kabisch, N.; Berry, P.; Breile, M.; Nita, M.R.; Geneletti, D.; Calfapietra, C. A framework for assessing and implementing the co-benefits of nature-based solutions in urban areas. *Environ. Sci. Policy* **2017**, *77*, 15–24. [CrossRef]
32. Cohen-Shacham, E.; Andrade, A.; Dalton, J.; Dudley, N.; Jones, M.; Kumar, C.; Maginnis, S.; Maynard, S.; Nelson, C.R.; Renaud, F.G.; et al. Core principles for successfully implementing and upscaling Nature-based Solutions. *Environ. Sci. Policy* **2019**, *98*, 20–29. [CrossRef]
33. United Nations World Water Assessment Programme (UN-Water). *The United Nations World Water Development Report 2018: Nature-Based Solutions for Water*; UNESCO: Paris, France, 2018.
34. Eggermont, H.; Balian, E.; Azevedo, J.M.N.; Beumer, V.; Brodin, T.; Claudet, J.; Fady, B.; Grube, M.; Keune, H.; Lamarque, P.; et al. Nature-based solutions: New influence for environmental management and research in Europe. *Gaia* **2015**, *24*, 243–248. [CrossRef]
35. Science for Environment Policy. *Future Brief: The Solution Is in Nature*; Future Brief 24; Brief Produced for the European Commission DG Environment; Science Communication Unit, UWE Bristol: Bristol, UK, 2021.
36. Miralles-Wilhelm, F. *Nature-Based Solutions in Agriculture: Sustainable Management and Conservation of Land, Water, and Biodiversity*; The United Nations Food and Agriculture Organization: Rome, Italy; The Nature Conservancy: Arlington, VA, USA, 2021.
37. Sonneveld, B.G.J.S.; Merbis, M.D.; Alfarra, A.; Ünver, O.; Arnal, M.A. *Nature-Based Solutions for Agricultural Water Management and Food Security*; FAO Land and Water Discussion Paper No. 12; FAO: Rome, Italy, 2018; p. 66.
38. Sartison, K.; Artmann, M. Edible cities—An innovative nature-based solution for urban sustainability transformation? An explorative study of urban food production in German cities. *Urban For. Urban Green.* **2020**, *49*, 126604. [CrossRef]
39. Di Crosta, A.; Ceccato, I.; Marchetti, D.; La Malva, P.; Maiella, R.; Cannito, L.; Cipi, M.; Mammarella, N.; Palumbo, R.; Verrocchio, M.C.; et al. Psychological factors and consumer behavior during the COVID-19 pandemic. *PLoS ONE* **2021**, *16*, e0256095. [CrossRef] [PubMed]
40. Durante, K.M.; Laran, J. The effect of stress on consumer saving and spending. *J. Mark. Res.* **2016**, *53*, 814–828. [CrossRef]
41. Hayek, M.; Harwatt, H.; Ripple, W.; Mueller, N. The carbon opportunity cost of animal-sourced food production on land. *Nat. Sustain.* **2021**, *4*, 21–24. [CrossRef]
42. Hartig, T.; van den Berg, A.E.; Hagerhall, M. Health Benefits of Nature Experience: Psychological, Social and Cultural Processes. In *Forests, Trees and Human Health*; Nilsson, K., Sangster, M., Gallis, C., Hartig, T., de Vries, S., Seeland, K., Schipperijn, J., Eds.; Springer: Berlin, Germany, 2011; pp. 127–168. [CrossRef]
43. Loomis, D.; Grosse, Y.; Lauby-Secretan, B.; El Ghissassi, F.; Bouvard, V.; Benbrahim-Tallaa, L.; Guha, N.; Baan, R.; Mattock, H.; Straif, K. The carcinogenicity of outdoor air pollution. *Lancet Oncol.* **2013**, *14*, 1262–1263. [CrossRef] [PubMed]
44. Kopsieker, L.; Gerritsen, E.; Stainforth, T.; Lucic, A.; Costa Domingo, G.; Naumann, S.; Röschel, L.; Davis, M. *Nature-Based Solutions and Their Socio-Economic Benefits for Europe's Recovery: Enhancing the Uptake of Nature-Based Solutions across EU Policies*; Policy Briefing by the Institute for European Environmental Policy (IEEP) and the Ecologic Institute; Institute for European Environmental Policy: Brussels, Belgium, 2021.
45. International Labour Organization (ILO). *Nature Hires: How Nature-Based Solutions Can Power a Green Jobs Recovery*; WWF: Gland, Switzerland; ILO: Geneva, Switzerland, 2020. Available online: https://www.ilo.org/wcmsp5/groups/public/---ed_emp/documents/publication/wcms_757823.pdf (accessed on 22 November 2021).
46. Charveriat, C.; Brzeziński, B.; Filipova, T.; Ramírez, O. Mental Health and the Environment: Bringing Nature Back into People's Lives. 2021. Available online: https://ieep.eu/uploads/articles/attachments/c2cc2d58-d8a0-4dee-b45e-57a7dfa2620d/Mental%20health%20and%20environment%20pol-icy%20brief%20(IEEP%20&%20ISGLOBAL%202021).pdf?v=63778955421 (accessed on 12 October 2021).

47. Xu, X.; Chen, M.; Yang, G.; Jiang, B.; Zhang, J. Wetland ecosystem services research: A critical review. *Glob. Ecol. Conserv.* **2020**, *22*, e01027. [CrossRef]
48. Metcalfe, C.D.; Nagabhatla, N.; Fitzgerald, S.K. Multifunctional Wetlands: Pollution Abatement by Natural and Constructed Wetlands. In *Multifunctional Wetlands, Environmental Contamination Remediation and Management*; Nagabhatla, N., Metcalfe, C.D., Eds.; Springer International Publishing: Cham, Switzerland, 2018; pp. 1–14. [CrossRef]
49. European Commission (EC). *Ecosystem Services and Green Infrastructure*; European Commission: Brussels, Belgium, 2020; Available online: http://ec.europa.eu/environment/nature/ecosystems/index_en.htm (accessed on 20 December 2021).
50. Muñoz Castillo, R.; Crisman, T. *The Role of Green Infrastructure in Water, Energy and Food Security in Latin America and the Caribbean: Experiences, Opportunities and Challenges*; IDB Discussion Paper 693; IDB Water and Sanitation Division, Inter-American Development Bank: Washington, DC, USA, 2019.
51. Kabisch, N.; KornJutta, H.; Stadler, J.; Bonn, A. (Eds.) *Nature-Based Solutions to Climate Change Adaptation in Urban Areas: Linkages between Science, Policy and Practice*; Springer International Publishing: Cham, Switzerland, 2017.
52. European Commission (EC). *Nature-Based Solutions*; European Commission: Brussels, Belgium, 2020; Available online: https://ec.europa.eu/info/research-and-innovation/research-area/environment/naturebased-solutions_en (accessed on 10 October 2021).
53. Woodward, B.; Tanner, C.C.; McKergow, L.; Sukias, J.P.S.; Matheson, F.E. *Diffuse Source Agricultural Sediment and Nutrient Attenuation by Constructed Wetlands: A Systematic Literature Review to Support Development of Guidelines*; NIWA report to DairyNZ; NIWA: Hamilton, New Zealand, 2020.
54. Plakas, K.; Karabelas, A.; Takavakoglou, V.; Chatzis, V.; Oller, I.; Polo-López, M.I.; Al-Naboulsi, T.; El Moll, A.; Kallali, H.; Mensi, K.; et al. Development and demonstration of an eco-innovative system for sustainable treatment and reuse of municipal wastewater in small and medium size communities in the Mediterranean region. In Proceedings of the 17th International Conference on Environmental Science and Technology (CEST 2021), Athens, Greece, 1–4 September 2021.
55. Liu, D.; Wu, X.; Chang, J.; Gu, B.; Min, Y.; Ge, Y.; Shi, Y.; Xue, H.; Peng, C.; Wu, J. Constructed wetlands as biofuel production systems. *Nat. Clim. Chang.* **2012**, *2*, 190–194. [CrossRef]
56. Wang, Y.; Zhao, Y.; Xu, L.; Wang, W.; Doherty, L.; Tang, C.; Ren, B.; Zhao, J. Constructed wetland integrated microbial fuel cell system: Looking back, moving forward. *Water Sci. Technol.* **2017**, *76*, 471–477. [CrossRef] [PubMed]
57. Hultberg, M.; Prade, T.; Bodin, H.; Vidakovic, A.; Asp, H. Adding benefit to wetlands—Valorization of harvested common reed through mushroom production. *Sci. Total Environ.* **2018**, *637–638*, 1395–1399. [CrossRef]
58. Chen, R.Z.; Wong, M.H. Integrated wetlands for food production. *Environ. Res.* **2016**, *148*, 429–442. [CrossRef]
59. Avellan, C.T.; Ardakanian, R.; Gremillion, P. The role of constructed wetlands for biomass production within the water-soil-waste nexus. *Water Sci. Technol.* **2017**, *75*, 2237–2245. [CrossRef] [PubMed]
60. Zhang, C.; Wen, L.; Wang, Y.; Liu, C.; Zhou, Y.; Lei, G. Can Constructed Wetlands be Wildlife Refuges? A Review of Their Potential Biodiversity Conservation Value. *Sustainability* **2020**, *12*, 1442. [CrossRef]
61. Ionescu, D.T.; Hodor, C.V.; Petritan, I.C. Artificial Wetlands as Breeding Habitats for Colonial Waterbirds within Central Romania. *Diversity* **2020**, *12*, 371. [CrossRef]
62. Belle, J.A.; Collins, N.; Jordaan, A. Managing wetlands for disaster risk reduction: A case study of the eastern Free State, South Africa. *Jamba* **2018**, *10*, 400. [CrossRef] [PubMed]
63. Tanner, C.; Sukias, J.; Woodward, B. *Technical Guidelines for Constructed Wetland Treatment of Pastoral Farm Run-Off*; NIWA Report No: 20200208.120200208.1HN for DairyNZ; National Institute of Water & Atmospheric Research Ltd.: Hamilton, New Zealand, 2021.
64. Stefanakis, A. The Role of Constructed Wetlands as Green Infrastructure for Sustainable Urban Water Management. *Sustainability* **2019**, *11*, 6981. [CrossRef]
65. Petitjean, A.; Forquet, N.; Choubert, J.M.; Coquery, M.; Bouyer, M.; Boutin, C. Land characterisation for soil-based constructed wetlands: Adapting investigation methods to design objectives. *Water Pract. Technol.* **2015**, *10*, 660–668. [CrossRef]
66. Sehar, S.; Nasser, H.A.A. Wastewater treatment of food industries through constructed wetland: A review. *Int. J. Environ. Sci. Technol.* **2019**, *16*, 6453–6472. [CrossRef]
67. Addo-Bankas, O.; Zhao, Y.; Vymazal, J.; Yuan, Y.; Fu, J.; Wei, T. Green walls: A form of constructed wetland in green buildings. *Ecol. Eng.* **2021**, *169*, 106321. [CrossRef]
68. Riggio, V.A.; Ruffino, B.; Campo, G.; Comino, E.; Comoglio, C.; Zanetti, M. Constructed wetlands for the reuse of industrial wastewater: A case-study. *J. Clean. Prod.* **2018**, *171*, 723–732. [CrossRef]
69. Coletta, V.R.; Pagano, A.; Pluchinotta, I.; Fratino, U.; Scrieciu, A.; Nanu, F.; Giordano, R. Causal Loop Diagrams for supporting Nature Based Solutions participatory design and performance assessment. *J. Environ. Manag.* **2021**, *280*, 111668. [CrossRef]
70. Yadav, A.K.; Srivastava, P.; Kumar, N.; Abbassi, R.; Mishra, B.K. Constructed Wetland-Microbial Fuel Cell: An Emerging Integrated Technology for Potential Industrial Wastewater Treatment and Bio-Electricity Generation. In *Constructed Wetlands for Industrial Wastewater Treatment*; Stefanakis, A., Ed.; John Wiley & Sons: Chichester, UK, 2018. [CrossRef]
71. Shutes, R.B.E.; Ellis, J.B.; Revitt, D.M.; Forshaw, M.; Winter, B. Chapter 20—Urban and Highway Runoff Treatment by Constructed Wetlands. In *Developments in Ecosystems, Wetlands Ecosystems in Asia*; Wong, M.H., Ed.; Elsevier: Amsterdam, The Netherlands, 2004; Volume 1, pp. 361–382. ISBN 9780444516916. [CrossRef]

72. Rosli, F.A.; Lee, K.E.; Choo, T.G.; Mokhtar, M.; Latif, M.T.; Goh, T.; Simon, N. The use of constructed wetlands in sequestrating carbon: An overview. *Nat. Environ. Pollut. Technol.* **2017**, *16*, 813–819.
73. Jeroen, J.M.; de Klein, A.; van der Werf, K. Balancing carbon sequestration and GHG emissions in a constructed wetland. *Ecol. Eng.* **2014**, *66*, 36–42. [CrossRef]
74. Were, D.; Kansiime, F.; Fetahi, T.; Cooper, A.; Jjuuko, C. Carbon Sequestration by Wetlands: A Critical Review of Enhancement Measures for Climate Change Mitigation. *Earth Syst. Environ.* **2019**, *3*, 327–340. [CrossRef]
75. Webb, B.; Dix, B.; Douglass, S.; Asam, S.; Cherry, C.; Buhring, B. *Nature-Based Solutions for Coastal Highway Resilience: An Implementation Guide*; Report No. FHWA-HEP-19-042; US Department of Transportation-Federal Highway Administration: Washington, DC, USA, 2019.
76. Sandström, U.G.; Elander, I. Biodiversity, road transport and urban planning: A Swedish local authority facing the challenge of establishing a logistics hub adjacent to a Natura 2000 site. *Prog. Plann.* **2021**, *148*, 100463. [CrossRef]
77. Takavakoglou, V.; Georgiadis, A.; Pana, E.; Georgiou, P.E.; Karpouzos, D.K.; Plakas, K.V. Screening Life Cycle Environmental Impacts and Assessing Economic Performance of Floating Wetlands for Marine Water Pollution Control. *J. Mar. Sci. Eng.* **2021**, *9*, 1345. [CrossRef]
78. Asian Development Bank. *From Toilets to Rivers: Experiences New Opportunities, and Innovative Solutions*; Asian Development Bank: Mandaluyong City, Philippines, 2014; p. 100. ISBN 978-92-9254-460-7.
79. Bakhshoodeh, R.; Alavi, N.; Oldham, C.; Santos, R.M.; Babaei, A.A.; Vymazal, J.; Paydary, P. Constructed wetlands for landfill leachate treatment: A review. *Ecol. Eng.* **2020**, *146*, 105725. [CrossRef]
80. World Bank. East Asia and the Pacific Region Urban Sanitation Review. No. 84290. Available online: https://documents1.worldbank.org/curated/en/771821468036884616/pdf/842900WP0P12990Box0382136B00PUBLIC0.pdf (accessed on 29 December 2021).
81. Masi, F.; Rizzo, A.; Regelsberger, M. The role of constructed wetlands in a new circular economy, resource oriented, and ecosystem services paradigm. *J. Environ. Manag.* **2018**, *216*, 275–284. [CrossRef] [PubMed]
82. Volkan Oral, H.; Carvalho, P.; Gajewska, M.; Ursino, N.; Masi, F.; van Hullebusch, E.D.; Kazak, J.K.; Expositoh, A.; Cipolletta, G.; Andersen, T.R.; et al. A review of nature-based solutions for urban water management in European circular cities: A critical assessment based on case studies and literature. *Blue-Green Syst.* **2020**, *2*, 112–136. [CrossRef]
83. Alikhani, S.; Nummi, P.; Ojala, A. Urban Wetlands: A Review on Ecological and Cultural Values. *Water* **2021**, *13*, 3301. [CrossRef]
84. Rahmani, D.R.; Wahyunah, W. Urban Floating Farming: The Alternative of Valuable Private Green Space for Urban Communities in the Wetland Area. *ESE Int. J. Environ. Sci. Eng.* **2019**, *2*, 1–3.
85. Hettiarachchi, M.; Morrison, T.H.; McAlpine, C. Forty-three years of Ramsar and urban wetlands. *Glob. Environ. Change* **2015**, *32*, 57–66. [CrossRef]
86. White, M.P.; Elliott, L.R.; Gascon, M.; Roberts, B.; Fleming, L.E. Blue space, health and well-being: A narrative overview and synthesis of potential benefits. *Environ. Res.* **2020**, *191*, 110169. [CrossRef]
87. IPBES. Summary for Policymakers of the Global Assessment Report on Biodiversity and Ecosystem Services, Intergovernmental Science-Policy Platform on Biodiversity and Ecosystem Services. 2019. Available online: https://zenodo.org/record/3553579 (accessed on 6 October 2020).
88. Seddon, N.; Chausson, A.; Berry, P.; Girardin, C.A.J.; Smith, A.; Turner, B. Understanding the value and limits of nature-based solutions to climate change and other global challenges. *Philos. Trans. R. Soc.* **2020**, *375*, 20190120. [CrossRef] [PubMed]
89. European Commission (EC). *Horizon Europe Strategic Plan 2021–2024*; Publications Office of the European Union: Luxembourg, 2021; p. 101.
90. European Commission (EC). List of Potential Agricultural Practices that Eco-Schemes Could Support. 2021. Available online: https://ec.europa.eu/info/sites/default/files/food-farming-fisheries/key_policies/documents/factsheet-agri-practices-under-ecoscheme_en.pdf (accessed on 30 November 2021).
91. Dale, P.; Sporne, I.; Knight, J.; Sheaves, M.; Eslami-Andergoli, L.; Dwyer, P. A conceptual model to improve links between science, policy and practice in coastal management. *Mar. Policy* **2019**, *103*, 42–49. [CrossRef]
92. Zabaniotou, A. A systemic approach to resilience and ecological sustainability during the COVID-19 pandemic: Human, societal, and ecological health as a system-wide emergent property in the Anthropocene. *Glob. Transit.* **2020**, *2*, 116–126. [CrossRef] [PubMed]
93. Gordon, L. The COVID-19 pandemic stress the need to build resilient production ecosystems. *Agric. Hum. Values* **2020**, *37*, 645–646. [CrossRef] [PubMed]
94. Vivid Economics. Green Stimulus Index. An Assessment of the Orientation of COVID-19 Stimulus in Relation to Climate Change, Biodiversity and Other Environmental Impacts. 2021. Available online: https://www.vivideconomics.com/wp-content/uploads/2021/07/Green-Stimulus-Index-6th-Edition_final-report.pdf (accessed on 2 September 2021).
95. Guerriero, C.; Haines, A.; Pagano, M. Health and sustainability in post-pandemic economic policies. *Nat. Sustain.* **2020**, *3*, 494–496. [CrossRef]
96. Rosenbloom, D.; Markard, J. A COVID-19 recovery for climate. *Science* **2020**, *368*, 447. [CrossRef]

 sustainability

Article

Trends in Food Innovation: An Interventional Study on the Benefits of Consuming Novel Functional Cookies Enriched with Olive Paste

Olga Papagianni [1], Iraklis Moulas [1], Thomas Loukas [2], Athanasios Magkoutis [2], Dimitrios Skalkos [3], Dimitrios Kafetzopoulos [4], Charalampia Dimou [1], Haralabos C. Karantonis [5] and Antonios E. Koutelidakis [1,*]

1 Laboratory of Nutrition and Public Health, Unit of Human Nutrition, Department of Food Science and Nutrition, University of the Aegean, 81400 Myrina, Greece; olga3_pap@yahoo.gr (O.P.); fns17077@fns.aegean.gr (I.M.); chadim@aegean.gr (C.D.)
2 Outpatient Clinic, 81400 Myrina, Greece; tloukas2002@yahoo.com (T.L.); tmagoutis@gmail.com (A.M.)
3 Laboratory of Food Chemistry, Department of Chemistry, University of Ioannina, 45110 Ioannina, Greece; dskalkos@uoi.gr
4 Department of Business Administration, University of Macedonia, 54636 Thessaloniki, Greece; dimkafe@uom.edu.gr
5 Laboratory of Food Chemistry, Biochemistry and Technology, Department of Food Science and Nutrition, University of the Aegean, 81400 Myrina, Greece; chkarantonis@aegean.gr
* Correspondence: akoutel@aegean.gr

Citation: Papagianni, O.; Moulas, I.; Loukas, T.; Magkoutis, A.; Skalkos, D.; Kafetzopoulos, D.; Dimou, C.; Karantonis, H.C.; Koutelidakis, A.E. Trends in Food Innovation: An Interventional Study on the Benefits of Consuming Novel Functional Cookies Enriched with Olive Paste. *Sustainability* **2021**, *13*, 11472. https://doi.org/10.3390/su132011472

Academic Editor: Mariarosaria Lombardi

Received: 18 July 2021
Accepted: 13 October 2021
Published: 17 October 2021

Publisher's Note: MDPI stays neutral with regard to jurisdictional claims in published maps and institutional affiliations.

Abstract: Olive paste may exert bioactivity due to its richness in bioactive components, such as oleic acid and polyphenols. The present interventional human study investigated if the fortification of cookies with olive paste and herbs may affect postprandial lipemia, oxidative stress, and other biomarkers in healthy volunteers. In a cross-over design, 10 healthy volunteers aged 20–30 years, consumed a meal, rich in fat and carbohydrates (50 g cookies). After a washout week, the same volunteers consumed the same cookie meal, enhanced with 20% olive paste. Blood sampling was performed before, 0.5 h, 1.5 h, and 3 h after eating. Total plasma antioxidant capacity according to FRAP, ABTS, and resistance to copper-induced plasma oxidation, serum lipids, glucose, uric acid, and antithrombotic activity in platelet-rich plasma were determined at each timepoint. There was a significant decrease in triglycerides' concentration in the last 1.5 h in the intervention compared to the control group ($p < 0.05$). A tendency for a decrease in glucose levels and an increase in the plasma antioxidant capacity was observed 0.5 h and 1.5 h, respectively, in the intervention compared to the control group. The remaining biomarkers did not show statistically significant differences ($p > 0.05$). More clinical and epidemiological studies in a larger sample are necessary in order to draw safer conclusions regarding the effect of olive paste on metabolic biomarkers, with the aim to enhance the industrial production of innovative functional cookies with possible bioactivity.

Keywords: postprandial bioactivity; bioactive compounds; metabolic biomarkers; functional cookies; olive paste

1. Introduction

COVID-19 pandemic has generated opportunities and challenges for the production of innovative functional foods and nutraceuticals containing bioactive compounds, and highlighted the advances of nutrition to boost consumers' immune system and improve their overall health [1]. Concerning that food companies are seeking new innovative products to cover consumer needs, the production and commercialization of food such as cookies rich with bioactive compounds will increase the interest of health-conscious individuals. Inflammation and oxidative stress are related to diet and play an essential role in the pathogenesis of chronic diseases such as diabetes, cardiovascular disease, neurodegenerative diseases, and cancer. The modern lifestyle associated with COVID-19 pandemic and unhealthy dietary patterns, exposure to a wide range of chemicals and lack of physical activity, seems to play a significant role in the induction of oxidative stress and inflammation [2].

The postprandial state is a normal, metabolic process that takes place throughout the day and involves multiple mechanisms. It refers to the state after meal consumption, when the digestion and absorption of nutrients are completed (duration 6–12 h) [3]. Numerous factors related to dietary intake have been associated with the activation of the endogenous or innate immune system, which is followed by the induction of a mild inflammatory response. Consumption of a meal, rich in fats or carbohydrates or their complex, may lead to the promotion of oxidative stress [3]. Postprandial lipemia and hyperglycemia may cause vascular damage by molecular mechanisms, including endothelial dysfunction, activation of molecules' adhesion, hemostasis activation, oxidative stress, or genetic polymorphisms that affect genes involved in lipoprotein metabolism [3].

After a meal intake and under normal conditions, there is a rapid increase in plasma glucose concentration, while the rate of glucose absorption is higher than the rate of endogenous glucose production. Important factors affecting the postprandial glycemia, are the amount and duration of food intake, carbohydrate content, glucose uptake rate and insulin resistance. Glucose levels 1–2 h after ingestion is an important factor in predicting the risk of cardiovascular disease, as concentrations higher than 7.8 mmol/L mean hyperglycemia [4,5]. In addition, after a meal consumption, plasma is enriched with triglyceride lipoproteins (TRL) of intestinal (chylomicron) or hepatic (very low-density lipoprotein, VLDL) origin [6].

Both postprandial hyperglycemia and hyperlipidemia are involved in inhibiting oxidative phosphorylation of mitochondria, allowing the passage of peroxide anions into the circulation and leading to increased ROS production by leukocytes. The main factor that affects the degree of oxidative stress after eating a meal is the amount of caloric intake [7]. Scientific evidence suggests that low-grade inflammation and endothelial dysfunction, combined with insulin resistance, are associated with increased risk of metabolic syndrome (MS), cardiovascular disease (CVD), and type 2 diabetes. Scientific data show that oxidation, inflammation, and thrombosis are key mechanisms that coexist and underly to the onset of atherogenesis [8].

Epidemiological and clinical studies suggest that the adoption of a diet rich in fruits, vegetables, raw cereals, fish and dairy products, low in saturated fat, such as the Mediterranean diet, reduces the risk of developing cardiovascular disease. However, the achievement of such a diet requires significant modifications to the wider dietary behavior of the population [9]. The development of innovative foods, such as cookies, enriched with components from the Mediterranean diet, could be an alternative way for diet improvement in parallel with the modern lifestyle. Recent scientific evidence suggests that the consumption of functional foods may have a beneficial effect on reducing the risk of chronic diseases such as obesity, cardiovascular disease, and diabetes [10]. Functional foods are defined as natural or processed foods, which have been shown to contribute to the achievement of specific functional goals within the human body, contributing to the possible prevention of diseases and to general health promotion [11–13]. The enrichment of foods in order to produce functional foods is achieved by adding bioactive compounds, such as phytosterols and stanols, antioxidants (carotenoids, polyphenols), dietary fiber (β-glucans), oligosaccharides (fructans), n-3 long chain fatty acids, etc., as they have been shown to play an important role in both inhibiting intestinal cholesterol absorption and in reduced glycemic response and oxidative stress [9,10]. However, limited number of scientific data are available on the potential postprandial bioactivity of innovative foods, such as cookies, enhanced with bioactive compounds from olive products.

Olive products may reduce the pre-inflammatory status as well as the oxidative damage, which are caused by the LDL cholesterol and free radicals' oxidation, respectively. The antioxidant activity possessed by olive phenolic compounds has been supported by an increase in plasma antioxidant capacity, modifying the lipid profile, and preventing oxidative damage in a group of young and middle-aged healthy volunteers. These results are mainly attributed to the redox properties of phenolic compounds, which allow to act as reducing agents and hydrogen donors [14]. The main bioactive ingredients found in olives are phenolic alcohols, 3,4-DHPEA (hydroxytyrosol) and *p*-HPEA (tyrosol), as well as the secoiridoid derivatives 3,4-DHPEA-EA (an oleuropein aglycon), *p*- HPEA-EA (ligstroside aglycon), 3,4-DHPEA-EDA (oleuropein aglycon di-aldehyde), *p*-HPEA-EDA (ligstroside aglycon de-aldehyde) and oleuropein [15]. Additionally, the oleic acid contained in olive products has been shown to possess a protective effect against insulin resistance, and contribute to the improvement of endothelial dysfunction in response to pro-inflammatory stimuli. Likewise, it has been reported that oleic acid may reduce blood lipids levels, mainly cholesterol, LDL, and triglycerides [15].

The plasma antioxidant levels reflect the general body antioxidant capacity. Total plasma antioxidant capacity is primarily credited to ascorbic acid, α-tocopherol, glutathione, lipoic acid, uric acid and urea, β-carotene, ubiquinone, and bilirubin. Low antioxidants levels in blood have been linked to the pathogenesis of various diseases such as cancer, Alzheimer, Parkinson, diabetes, rheumatoid arthritis, hypertension, heart disease, and aging. In contrast, high concentrations of antioxidants in the blood strengthen the body's defenses against degenerative diseases. The advantages acquired from antioxidants' dietary intake is attributed to the food content of antioxidants with various antioxidant potential that can participate in numerous reactions, neutralizing effectively the free radicals [16]. The nutritional biomarkers' levels, which are determined in biological samples (urine, blood, and other tissues) may reflect either dietary intake levels, the metabolic rate of food components, or nutritional status and could be used as indicators of a person's normal functioning or of a clinical condition [15].

Thus, the development of strong biomarkers is a matter of necessity, as it is presumed that it will contribute to the achievement of an optimal classification of dietary intake or evaluation of the interdependence of nutrition with chronic diseases. Scientific interest in the application of metabolic biomarkers has intensified in recent years, with the aim of discovering further biological elements of dietary intake. Existing findings from dietary interventions in clinical trials suggest that the consumption of refined wheat flour of cookies may cause hyperglycemia, as the source of rapidly digestible starch [17]. Furthermore, it is noteworthy that the lack of scientific data observed, regarding the possible effect of a meal containing cookies enhanced with bioactive compounds, derived from natural food functional sources, on the postprandial state.

The aim of the present interventional study was to investigate if the fortification of cookies with olive paste and herbs may affect postprandial lipemia, oxidative stress, and other biomarkers in healthy volunteers.

2. Materials and Methods

2.1. Subjects

The study protocol was approved by the Ethics Committee at the University of the Aegean and performed in accordance with the Declaration of Helsinki. The study duration was from 1 May to 30 June 2021. All volunteers signed an updated consent form and were informed about the ultimate purpose of the study, the confidentiality of the data obtained, and the voluntary nature of participation. All participants were initially tested using a medical history questionnaire that also included demographic characteristics, level of physical activity, frequency of polyphenol-rich foods, and general habits, as smoking, and alcohol consuming, referring to the last 6 months. Anthropometric measurements, especially measurement of height, weight, and body composition was performed with the use of a suitable body composition analyser (Tanita SC 330).

After an initial screening of 16 potential participants, 11 volunteers were selected according to the inclusion and exclusion criteria and finally 10 healthy volunteers, 4 men and 6 women, aged 20–30 years were recruited to the study. The recruitment of the participants was carried out in a random way via social media and online announcements at Lemnos Island and Lemnos University Department. The study excluded people over the age of 30, in order to achieve the participation of a homogeneous sample of volunteers, those who had received dietary supplements in the last two months, those with a history of chronic diseases including type I and II diabetes (HbA1c > 5%), those who showed heavy smoking (>10 cigarettes/day), abnormal BMI (>25 kg/m^2), and alcohol overdose (>40 g alcohol/day), factors that could lead to unstable conclusions. Prior to the start of the study, participants underwent biochemical blood tests, in collaboration with external physicians, in order to exclude cases with hematological and biochemical profile beyond normal values (cholesterol > 6.8 mM, triglycerides > 2.8 mM, glucose > 6.11 mM).

Subjects were asked to come after a 12 h fast, as well as to avoid taking medication on the day of the study as well as any dietary supplements. They were also asked to abstain from foods high in antioxidants and alcohol for 24 h before the study.

2.2. Treatments

Cookies were supplied by Greek olive company AMALTHIA SA. Each functional cookie weighed a total of 12.5 g, contained 5.4 g of soft wheat flour, 1.6 g of oat flour, 0.08 g of soda, 2.04 g of vegetable margarine, 0.72 g of sugar, 2.5 g of olive paste, as well as 0.12 g garlic, 0.06 g of oregano, and 0.06 g of thyme. The functional meal weighed a total of 50 g, and contained 4 functional cookies, enhanced with 20.0% olive paste, 1.0% garlic, and thyme and oregano at a percentage of 0.5 % for each.

Each control cookie, which weighed a total of 12.5 g, contained 5.4 g of soft wheat flour, 1.6 g of oat flour, 0.08 g of soda, 4.7 g of vegetable margarine, and 0.72 g of sugar. The difference from the functional ones lies in the replacement of the fats, derived from the olive paste from an additional amount of fats, derived from vegetable margarine, while in the control cookies no herbs and spices were added. The control meal weighed a total of 50 g and contained 4 control cookies. Dietary composition of the test meals is shown in Table 1.

Table 1. Nutritional composition of each meal cookies.

Meals' Dietary Composition	Control	Functional
Energy (kcal)	242.32	185.1
Carbohydrates (g)	23.68	24.27
Fat, total (g)	15.04	8.37
Protein (g/kg)	3.2	3.28
Saturated fat (g)	2.84	1.52
Unsaturated fat (g)	6.38	6.52
Cholesterol (mg)	0	0
Dietary fiber, total (g)	1.24	1.59
Sugar, total (g)	2.96	2.96

Figure 1 presents control coolies (a) and functional cookies (b), prior to the baking.

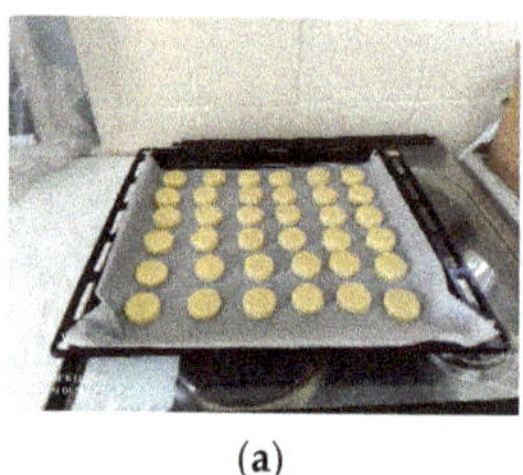
(**a**)

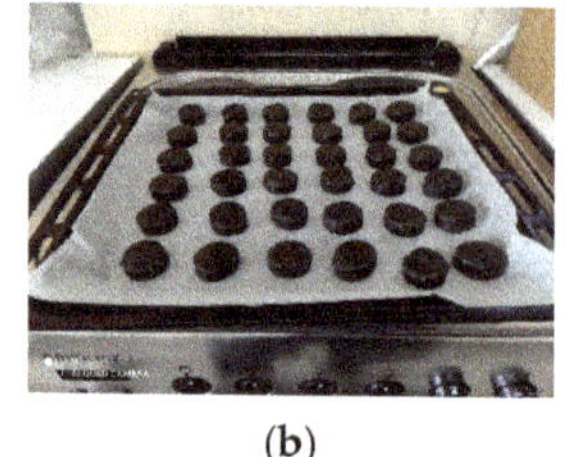
(**b**)

Figure 1. Control (**a**) and Functional (**b**) cookies prior to the baking.

In vitro preliminary studies for the tested cookies have been performed in order to test sensorial acceptability, total phenolics, and antioxidant activity by DPPH, ABTS, FRAP, and CUPRAC assays, and the results have been recorded by Argyri et al. [18].

2.3. Study Design

It was an acute cross-over and two-period, interventional study. All participants on enrollment were randomly assigned to group C:F (Control:Functional-Control:Functional) or to group F:C (Functional:Control-Functional:Control). Individuals who joined the C:F group during the first period received the control meal and during the second trial period ate the functional meal, while participants in the F: C group received the functional meal during the first period and the control meal in the second period. Figure 2 shows the crossover design study illustration.

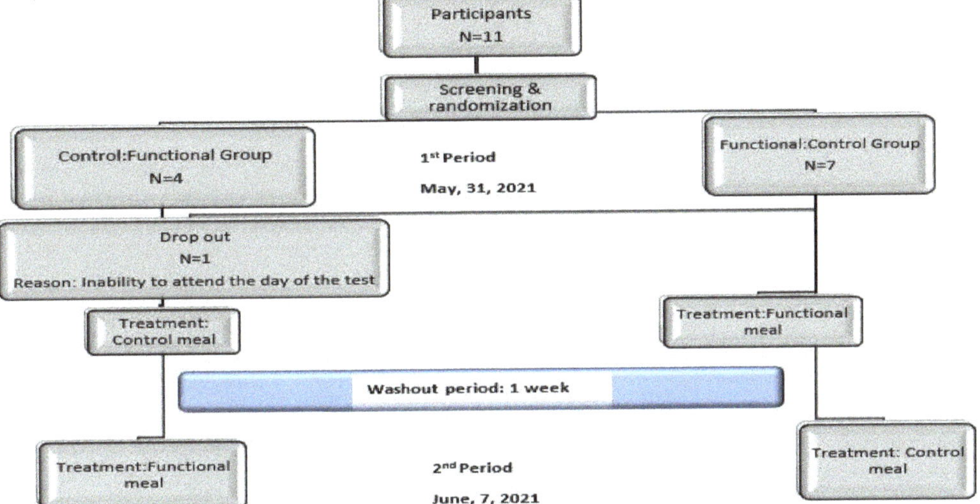

Figure 2. Illustration of study crossover design.

The volunteers arrived at the Human Nutrition Unit at 9 a.m., after a 12 h fast and abstinence from dietary supplements and any medication. The participants were asked to complete a short 24 h recall questionnaire, which recorded all meals eaten in the last 24 h.

A meal consisting of 4 biscuits (50 g) was then offered for consumption, while a glass of water (250 mL) was available for each participant. For the first trial period, volunteers in the C:F group consumed the control cookie meal, while individuals in the F:C group received the functional cookie meal. Respectively, during the second trial period, the participants in the C:F group received the functional cookie meal and those in the F:C group received the control cookie meal.

Ten mL of blood was drawn from all volunteers, shortly before the meal (baseline) and 30 min, 1.5 h and 3 h (Figure 3) after meal consumption. Blood samples were collected in EDTA and citric acid tubes for plasma separation or in heparin tubes for serum separation. Plasma and serum of each volunteer and for each time point were separated by 10-min centrifugation at 20,000× *g* and cooled to −4 °C in a tabletop high speed refrigerated centrifuge immediately after blood collection. Aliquots of plasma or serum were stored at –40 °C until analysis.

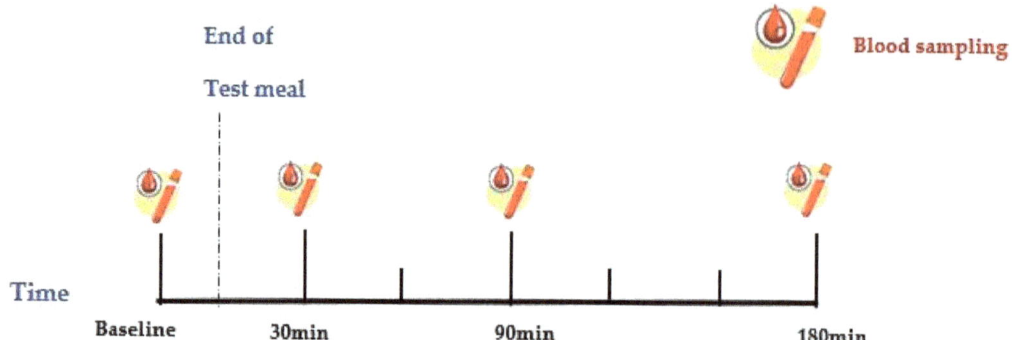

Figure 3. Test visit flow diagram.

2.4. Biochemical Analyses in the Blood Samples

Total Antioxidant Capacity (TAC) was evaluated in plasma by Ferric Reducing Antioxidant Power (FRAP assay), as described by Argyri et al. [18,19]. Antiradical activity was determined by Trolox Equivalent Antioxidant Capacity (TEAC) assay, as previously described in deproteinized plasma according to Prymont-Przyminska et al. [20]. Resistance to copper-induced plasma oxidation was determined according to Sakka and Karantonis [21].

Inhibition of platelet activating factor (PAF)—induced thrombosis in platelet-rich plasma was determined according to Antonopoulou et al. [22].

Total, HDL and LDL cholesterol, triglycerides, glucose and uric acid were measured in serum with an automated analyzer (COBAS c111, Roche, Basel, Switzerland).

2.5. Statistical Analysis

The statistical analysis was performed using SPSS (SPSS V210). The computational power of each sample was calculated for the outcome, as well as the venous plasma antioxidant capacity (TAC) using Statmate version 2.0 (GraphPad Software, Inc., San Diego, CA, USA). Taking $\alpha = 0.01$, the sample of 10–14 individuals allows the detection of a difference of 0.21 mmol TAC/L between the control group and the intervention group, calculated from the expected SD between the differences of the meal group 0.21 mmol/L. The level of statistical significance was at $p < 0.05$. Prior to any statistical analysis, all variables were tested for normal distribution. For the variables that followed normal distribution, repeated post-hoc test ANOVA and Bonferroni test, as well as Wilcoxon sign rank test were performed for the differences between plasma and serum samples at 0.5 h, 1.5 h, and 3 h for each meal group and by change from baseline for venous plasma TAC, TEAC, and copper-induced oxidation, platelet rich plasma thrombosis, and serum biomarkers. Differences between the two treatment groups at any time and time period from the baseline were also examined by the paired *t*-test. For variables that did not follow a normal distribution, Wilcoxon sign rank tests were performed, both for changes in each biomarker and for each treatment group, and for significant differences between the two treatments, at different times, at intervals between blood sampling and at intervals for changes from baseline. The variables LDL and HDL cholesterol, antioxidants, uric acid, ABTS, copper-induced plasma oxidation, and platelet rich plasma thrombosis did not follow a normal distribution. The variables cholesterol, triglycerides, and glucose followed a normal distribution. All data were taken into account for statistical analyses.

3. Results

3.1. Baseline Characteristics

Ten participants completed the study, while one volunteer was unable to attend the study appointments. The volunteers' baseline characteristics during the initial screening are presented in Table 2. No difference was observed between men and women in all tested parameters. The analysis of food frequency questionnaires showed that the majority of the participants consumed fruits 1–2 times a week and vegetables 1–2 times a week, while they stated that in their diet, they included starchy foods 1–2 times a week (data not shown).

Table 2. Volunteers' characteristics at baseline.

	N
Volunteers	11
Men	4
Women	7
Smokers	8
Dietary Supplementation	1
Physical Activity	
Low	5
Medium	4
High	2
	Mean ± SD
Age (years)	22.8 ± 1.6
Weight (kg)	75.2 ± 10.3
Height (cm)	168 ± 6.7
BMI	25.8 ± 6

3.2. *Plasma Total Antioxidant Capacity and Oxidation Resistance*

3.2.1. Plasma Total Antioxidant Capacity

Plasma TAC differed between the two groups for value changes from 1.5 to 3 h after the meal consumption. Venous plasma TAC increased from 1.5 h to 3 h after the functional meal (MD = 0.02, 6.5%) compared to the control meal, where TAC decreased significantly (MD = −0.13, 27.5%, p = 0.05) as shown in Table 3. No other statistically significant differences were observed between the control group and the treatment group.

Table 3. Effects of functional cookies on plasma TAC, serum glucose, and triglycerides levels.

	a. Significant Changes of Each Treatment over Time			
				Mean Difference
Biomarker	p Value [a]	Timepoints' Difference	P (Bonferroni Test) [b]	P (Wilcoxon Test) [c]
Antioxidant Capacity (TAC) (mmol/L)				
Control		30 min–Baseline		0.109
		1.5 h–30 min		0.285
		3 h–1.5 h		0.05 *
Functional		30 min–Baseline		0.109
		1.5 h–30 min		0.109
		3 h–1.5 h		0.285
Glucose(mg/dL)				
Control	0.024 *	Δbaseline–Δ30 min		0.012
		Δ30 min–Δ1.5 h		0.05
		Δ1.5 h–Δ3 h		0.035
	0.001 *	Δ1.5 h–30 min	−23	
Functional	0.896			
Triglycerides (mg/dL)				
Control	0.130			
Functional	0.05 *	3 h–1.5 h	−19.8	

[a] p values represents the statistical significance over time for each treatment, and for the variables that follow a normal distribution. [b] p indicates the mean difference obtained from the Bonferroni test for the baseline time periods for each treatment, and for the variables following a normal distribution. [c] p values represent the statistical significance over time for each treatment, and for variables that do not follow a normal distribution. * correspond statistical significant differences (p < or = 0.05).

3.2.2. Plasma Antiradical Activity Based on TEAC Assay

A gradual increase in plasma antiradical activity expressed as % scavenging activity of the ABTS radical cation was observed in the functional group, reaching an increase of MD = 35.0 (92.1%), 3 h after consumption in comparison with baseline values, when for the control group a decrease of MD = 6.0 (12.5%) was observed at 0.5 h compared to baseline and an increase of MD = 18 (37.5%) and 8 (16.7%) at 1.5 h and 3 h, respectively. All differences failed to reach statistical significance.

3.2.3. Coper-Induced Plasma Oxidation Resistance

A non-significant, gradual increase in oxidation resistance was observed in the functional group, reaching an increase of MD = 9.06 (65.6%), 3 h after consumption in comparison with baseline values, when for the control group a gradual decrease in oxidation resistance was observed, especially in the last 1.5 h (1.5 h−3 h), reaching a decrease from the baseline of MD = −1.4 (−28.5%). No statistically significant difference was observed between different fractions for oxidation resistance measurements.

3.3. *Antithrombotic Activity*

PAF-induced platelet rich plasma thrombosis was not affected by the intervention, since no differences were noticed in the intervention compared to the control group.

3.4. *Serum Lipids, Glucose, and Uric Acid Concentrations*

Serum triglyceride concentrations differed significantly between the two treatments 3 h after each meal intake (paired sample *t*-test, $p = 0.041$ and treatment * time, $p = 0.02$) (Table 4). In addition, a statistically significant difference was found between the two treatments in the different time intervals ($p = 0.003$) and between each timepoint and baseline ($p = 0.004$). Although, triglyceride levels 3 h after the control meal are represented as not significant increase, while triglyceride values were decreased significantly in the last 1.5 h (the period from 1.5 h to 3 h) after functional meal consumption (MD = −19.87, 34.9%, $p = 0.05$) (Table 3).

Serum glucose concentration decreased to a non-significant, greater degree 3 h after the consumption of the functional meal (18.7%, MD = −16.1) compared to the degree of decrease observed 3 h after the control meal consumption (2.1 %, MD = −1.6). A statistically significant difference was observed for changes in glucose values over time in the control group for changes from 30 min to 3 h ($p = 0.001$, $P_{Bonferroni} = -23$), and for all time intervals between blood sampling (Δ30-baseline $p = 0.012$, Δ90–30 $p = 0.051$, Δ180–90 $p = 0.035$) (Σφάλμα! Το αρχείο προέλευσης της αναφοράς δεν βρέθηκε). Serum glucose concentration differed significantly in terms of sampling timepoints ($p = 0.022$) and for all changes observed from baseline ($p = 0.046$), after consuming both meals (Table 4).

No statistically significant differences were observed for any interaction regarding the remaining biomarkers tested (total, HDL and LDL cholesterol and uric acid), as there was a similar response to the levels of these biomarkers after consuming both meals.

Table 4. Significant differences between control and functional cookies consumption, over time and for changes from baseline, regarding plasma TAC concentrations as well as serum glucose and triglycerides.

		a. Significant Difference for Each Time Point		**Treatment Comparison**		
	Treatment	Time	Treatment * Time	Mean Difference	Paired-Samples *t*-Test	Wilcoxon Sign Rank Test
Biomarker	*p* Value [a]	*p* Value [a]	*p* Value [a]	Timepoint	*p* Value [b]	*p* Value [c]
Glucose (mg/dL)	0.114	0.022 *	0.450			
Triglycerides (mg/dL)	0.171	0.297	0.02*	3 h	0.041	
Antioxidant capacity (TAC) (mmol/L)				30 min		0.782
				1.5 h		0.312
				3 h		0.153
		b. Significant difference for baseline changes				
	Treatment	Time	Treatment * Time period		Paired-Samples *t*-test	Wilcoxon sign rank test
Biomarker	*p* value [d]	*p* value [d]	*p* value [d]	Time period	*p* value	*p* value [e]
Glucose (mg/dL)	0.146	0.046 *	0.777			
Triglycerides (mg/dL)	0.004 *	0.182	0.297			
Antioxidant capacity (TAC) (mmol/L)				Δ30 min-baseline		0.875
				Δ1.5 h-baseline		0.381
				Δ3 h-baseline		0.081

* Indicates the statistical significance. [a] *p* values indicate the statistical significance for the effect of treatment, the effect of time, and the effect of treatment x time interaction, obtained by repeated ANOVA tests, for the changes between the time points, and for the variables that follow a normal distribution. [b] *p* value indicates the statistical significance for the effect of treatment on the time points obtained with Paired samples *t*-test for the changes between the time points, and for the variables that follow a normal distribution. [c] *p* value indicates the statistical difference between the two treatments (Control-Functional) in the time points obtained with Wilcoxon sign rank test, and for the variables that do not follow a normal distribution. [d] *p* value indicates the statistical significance for the effect of treatment, the effect of time and the effect of therapy x interaction time, obtained by repeated ANOVA tests for changes up to 3 h from the baseline (fasting, 0 h), and for the variables that follow a normal distribution. [e] *p* value indicates the statistical difference between the two treatments (Control-Functional) for the changes up to 3 h from the baseline (fasting, 0 h), obtained with Wilcoxon sign rank test, and for the variables that do not follow a normal distribution.

4. Discussion

The significance of the findings of the acute effect of this dietary intervention on the metabolic, postprandial biomarkers lies in the scientific data which suggest that postprandial hyperlipidemia, hyperglycemia, and induction of oxidative stress are associated with vascular dysfunction [23]. The main finding of this nutritional intervention study was that serum triglyceride levels gradually decreased after consuming the meal containing cookies enriched with olive paste and herbs, as presented in Table 3. In particular, serum triglyceride concentration was significantly reduced at 3 h after the intake of the functional meal. In contrast, for the control group, it was observed that 3 h after consuming the meal containing the control cookies, serum triglyceride values were significantly higher than the baseline values (Tables 3 and 4). The development of innovative functional foods rich in bioactive compounds from the Mediterranean diet could be an alternative way to enhance nutrients intake, improve dietary habits, and decrease the risk of chronic diseases and infections, which is of high importance in the period after COVID-19 pandemic.

The possible beneficial effect of the meal contained olive paste enriched cookies, in the concentrations of serum triglycerides, could be attributed to oleic acid, the basic monounsaturated fatty acid contained in olive paste. Oleic acid consumption has been associated with an improved or unchanged lipid profile, mainly due to a reduction in total cholesterol and LDL cholesterol levels. Data on the effect of postprandial consumption of oleic acid-rich foods on triglyceride levels are conflicting, when some interventional studies showed that their values decrease, and in other studies, HDL cholesterol concentrations appear an increase. Replacing trans fats with oleic acid has been suggested to increase HDL cholesterol and lower triglyceride levels [24].

The fatty acid composition of a meal greatly affect the subsequent responses, as the first chylomicrons enriched with triacylglycerol, appearing in the post-meal circulation contain the triacylglycerols from the previous meal. In a study examining the responses of a control meal, a butter-fortified meal, and an olive oil-fortified meal, it was found that butter tended to cause significantly higher triacylglycerol concentrations during the 8-h test period than olive oil. In addition, a meal fortified with butter led to a reduction in HDL-cholesterol levels, which was not observed after eating a meal fortified with olive oil. These results are probably attributed to the fact that the administration of triglyceride emulsions leads to increased concentrations of fatty acids, which lies in the activity of the enzyme lipoprotein lipase [25]. These findings are in consistence with the findings of the presented study, where after consuming the meal containing the enriched with olive paste cookies, there was a decrease in the concentration of volunteers' plasma triacylglycerols, in contrast to the control meal. However, no significant difference was found in the response of HDL cholesterol between the control meal and the functional meal. In an intervention study, in which 10 volunteers with high fasting glucose levels consumed a meal with olive oil or not, it was found that the meal containing olive oil led to a statistically significant reduction in triglyceride and Apo B-48 levels and glucose levels, compared to the meal that did not contain olive oil. This beneficial postprandial effect of olive oil was probably attributed to the regulation of incretin secretion [26]. In addition, Gilmore et al. observed that a 5-week diet intervention, with a high content of monounsaturated fats, led to an increase in the concentration of HDL cholesterol and a decrease in the ratio of LDL:HDL cholesterol [27].

The increase in plasma TAC observed 3 h after the consumption of the functional cookie meal could be attributed to increased concentrations of polyphenolic metabolites, such as tyrosol, hydroxy–tyrosol, and oleuropein. The combination of these phenolics has been shown to be potent against free radical oxidation, inhibiting the oxidation of low-density lipoproteins. It is speculated that the metabolites of the aforementioned polyphenols may be transported into the blood plasma [26]. However, research shows that this change can also be attributed to the presence of uric acid, which can contribute up to 60% of antioxidant activity as an endogenous, postprandially increasing antioxidant [28]. Numerous studies confirm the above finding, regarding the antioxidant effect of a meal containing olive paste-enriched cookies, as it has been observed that consuming a meal, high in monounsaturated fatty acid may lead to increased plasma concentrations of antioxidants, when compared with meals, rich in polyunsaturated or saturated fats. This could be explained by the fact that the TRL particles provided by the monounsaturated fatty acid-rich meal could have a greater affinity for the hepatic receptor involved in metabolism, leading to faster and more efficient neutralization of TRLs compared to other fat types [26]. The tendency for an increase of plasma antiradical activity 3 h after consumption of functional meal is in accordance with the increase in plasma TAC (Tables 3 and 4). This tendency failed to reach statistical significance possibly due to the limited number of volunteers, which is a limitation of the study.

The fact that PAF-induced platelet rich plasma thrombosis was not affected in the intervention compared to the control group shows that after digestion of olive paste, garlic, thyme, and oregano constituents do not exert antithrombotic effects at the levels relevant to the present study. To the best of our knowledge, nutritional interventions with olive paste or olives have not been performed until now. A previous study has been referring an effect on PAF-induced platelet thrombosis in human volunteers who consumed 15 g of olive oil (1 g capsules of olive oil three times per day with meals). Taking into account that olive paste in cookies contained 20.3 ± 0.8 olive oil [18], volunteers in this study consuming 50 g of cookies received about 2 mL olive oil, which is much lower than the study with olive oil to have an effect [29].

The latest finding of this study was the trend toward a more pronounced reduction in volunteer blood glucose levels, 30 min after the consumption of the meal, which included olive paste-enhanced cookies (Tables 3 and 4). This result is confirmed by a clinical study in healthy volunteers, which found that oleuropein, which is included in the composition of olive oil, is believed to reduce postprandial lipemia. This beneficial effect is probably due to the increased activity of the incretins GLP1 and GIP, inhibiting the activity of the enzyme DPP-4 (dipeptidyl peptidase). Incretins are secreted by the peripheral small intestine in response to its stimulation by binding to receptors in the endocrine pancreas and causing insulin secretion and a decrease in postprandial blood glucose [26].

Finally, it is worth mentioning that oregano and thyme added to the cookies probably contribute to a lesser extent to the beneficial effect that the enrichment of the cookies with a mixture of olive paste may possess. The possible bioactivity of these aromatic plants in postprandial lipemia, glycemia, and oxidative stress has been previously investigated, and lies mainly in the bioactive ingredients that they contain such as thymol, carvacrol etc. [30].

During the COVID-19 pandemic, the need for development by the food industry of innovative functional food with increased nutritional value and possible health effects is constantly increasing. Consumers' demands are also focused on novel foods based on both traditional and nutritious raw materials, with a parallel positive environmental impact. The present study indicates that the development of functional cookies enriched with olive bioactive compounds and herbs may contribute to biomarkers which link to possible bioactivity and potential health effects. The use of second sort olives, as well as future use of olive leaves and olive pomace extracts, as olive oil by-products, may help solve a huge environmental problem, by valorization of industrial pollutant wastes, and in parallel contribute to the isolation of valuable bioactive compounds to create innovative food products. Furthermore, the use of traditional natural functional foods, such as olives, olive oil, and herbs could enhance the trend on food industry innovation, via development of processed novel biofunctional foods with nutritional claims and possible bioactivity, such as cookies, cheeses, yogurts, etc. [1,2,10,31,32].

Nevertheless, the study has some limitations. Initially, although the adequacy of the sample size used was statistically calculated and is similar to other studies [32], the small sample size may have influenced the lack of statistical significance for most of the biomarkers tested. Studies with a higher number of participants should be performed to determine if a novel cookie with olive paste has more pronounced effects in the postprandial state. Moreover, the study assessed the total antioxidant capacity and oxidation resistance of plasma, but individual polyphenols detection was not performed. Furthermore, the possible functional effect of the novel cookie was investigated in healthy volunteers; more interventional studies should be performed in order to evaluate its postprandial effect in patients with cardiovascular disease or metabolic syndrome. Finally, the present study did not detect differences between men and women in each biomarker tested, possibly due to the small sample size.

5. Conclusions

The findings of this nutritional intervention study indicate that the enhancement of cookies with mainly olive paste, but also with medicinal herbs, is likely to be beneficial, as there was a trend of bioactivity, which is mainly focused on reducing serum triglyceride levels, increasing antioxidant activity, and greater reduction in serum glucose levels, postprandially. Monounsaturated fatty acids, especially oleic acid, and polyphenols contained in olive paste, may have a beneficial effect on metabolic, postprandial biomarkers, such as blood lipids and glucose. These findings enhance the need for the development of novel industrial functional foods, with the aim to cover the increasing consumers' need for innovative food items, rich in nutrients and bioactive compounds. However, it is necessary to continue and expand clinical trials in a larger sample of the population, in order to draw safer conclusions about the effect of olive paste on postprandial and other biomarkers of lipemia, glycemia, and also oxidative stress, factors that significantly affect the risk of cardiovascular disease.

In addition, epidemiological studies examining consumers' views and acceptance of functional foods suggest that organoleptic characteristics play an important role in their purchasing decision. Thus, a next step could be the design of sensory analysis and organoleptic tests of the produced, functional cookies, fortified with olive paste and aromatic herbs, in order to investigate consumer attitudes and perform possible improvements of the final product.

Author Contributions: Conceptualization, A.E.K.; methodology, A.E.K., O.P. and H.C.K.; software, O.P.; validation, A.E.K. and H.C.K.; formal analysis, O.P. and C.D.; investigation, O.P., I.M., A.E.K., H.C.K. and C.D.; resources, O.P., I.M., C.D., D.S., D.K., T.L. and A.M.; data curation, O.P., A.E.K. and H.C.K.; writing—original draft preparation, O.P., A.E.K. and H.C.K.; writing—review and editing, A.E.K.; visualization, A.E.K. and H.C.K.; supervision, A.E.K.; project administration, A.E.K.; funding acquisition, A.E.K., H.C.K., D.S. and D.K. All authors have read and agreed to the published version of the manuscript.

Funding: This work was supported by the ERDF research regional program of Western Greece entitled "Production of innovative olive based biscuits with dietary added value" code number: DEP6-0022676, granted to the Greek olive company AMALTHIA S.A.

Institutional Review Board Statement: The study was conducted according to the guidelines of the Declaration of Helsinki, and approved by the Ethics Committee of University of the Aegean.

Informed Consent Statement: Informed consent was obtained from all subjects involved in the study.

Data Availability Statement: The data presented in this study are available within this article.

Acknowledgments: Our thanks to AMALTHIA S.A. for supply of the cookies used in the study.

Conflicts of Interest: The authors declare no conflict of interest.

References

1. Galanakis, C.M.; Rizou, M.; Aldawoud, T.M.S.; Ucak, I.; Rowan, N.J. Innovations and Technology Disruptions in the Food Sector within the COVID-19 Pandemic and Post-Lockdown Era. *Trends Food Sci. Technol.* **2021**, *110*, 193–200. [CrossRef]
2. Sharifi-Rad, M.; Anil Kumar, N.V.; Zucca, P.; Varoni, E.M.; Dini, L.; Panzarini, E.; Rajkovic, J.; Tsouh Fokou, P.V.; Azzini, E.; Peluso, I.; et al. Lifestyle, Oxidative Stress, and Antioxidants: Back and Forth in the Pathophysiology of Chronic Diseases. *Front. Physiol.* **2020**, *11*, 694. [CrossRef] [PubMed]
3. Moreno-Luna, R.; Villar, J.; Costa-Martins, A.F.; Vallejo-Vaz, A.; Stiefel, P. The Postprandial State and Its Influence on the Development of Atherosclerosis. *Immunol. Endocr. Metab. Agents Med. Chem.* **2011**, *11*, 1–9. [CrossRef]
4. Meessen, E.C.E.; Warmbrunn, M.V.; Nieuwdorp, M.; Soeters, M.R. Human Postprandial Nutrient Metabolism and Low-Grade Inflammation: A Narrative Review. *Nutrients* **2019**, *11*, 3000. [CrossRef]
5. Hiyoshi, T.; Fujiwara, M.; Yao, Z. Postprandial Hyperglycemia and Postprandial Hypertriglyceridemia in Type 2 Diabetes. *J. Biomed. Res.* **2019**, *33*, 1–16. [CrossRef]
6. Bozzetto, L.; Della Pepa, G.; Vetrani, C.; Rivellese, A.A. Dietary Impact on Postprandial Lipemia. *Front. Endocrinol. (Lausanne)* **2020**, *11*, 337. [CrossRef]
7. Diamanti-Kandarakis, E.; Papalou, O.; Kandaraki, E.A.; Kassi, G. Nutrition as a Mediator of Oxidative Stress in Metabolic and Reproductive Disorders in Women. *Eur. J. Endocrinol.* **2017**, *176*, R79–R99. [CrossRef]
8. Demopoulos, C.A.; Karantonis, H.C.; Antonopoulou, S. Platelet Activating Factor—A Molecular Link between Atherosclerosis Theories. *Eur. J. Lipid Sci. Technol.* **2003**, *105*, 705–716. [CrossRef]
9. Sirtori, C.R.; Galli, C.; Anderson, J.W.; Sirtori, E.; Arnoldi, A. Functional Foods for Dyslipidaemia and Cardiovascular Risk Prevention. *Nutr. Res. Rev.* **2009**, *22*, 244–261. [CrossRef] [PubMed]
10. Koutelidakis, A.; Dimou, C. The Effects of Functional Food and Bioactive Compounds on Biomarkers of Cardiovascular Diseases. In *Functional Foods Text Book*, 1st ed.; Martirosyan, D., Ed.; Functional Food Center: Dallas, TX, USA, 2017; pp. 89–117.
11. Martirosyan, D.; Miller, E. Bioactive Compounds: The Key to Functional Foods. *Bioact. Compd. Health Dis.* **2018**, *1*, 36. [CrossRef]

2. Bidinger, M. Nutraceutical and Functional Food Regulations in the United States and around the World. *J. Chem. Inf. Model.* **2019**, *53*, 1689–1699.
3. Birch, C.S.; Bonwick, G.A. Ensuring the Future of Functional Foods. *Int. J. Food Sci. Technol.* **2019**, *54*, 1467–1485. [CrossRef]
4. Gambino, C.M.; Accardi, G.; Aiello, A.; Candore, G.; Dara-Guccione, G.; Mirisola, M.; Procopio, A.; Taormina, G.; Caruso, C. Effect of Extra Virgin Olive Oil and Table Olives on the ImmuneInflammatory Responses: Potential Clinical Applications. *Endocr. Metab. Immune Disord.-Drug Targets* **2017**, *18*, 14–22. [CrossRef]
5. Flori, L.; Donnini, S.; Calderone, V.; Zinnai, A.; Taglieri, I.; Venturi, F.; Testai, L. The Nutraceutical Value of Olive Oil and Its Bioactive Constituents on the Cardiovascular System. Focusing on Main Strategies to Slow Down Its Quality Decay during Production and Storage. *Nutrients* **2019**, *11*, 1962. [CrossRef]
6. Drouin-Chartier, J.P.; Côté, J.A.; Labonté, M.É.; Brassard, D.; Tessier-Grenier, M.; Desroches, S.; Couture, P.; Lamarche, B. Comprehensive Review of the Impact of Dairy Foods and Dairy Fat on Cardiometabolic Risk. *Adv. Nutr.* **2016**, *7*, 1041–1051. [CrossRef] [PubMed]
7. Ahmad, J.; Khan, I.; Johnson, S.K.; Alam, I.; Din, Z. Effect of Incorporating Stevia and Moringa in Cookies on Postprandial Glycemia, Appetite, Palatability, and Gastrointestinal Well-Being. *J. Am. Coll. Nutr.* **2018**, *37*, 133–139. [CrossRef]
8. Argyri, E.; Piromalis, S.; Koutelidakis, A.; Kafetzopoulos, D.; Petsas, A.S.; Skalkos, D.; Nasopoulou, C.; Dimou, C.; Karantonis, H.C. Olive Paste-Enriched Cookies Exert Increased Antioxidant Activities. *Appl. Sci.* **2021**, *11*, 5515. [CrossRef]
19. Xiao, F.; Xu, T.; Lu, B.; Liu, R. Guidelines for Antioxidant Assays for Food Components. *Food Front.* **2020**, *1*, 60–69. [CrossRef]
20. Przyminska, A.P.; Zwolinska, A.; Sarniak, A.; Wlodarczyk, A.; Krol, M.; Nowak, M.; Johnson, J.D.G.; Padula, G.; Bialasiewicz, P.; Markowski, J.; et al. Isoflavone Intake Inhibits the Development of 7,12 Dimethylbenz(a)Anthracene(DMBA) Induced Mammary Tumors in Normal Andovariectomized Rats. *J. Clin. Biochem. Nutr.* **2014**, *54*, 31–38. [CrossRef]
21. Sakka, D.; Karantonis, H.C. In Vitro Health Beneficial Activities of Pumpkin Seeds from Cucurbita Moschata Cultivated in Lemnos. *Int. J. Food Stud.* **2015**, *4*, 221–237. [CrossRef]
22. Antonopoulou, S.; Fragopoulou, E.; Karantonis, H.C.; Mitsou, E.; Sitara, M.; Rementzis, J.; Mourelatos, A.; Ginis, A.; Phenekos, C. Effect of Traditional Greek Mediterranean Meals on Platelet Aggregation in Normal Subjects and in Patients with Type 2 Diabetes Mellitus. *J. Med. Food* **2006**, *9*, 356–362. [CrossRef]
23. Dimina, L.; Mariotti, F. The Postprandial Appearance of Features of Cardiometabolic Risk: Acute Induction and Prevention by Nutrients and Other Dietary Substances. *Nutrients* **2019**, *11*, 1963. [CrossRef] [PubMed]
24. Pastor, R.; Bouzas, C.; Tur, J.A. Beneficial Effects of Dietary Supplementation with Olive Oil, Oleic Acid, or Hydroxytyrosol in Metabolic Syndrome: Systematic Review and Meta-Analysis. *Free Radic. Biol. Med.* **2021**, *172*, 372–385. [CrossRef] [PubMed]
25. Thomsen, C.; Storm, H.; Holst, J.J.; Hermansen, K. Differential Effects of Saturated and Monounsaturated Fats on Postprandial Lipemia and Glucagon-like Peptide 1 Responses in Patients with Type 2 Diabetes. *Am. J. Clin. Nutr.* **2003**, *77*, 605–611. [CrossRef] [PubMed]
26. Carnevale, R.; Loffredo, L.; Del Ben, M.; Angelico, F.; Nocella, C.; Petruccioli, A.; Bartimoccia, S.; Monticolo, R.; Cava, E.; Violi, F. Extra Virgin Olive Oil Improves Post-Prandial Glycemic and Lipid Profile in Patients with Impaired Fasting Glucose. *Clin. Nutr.* **2017**, *36*, 782–787. [CrossRef]
27. Gilmore, L.A.; Walzem, R.L.; Crouse, S.F.; Smith, D.R.; Adams, T.H.; Vaidyanathan, V.; Cao, X.; Smith, S.B. Consumption of High-Oleic Acid Ground Beef Increases HDL-Cholesterol Concentration but Both High- and Low-Oleic Acid Ground Beef Decrease HDL Particle Diameter in Normocholesterolemic Men. *J. Nutr.* **2011**, *141*, 1188–1194. [CrossRef] [PubMed]
28. Tasaki, E.; Sakurai, H.; Nitao, M.; Matsuura, K.; Iuchi, Y. Uric Acid, an Important Antioxidant Contributing to Survival in Termites. *PLoS ONE* **2017**, *12*, e0179426. [CrossRef] [PubMed]
29. Mori, T.A.; Vandongen, R.; Mahanian, F.; Douglas, A. Plasma Lipid Levels and Platelet and Neutrophil Function in Patients with Vascular Disease Following Fish Oil and Olive Oil Supplementation. *Metabolism* **1992**, *41*, 1059–1067. [CrossRef]
30. Sekhon-Loodu, S.; Rupasinghe, H.P.V. Evaluation of Antioxidant, Antidiabetic and Antiobesity Potential of Selected Traditional Medicinal Plants. *Front. Nutr.* **2019**, *6*, 53. [CrossRef]
31. Dimou, C.; Karantonis, C.; Skalkos, D.; Koutelidakis, A.E. Valorization of fruits by-products to unconventional sources of additives, oil, biomolecules and innovative functional foods. *Cur. Pharm. Biotechnol.* **2019**, *20*, 776–786. [CrossRef]
32. Papagianni, O.; Argiri, K.; Loukas, T.; Magkoutis, A.; Biagki, T.; Dimou, C.; Karantonis, H.; Koutelidakis, A. Postprandial Bioactivity of Spread Cheese, Enhanced with Mountain Tea and Orange Peel Extract, in Healthy Volunteers. An interventional study. *Biomolecules* **2021**, *11*, 1241. [CrossRef]

MDPI
St. Alban-Anlage 66
4052 Basel
Switzerland
Tel. +41 61 683 77 34
Fax +41 61 302 89 18
www.mdpi.com

Sustainability Editorial Office
E-mail: sustainability@mdpi.com
www.mdpi.com/journal/sustainability

www.ingramcontent.com/pod-product-compliance
Lightning Source LLC
LaVergne TN
LVHW070948170726
843515LV00005B/941

* 9 7 8 3 0 3 6 5 4 1 8 8 4 *